The History of Correlation

After 30 years of research, the author of *The History of Correlation* organized his notes into a manuscript draft during the lockdown months of the COVID-19 pandemic. Getting it into shape for publication took another few years. It was a labor of love.

Readers will enjoy learning in detail how correlation evolved from a completely non-mathematical concept to one today that is virtually always viewed mathematically. This book reports in detail on 19th- and 20th-century English-language publications; it discusses the good and bad of many dozens of 20th-century articles and statistics textbooks in regard to their presentation and explanation of correlation. The final chapter discusses 21st-century trends.

Some topics included here have never been discussed in depth by any historian. For example: Was Francis Galton lying in the first sentence of his first paper about correlation? Why did he choose the word "co-relation" rather than "correlation" for his new coefficient? How accurate is the account of the history of correlation that is found in H. Walker's 1929 classic book, *Studies in the History of Historical Method*? Have 20th-century textbooks misled students as to how to use the correlation coefficient?

Key features of this book:

- Charts, tables, and quotations (or summaries of them) are provided from about 450 publications.
- In-depth analyses of those charts, tables, and quotations are included.
- Correlation-related claims by a few noted historians are shown to be in error.
- Many funny findings from 30 years of research are highlighted.

This book is an enjoyable read that is both serious and (occasionally) humorous. Not only is it aimed at historians of mathematics, but also professors and students of statistics and anyone who has enjoyed books such as Beckmann's *A History of Pi* or Stigler's *The History of Statistics*.

The History of Correlation

John Nicholas Zorich

CRC Press
Taylor & Francis Group
Boca Raton London New York

CRC Press is an imprint of the
Taylor & Francis Group, an **informa** business
A CHAPMAN & HALL BOOK

First edition published 2025
by CRC Press
2385 NW Executive Center Drive, Suite 320, Boca Raton FL 33431

and by CRC Press
4 Park Square, Milton Park, Abingdon, Oxon, OX14 4RN

CRC Press is an imprint of Taylor & Francis Group, LLC

ISBN: 9781032865249 (hbk)
ISBN: 9781032865041 (pbk)
ISBN: 9781003527893 (ebk)

DOI: 10.1201/9781003527893

Typeset in Palatino
by codeMantra

Contents

Preface

This book is the result of 30 years of enjoyable research. That research included many hours on many weekends in the Stanford University archive libraries as well as in the stacks of virtually all the used bookstores in the San Francisco Bay Area. Other source material was found in antiquarian bookshops in the cities and countries I visited while on vacations or business trips. Online used bookstores throughout the English-speaking world provided the remainder of my source for hard-copy books and papers. In the 21st century, I felt like I'd won the lottery when I discovered that many 19th- and 20th-century books and journals were now available on websites such as Archive.org, Gutenberg.org, Jstor.org, and Galton.org.

I started my research without intending to publish a book; rather, I was hoping to better understand the history of statistics, especially that of correlation, and to possibly someday publish a paper on the topic. The majority of my writing began in the early 2020 "lockdown" months of the COVID-19 pandemic, at the start of which I asked myself, "Years from now, what will I have to show for my time at home, besides bills from Zoom and Netflix?" My answer was: "I'll write a book on the history of correlation." More than half of the research for this book was conducted during that pandemic, and almost all of that was done online. I especially recall one day in mid-2020 when I spent several hours downloading about 600 pdf files, one by one, from Galton.org. Also, I recall emailing and then waiting for 15 months for the University College London's library staff to return to work (from their stay-at-home orders during the pandemic), so that they could fulfill my request for copies of pages from two books in their Francis Galton Library collection.

The research and writing of this book were fun and satisfying. I would sometimes spend 2 days writing a single-sentence comment about a quote, when the research on that quote would take me from reference to reference; such research allowed me to assess the accuracy of the original quote and of published statements about it. As justification for such labors, let me narrate an infamous example from the history of the concept of biological evolution: At the end of the 19th century, philosopher Herbert Spencer published a book-length collection of his previously individually published papers. That book included an 1852 essay titled, "The Development Hypothesis", which discussed what he called the "Theory of Evolution" of plants and animals from simpler life forms (Spencer, 1901, p.1). However, the term he'd originally used in that essay (in 1852) was not "Theory of Evolution" but rather "Theory of Lamarck" (Spencer, 1852, p. 280), which Charles Darwin's Theory of Evolution had (by 1901) famously supplanted.

During my research for this book, I examined thousands of papers and books. For inclusion in this book, I then chose quotations, figures, and tables that were historically significant or were either interesting or humorous to me.

I have certainly focused on historical minutia. If I am to be mocked for that minutia, that mockery will most likely center on my many-page discussion of why Francis Galton hyphenated the first word in the title of his December 1888 paper that introduced mathematical correlation.

As you will see, I have discovered mistakes galore in publications aplenty, my descriptions of which you may consider to be unkind. A very experienced author friend of mine assures me that it is impossible for a book of this length to be error-free, no matter how many times I and others proof-read it. Therefore, please write to *kindly* inform me of any errors that you find (send to johnzorich@yahoo.com). I will maintain a list of such errata at www.johnzorich.com/correlationbook.

THE HISTORY OF CORRELATION
by John Nicholas Zorich

Researching my book was very much fun,
although I admit, I'm happy it's done.
Old textbooks I hunted, and purchased a ton;
in Stanford Archives were many more ones.

Internet sites I found to be great
for finding old books and papers of late.
Thousands of pages I did read and quotate;
comments from me, to those I did mate.

Writing my book took many a year.
Would it be published? "Never, I fear!"
Twenty refusals were painful to hear;
the next was accepted, and it felt like a cheer.

The draft I submitted was far from a mess,
but "changes are needed" came the request
from Taylor & Francis and CRC Press;
not wanting to fight, I did acquiesce.

Now it's for sale on Amazon mobile,
the Routledge website, and Barnes & Noble,
with Access to Open and distribute it global;
maybe someday it will even be vocal.

Acknowledgments

I thank all those who provided help while I prepared this history, especially:

- My wife, Sylvia Zorich, for more than all the usual reasons.
- Family and friends who reviewed all or part of the drafts, especially for the enthusiastic encouragement given by my sister Monette Zorich and my friend Patrick Giuliano.
- Stephen Stigler (author, statistician, and historian) for publication advice.
- Melinda Baldwin (author and member of the Department of the History of Science, Harvard University) for advice regarding 19th-century journals.
- Alysoun Sanders (archivist at the journal *Nature*) for insight regarding 19th-century issues of *Nature*.
- Dan Mitchell (at University College London's Special Collections Library) for photographs of pages in books in the Francis Galton Library collection.
- Lizette Barton (reference archivist at the Drs. Nicholas and Dorothy Cummings Center for the History of Psychology) for a copy of a 1926 paper from a long-defunct journal.
- Joachim Dagg (historian of biology and evolution) for a copy of Herbert Spencer's original essay from 1852 titled, "The Development Hypothesis".
- My publication editors and their assistants at Chapman & Hall/CRC Press (Taylor & Francis Group).
- The following publishers and individuals for permitting the use of text and images from their publications:

  - The American Mathematical Association of Two-Year Colleges
  - Ian Johnston
  - BVT Publishing

Author

John Nicholas Zorich has bachelor's and master's degrees in botany from the University of California. He has worked in the biotechnology industry since 1979, in QA/QC, Mfg. and R&D. For the past 25 years, he has worked as a statistical consultant to that industry. His consulting clients have included several large multi-national biotechnology companies as well as many Silicon Valley startups. He designs statistical application software programs that have been purchased by more than 140 companies worldwide. For almost 20 years, he has been the in-house subject-matter expert in statistics for one of the EU medical-device Notified Bodies. For 20 years, he's been an instructor in applied industrial statistics for the Biotechnology Center of Ohlone College and is an annual guest lecturer in applied statistics at the Graduate Department of Biomedical Engineering at San Jose State University. In recent years, he's published three papers on aspects of teaching statistics and published one paper on how to extend the shelf-life of pharmaceuticals simply by improving the statistical method of calculation. The research for this book on correlation has been an ongoing passion of his for the past 30 years. He lives in Houston, Texas, with his wife Sylvia. When not involved in statistical pursuits, he enjoys gardening, reading, and traveling.

1

Essential Terms, Definitions, and Background Information

This book is primarily a history of the word "correlation" and of the various meanings it was given in publications by scientists during the past two centuries, mostly in regard to linear correlation. It is also a history of how the linear correlation coefficient was interpreted during that time and of how the explanation of its meaning evolved in textbooks and journal articles. It is not a history of the formulas used to calculate linear correlation, although part of that history is included.

This history is about 50% quotations and 50% my comments and analyses of those quotations. I added text within brackets within some quotations (e.g., "... [xxx]...") in order to correct grammar or punctuation (or lack of it) that interferes with comprehension; in some cases, that was necessary or in order to clarify the intent as determined from other text given by the original author but which I do not provide in this book. Lengthy quotations are typically shown indented and italicized; short quotations are typically shown within quotation marks, in regular font, and in-line with non-quotation text.

This first chapter focuses on background information that is necessary for the reader to understand what follows. The topics discussed in this chapter are:

- Summary of the history of correlation.
- Overview of linguistics.
- History of the symbols used to represent correlation.
- History of the nomenclature used for linear correlation.
- Tables, plots, and charts used for correlation analysis.
- Examples of formulas that have been used to calculate linear correlation.
- Definitions of correlation "classes", which are new terms introduced in this book.
- Definitions of "relation", "co-relation", and "correlation" in the Oxford English Dictionary.

DOI: 10.1201/9781003527893-1

Brief History of Correlation

The history of the word correlation is similar to that of the word "statistics". Both words (correlation and statistics) began the 19th century with meanings that had nothing to do with mathematical analysis. Statistics originally referred to state-secrets data (e.g., the number of men who could be drafted into the army in time of war). Subsequently, it was applied to *any* type of survey or census data; evaluation of such data primarily involved summaries, tables, and charts. The methods used to analyze those data were at first called "political arithmetic" (Porter, 1986, p. 18), a term that by the end of the 19th century was supplanted by "statistical methods" (e.g., C. B. Davenport, 1899, p. iii) or the "methods of statistics" (e.g., Bowley, 1901, p. v), or similar wording. In the 20th century, the word statistics slowly began to be used primarily in reference to analytic methods such as t-tests, ANOVA, and correlation; that transition was sign-posted by a 1920 paper titled "The Conception of Statistics as a Technique", in which the author proposed something new, namely that...

> *It is the statistician's function to measure social tendencies precisely, not to reason or philosophize about hypothetic al tendencies. It is his function to determine precisely rates and frequencies and <u>correlations</u> and probabilities...*
>
> *(Cummings, 1920, p. 176; underlining added)*

During the few decades when the meaning of statistics was transitioning from methods of data summary to methods of data analysis, the word "mathematical" was sometimes added to a book's title if its subject matter was primarily mathematical analysis rather than descriptive tables and charts (e.g., West, 1918, *Introduction to Mathematical Statistics*; Rietz, 1927, *Mathematical Statistics*; Burgess, 1927, *Introduction to the Mathematics of Statistics*).

In 1915 in the United States, it could be said that "as a rule, our professional statisticians are self-taught or get their training, which often may be imperfect and one-sided enough, by the aid of others whose only school-master had been experience"; in contrast, the training process in Germany was relatively formal (Koren, 1915, p. 353). After Francis Galton died in 1911, his will provided funds to University College London (UCL) that enabled Karl Pearson to become "the first head of the first [British] university department in which statistical theory was a major concern"; his title was "Professor of Eugenics" in the newly formed "Department of Applied Statistics", which encompassed the "Biometric and Eugenics Laboratories" (Mackenzie, 1981, pp. 10, 104–105). In 1933, UCL "severed the link of [sic] statistics and eugenics" by assigning R. A. Fisher to be chair of Eugenics and by assigning Egon Pearson (son of Karl) to be chair of Statistics (Mackenzie, 1981, p. 118).

The word correlation originally referred to a close association of characteristics (e.g., blond hair and blue eyes); subsequently, it was applied to sets of survey data (e.g., number of immigrants vs. city). Then, it was applied to data sets involving two or more characteristics that could be represented numerically (e.g., number of immigrants vs. population of city). After the turn of the century, the word correlation began to be used primarily in reference to a very specialized analytical method.

The word correlation has been in general use in the English-speaking world for more than two centuries. That time frame can be split into two eras whose mutual boundary is December 20, 1888. On that date, Francis Galton gave an oral presentation to London's Royal Society based upon a not-yet-published paper that he'd delivered in hard copy to the Society earlier that month (Galton, 1888c, p. 135). In that paper, he stated that...

> *"Co-relation or correlation of structure" is a phrase much used in biology... but I am not aware of any previous attempt to define it clearly, to trace its mode of action in detail, or to show how to measure its degree.... the closeness of co-relation in any particular case admits of being expressed by a simple number.*
>
> *(Galton, 1888c, pp. 135–136)*

Eventually, that simple number came to be called the correlation coefficient. Karl Pearson, Galton's close friend and first major biographer, pronounced that "The twentieth of December is therefore the birthday of the conception of [mathematical] correlation in biometric data" (K. Pearson, 1930a, p. 50). Historian T. M. Porter remarked that "The almost simultaneous appearance of Galton's book *Natural Inheritance* and his [December 1888] method of correlation in 1889 [the year that they were both formally published] marks the beginning of the modern period of statistics" (Porter, 1986, p. 296).

As a result of Galton's co-relation paper, the number of ways in which the word "correlation" was used in scientific publications increased dramatically. For example, the following is a list of words and syllables that I've found to precede the word correlation in post-1888 literature:

actual, anti-, average, biserial, close, correct, cross-, curvi-linear, direct, dynamical, entire, grade, index, individual, inter-, intimate, inverse, linear, multiple, natural, negative, non-, non-linear, nonsense, normal, ordinary, organic, partial, perfect, positive, rank, raw, re-, real, simple, skew, spurious, static, statical, statistical, true, un-.

As we shall see in the following chapters, many post-1888 authors applied the word correlation very restrictively. However, many others used it so generally as to be almost superfluous; for example:

> ...*two sets of measures are <u>related</u> if knowledge of an individual's score on one of them reduces the range of possibilities for his position on the other one of them.... When two measures are so <u>related</u>, the term <u>correlation</u> is usually used to describe the fact.*
>
> *(Willemsen, 1974, p. 64; underlining added)*

In 1984, the journal *Science* published an issue in which 20 authors wrote 20 articles about "20 Discoveries that Changed our [20th century] World". One of those authors was Ian Hacking, whose article traced the history of statistical tests and then claimed that… "Statistics has not yet aged into a stable discipline with complete agreement on foundations" (p. 70). As we shall see in later chapters of this book, that same claim can be applied to correlation.

Logistics

Almost all of the source materials for this book were written in English; hopefully, someday, someone will research the history of non-English sources. My sources were derived primarily from the following:

- Leland Stanford Junior University (a.k.a. Stanford) libraries.
- Journal Storage (https://www.jstor.org).
- Sir Francis Galton FRS (https://galton.org).
- Internet Archive (https://archive.org/index.php).
- Project Gutenberg (http://www.gutenberg.org).
- Bookfinder (https://www.bookfinder.com).
- Random selection of used-book stores in England, Scotland, Ireland, Canada, and the United States, perused while visiting various cities on business trips and vacations, 1992–2019.

According to the United Nations Educational, Scientific and Cultural Organization (www.unesco.org), artificial intelligence ("AI") is the future of teaching and learning; but recently the answers that an AI site gave to my questions about the history of correlation were all very wrong (e.g., wrong dates, wrong journals, and completely fictitious quotations).

I have focused on the concept of *linear* correlation. To include a full history of other types of correlation (e.g., curvi-linear, rank, multiple) is exhausting just to contemplate. My focus is coincidentally in sync with 21st-century author Nassim Taleb, who in his book *Fooled by Randomness: The Hidden Role of Chance in Life and in the Markets* advised those who trade stocks on Wall Street…

> *Don't take "correlation" and those who use the word seriously… The reader should do himself a favor by not taking the notion of correlation seriously except in very narrow matters where <u>linearity</u> is justified.*

> *(Taleb, 2004, p. 283; underlining added)*

This book is arranged semi-chronologically. Word usage in the 19th century is examined, first in the works of prominent scientists and then in works by Francis Galton. The remainder of the book focuses on how the word has been used by scientists, historians, and textbook authors in the 20th century.

I performed all word counts given herein; unless stated otherwise, such counts ignored words in titles of chapters and of chapter sub-sections. Counts were conducted while reading the entire document or while searching an electronic version of it; those electronic searches were conducted on files after they were converted to "Searchable PDF" format using "PDF Converter Professional 8.1 by Nuance Communications".

I am responsible for all translations, unless another author is cited.

Unless otherwise indicated, wherever it is stated that a document does or does not contain an instance of the noun correlation, it is meant that the document also does or does not contain instances of the associated parts of speech, such as correlate, correlated, correlative, and correlatively. The same is true for statements about the nouns relation and co-relation.

Linguistics

Words matter because meaning matters; otherwise, what is the purpose of writing and reading? An author of a book on grammar concurs:

> *Words are powerful. As a writer you can use to advantage the power that words hold to call up images in the mind of the reader. But to use words effectively, you have to understand as far as possible the meanings that are built into the reader's lexicon.*

> *(Kolln, 1996, p. 53)*

To that, I might add: "Conversely, as a reader you have to understand as far as possible the meanings that are built into the *writer's* lexicon."

The authors of a famous book on language echoed some of the thoughts that I had while writing this book: "To learn to think more clearly, to speak and to write more effectively, and to listen and to read with greater understanding – these are the goals of the study of language" (Hayakawa and Hayakawa, 1990, p. x). To that, I might add: "and these are the goals of the study of a single word".

Those same two linguistic scientists discussed at length how "Language… makes progress possible" (Hayakawa and Hayakawa, 1990, p. 7f); it does that by documenting history, not only the history of nations but of scientific discoveries (such as the discovery of mathematical correlation). Such discoveries are documented in published papers, books, and similar "reports". The value of language (the Hayakawas claim) is that the "accuracy of reports can be checked and rechecked by successive generations of observers". However, that is possible only if successive generations understand the meaning of the words in those reports *as intended by the original author.*

Linguistic scientists have long used the simile that words are like maps, maps to ideas (Hayakawa and Hayakawa, 1990, pp. 20–21); however, readers may end up in the wrong place if they misread the map (i.e., they may get the wrong idea if they interpret the words differently than the author intended). Dictionaries are relied upon to help in the interpretation of such maps. Although the average person considers dictionaries to be the highest authority for the meaning of a word, dictionaries actually are histories of how words have been used up to when the dictionary was compiled:

> *The writer of the dictionary is a historian, not a lawgiver.… In choosing our words when we speak or write, we can be guided by the historical record afforded us by the dictionary, but we cannot be bound by it, because new situations, new experiences, new inventions, new feelings, are always compelling us to give new uses to old words.… The way dictionary writers arrive at definitions is merely the systematization of the way we all learn the meaning of words, beginning at infancy and continuing for the rest of our lives.… [W]e learn from verbal context, arriving at a workable definition by understanding one word in relation to the others with which it appears.… [W]e learn by physical and social context.*
>
> (Hayakawa and Hayakawa, 1990, pp. 34–35)

Words have different *types* of meanings, and those meanings are contextually fluid:

> *The <u>extensional meaning</u> of an utterance is that to which it points in the extensional (physical) world.… An easy way to remember this is to put your hand over your mouth and point whenever you are asked to give an extensional meaning.… The <u>intensional meaning</u> of a word or expression is that which is suggested (connoted) inside one's head. Roughly speaking, whenever we express the meaning of words by uttering other words, we are giving connotations or intensional meaning.… Everyone, of course, who has ever given any thought to the meanings of words has noticed that words are always shifting and changing in meaning.… The situation, therefore, appears hopeless. Such an impasse is avoided when we start with a new premise altogether – <u>one of the premises upon which modern linguistic thought is based: namely, that no word ever has exactly the same meaning twice</u>.… We cannot know what a word means before it is uttered. All we can know in advance of its utterance is approximately what it will mean. After the utterance, we interpret what has been said in the light of both verbal and physical contexts, then act or understand according to our interpretation.*
>
> (Hayakawa and Hayakawa, 1990, pp. 36–39; underlining added)

As we shall see, some authors of statistical papers and textbooks preferred to define the word correlation extensionally only (i.e., they define correlation by using only a formula – no words), some authors preferred to define it intensionally only (i.e., they define correlation by using words only – no formula), whereas most authors provided both types of definitions.

In light of the preceding linguistic discussion, the word correlation is seen to have had two major problems for at least the first few decades after the 1888 invention/discovery of mathematical correlation, which problems slowly lessened in severity as the 20th century progressed; those problems were:

- Almost every author on the subject of mathematical correlation gave the word correlation a slightly to significantly different meaning. Even in more recently published textbooks, there is no universal agreement on what is "correlation". Authors have defined correlation and/or its coefficient as a measure of variation, an estimate of causes, a probability level of co-relation, a confidence level in a statement, a significance level in a test, etc.

- Dictionaries did not include a definition of mathematical correlation. When definitions did finally appear in dictionaries, it was impossible for dictionary writers to coalesce all the just-mentioned historical uses into a single clear statement, and so, they chose one or more based upon their own view of which authors were most authoritative.

Symbols

Starting in Galton's December 1888 paper, the symbol for the sample linear correlation coefficient has virtually always been r, even though Galton had previously used that letter to symbolize what we now call the regression coefficient (Galton, 1877a, p. 299). Many authors have added subscripts as well such as xy or yx, the order of letters in which was intended to indicate which of the variables (x or y) is the independent or fixed-value variable; unfortunately, the literature in that regard is not consistent. One author melodramatically recommended that r be used (instead of r_{xy}) only "when there is no danger of misinterpretation" (Kozelka, 1961, p. 131; underlining added).

Another author wrote in such an unclear manner that it discredited Galton for having assigned r to the correlation coefficient:

> r is Bravais' value of the coefficient of correlation.
>
> (Yule, 1897a, p. 481)

That is a complete sentence; in Yule's entire 13-page paper, his only mention of Galton is on page 482, in a footnote in which the symbol r is not mentioned.

Some authors have preferred to use R instead of r (e.g., Davenport and Bullard, 1896, p. 94; Brown and Thomson, 1921, pp. 204–205; Ostle, 1963, p. 222). Additionally, R has been used to represent the correlation coefficient of multiple correlation (Yule, 1897a, p. 485; Holland, 1998, p. 47). However, most authors have reserved R for the sample curvi-linear correlation coefficient; that coefficient was invented by Karl Pearson, who gave its population parameter the symbol η (the lower-case Greek letter *eta*) and called it the "correlation ratio" (Pearson, 1905, p. 484), a name that was widely adopted (Walker, 1929, p. 178). Amazingly, R has also been used to represent the regression coefficient (e.g., Pearson, 1896a, p. 277). The most common correlation-related use for R today is in R^2, which is the square of either the correlation coefficient or the correlation ratio, as applicable. Applied to linear correlation, R^2 is called the Coefficient of Determination, and its formula always results in a positive number; $_sR^2$ is called the Signed Coefficient of Determination because its formula can result in a positive or negative value (Zorich, 2018).

From the time it was introduced by Edgeworth (1892, p. 190), the symbol ρ (the lower-case Greek letter *rho*) has been used almost universally for the population parameter linear correlation coefficient. Unfortunately, ρ was later also used for the rank correlation coefficient and for the curvi-linear correlation coefficient (more about this, later). Pearson reported that Galton later in life wished to change the symbol r to ρ (Pearson, 1930a, p. 55n).

Nomenclature

The following is a discussion of the names for r that are historically important:

Index of Co-Relation

Based on the writings of some historians, it might seem reasonable to infer that Galton in his December 1888 paper intentionally coined the term "index of co-relation" or "index of correlation". For example:

- [Galton] *determines seven correlations which he here* [in Galton's December 1888 paper] *terms 'indices of correlation.'* (Pearson, 1920, p. 39)
- *Index of correlation… Galton, 1888 "Co-relations and their Measurements…."* (Walker, 1929, p. 181, in a section titled "The Origin of Certain Technical Terms Used in Statistics")
- *Galton viewed his "index of co-relation" (Galton, 1888c, p. 143) within the context of his earlier work on regression…. Edgeworth introduced this term* [coefficient of correlation]… *although Galton used "index of correlation" in 1888* (Stigler, 1986, pp. 297, 319n).

However, such an inference seems invalid, given the following facts and discussions:

- Despite Pearson's claim (just given), neither the term "indices of correlation", nor "indices of co-relation", nor even the word "indices" appears anywhere in Galton's 10-page, December 1888 paper.
- In that paper, Galton used the phrase "index of co-relation" only once (shown here next); the word index appeared nowhere else, not even in the paper's final summary paragraph:

When the deviations of the subject and those of the mean of the relatives are severally measured in units of their own Q [i.e. his version of probable error], there is always a regression in the value of the latter. This is precisely analogous to what was observed in kinship, as I showed in my paper read before this Society on "Hereditary Stature" (' Roy. Soc. Proc.,' vol. 40, 1886, p. 42). The statures of kinsmen are co-related variables; thus, the stature of the father is correlated to that of the adult son, and the stature of the adult son to that of the father; the stature of the uncle to that of the adult nephew, and the stature of the adult nephew to that of the uncle, and so on; but <u>the index of co-relation, which is what I there called "regression,"</u> is different in the different cases.

(Galton, 1888c, pp. 142–143, underlining added)

In that text, notice that Galton put quotation marks around his special term "regression" but not around the words "index of co-relation", as if that were *not* a special term; nor were those words underlined, capitalized, or italicized in the original.

- An abstract of Galton's December 1888 paper appeared in *Nature* early the next month; it provided no name for r and did not use the word "index" at all (Galton, 1889b). The same can be said for the published version of his "President's Address" that he gave at the Anthropological Institute a few weeks later, in which he discussed r at length (Galton, 1889e). Nor does the word index appear in the modified version of that Address that was soon-after published in *Nature* (Galton, 1889d). Later that year (per Stigler, 1989), Galton wrote another paper, in which he provided the history of his discovery of mathematical correlation; however, it too contained no explicit name for r, and its only use of the word index was this:

There is no numerical reciprocity in these figures, because the scales of dispersion of the lengths of the finger and of the stature differ greatly, being in the ratio of 15 to 175. But the 6 hundredths multiplied into the fraction of 175 divided by 15, and the 819 hundredths multiplied into that of 15 divided by 175, concur in giving the identical value of 7 tenths, which is <u>the index of their correlation</u>.

(Galton, 1890a, p. 430; underlining added)

- The manner in which Galton paired the word index with the words co-relation or correlation was not substantially different from other relevant pairings in the papers just mentioned. For example, in addition to being described as the index of co-relation, *r* was said to either be or represent…

    In Galton (1888c):

    o   *the closeness of the co-relation* (pp. 135–136).

    o   *the exactness of the co-relation* (p. 136).

    o   *the measure of the closeness of the co-relation* (p. 140).

    In Galton (1889b):

    o   *a measure of the closeness of correlation* (p. 238).

    In Galton (1889e):

    o   *the measure of correlation* (p. 403).

    o   *a measure of the closeness of correlation* (p. 408n).

    In Galton (1890a):

    o   *an exact measure of the weakness of the correlation* (p. 427).

Therefore, Galton's phrase (previously mentioned)…

    the <u>index</u> of co-relation, which is what I there called "regression,".

could have, without any loss of meaning, been written as…

    the <u>measure</u> of co-relation, which is what I there called "regression,".

In the coming years, Galton referred to *r* as the index of correlation only three times in print; in the first case (in 1894) he used neither capitals nor quotation marks, whereas in the other two cases (in 1907 and 1908) he used both:

- *I have worked out the theory of correlation in a Memoir read before the Royal Society, in 1888, where I showed that there exists what may be called an index of correlation* (Galton, quoted in Troup et al., 1894, p. 59).
- *It follows that the average deviation of a B value bears a constant ratio to the deviation of the corresponding A value. This ratio is called the 'Index of Correlation', and is expressed by a single figure* (Galton, 1907, pp. 22–23).
- *It had appeared from observation, and it was fully confirmed by this theory, that such a thing existed as an "Index of Correlation"; that is to say, a fraction, now commonly written r…* (Galton, 1908b, p. 303).

The first of those texts (in 1894) was written by an interviewer or stenographer who was quoting Galton, whereas Galton himself wrote the second and third texts (in 1907/1908).

Such evidence indicates that if Galton in December 1888 had intended "index of correlation" to be a special term, he would have encased it in quotation marks or at least capitalized it. After all, he'd capitalized every important term that he'd included in his book *Natural Inheritance*, which he'd written earlier that year. The following examples of capitalized terms in *Natural Inheritance* are found not in the title of a chapter or subchapter but rather in the middle of sentences (all page references are to Galton, 1889a):

> *Natural Selection* (p. 32).
> *Curve of Distribution* (p. 49).
> *Law of Error* (p. 55).
> *Probable Error* (p. 58).
> *Mid-Parent* (p. 89).
> *Co-Fraternals* (p. 94).
> *I call this ratio… "Filial Regression."* (p. 97; notice the quotation marks)
> *Ratios of Regression* (p. 101).

It seems reasonable to conclude that in 1888/1889 and for years thereafter, Galton himself did not in print assign *any* special nomenclature to what today we call the correlation coefficient. On the other hand, some of his contemporaries did use the term "Index of Correlation" (e.g., Davenport and Bullard, 1896, pp. 94, 96; notice the capitalization).

It is interesting that Galton has been credited with having discovered the "correlation-index" rather than the "index of correlation" (Davenport, 1900, p. 866; notice the hyphen). Surprisingly, the terms "index of correlation" and "correlation-index" have also been used to refer to the coefficient of *curvi*-linear correlation (e.g., Ezekiel, 1924, p. 434). In mid-20th-century, one author reserved the term "correlation index" for the *square* of the correlation coefficient, i.e., for what virtually everyone else was calling the Coefficient of Determination (Ostle, 1963, p. 223).

Galton's Function

Additional strong support for the just-given conclusion can be found in the published papers of W. F. R. Weldon, a young lecturer in invertebrate morphology at University College London. In 1889, Galton began mentoring Weldon in the mathematics of regression and correlation. In a paper published in early 1890, Weldon admitted…

> *My ignorance of statistical methods was so great that, without Mr. Galton's constant help… this paper would never have been written…. I have attempted to apply to the organs measured the test of correlation given by Mr. Galton* [in his December 1888 paper]….

> (Weldon, 1890, pp. 445, 453)

Galton continued his tutelage, which resulted in Weldon's papers titled "Certain Correlated Variations in *Crangon vulgaris*" and "On Certain Correlated Variations in *Carcinus moenas*"; relevant passages from both are:

In what follows [,] an attempt is made to apply Mr. Galton's method to the mea-
surement of the correlation between four organs of the common shrimp.

(Weldon, 1892, p. 2)

The method adopted to determine the degree of correlation between two organs
was that proposed by Mr. Galton.

(Weldon, 1893, p. 325)

If Galton had a name in mind for r, he certainly would have shared it with
Weldon; and then Weldon's 1892 paper would not have referred to r as "this
constant" and "the constant" (Weldon, 1892, p. 3); nor would Weldon have
proposed a name for it in his 1893 paper:

The importance of <u>this constant</u> *in all attempts to deal with the problems of ani-*
mal variation was first pointed out by Mr. Galton... and I would suggest that
<u>the constant</u> *whose changes he has investigated, and whose importance he has*
indicated, may fitly be known as "<u>Galton's function</u>*."*

(Weldon, 1893, p. 325; underlining added)

Although the term "Galton's function" was used in print by some of Weldon's
acquaintances (e.g., Thompson, 1894, p. 236; Yule, 1895, p. 604; Pearson, 1895,
p. 241, 1896a, p. 267), other terms came to be more widely adopted. Yule, in an
historical account written in 1909, referred to r as the "'coefficient of correlation,' as it is now termed, or 'Galton's function,' as it was called for some time"
(Yule, 1909, p. 722) – notice that he did not mention "index of correlation". In
Yule's textbook published in 1911, he spoke of the "coefficient of correlation...
[or] Galton's function, as it was termed at <u>first</u>" (Yule, 1911, p. 188; underlining added) – in other words, Yule did not consider "index of correlation" to
have been the "first" term ever used. Karl Pearson in 1920 wrote a history of
correlation in which he quoted that 1911 statement by Yule; Pearson did not
contradict Yule regarding it, although Pearson did contradict other historical
statements by Yule (Pearson, 1920, p. 28). Even the anonymous author of the
Royal Statistical Society's Galton obituary mentioned Weldon's term but not
any term coined by Galton (Anonymous, 1911).

Coefficient of Correlation

Karl Pearson's paper titled "Notes on the History of Correlation" referenced
an 1893 paper by Weldon as being titled "Correlated Variations in Naples and
Plymouth shore [sic] Crabs" (Pearson, 1920, p. 42) – however, the correct title
was "On certain Correlated Variations in *Carcinus moenas*" (Pearson seems
to have obtained the words for his version of the title from the last sentence
in the first paragraph in Weldon's paper). Pearson was equally creative when
referencing *terms* coined by others, as evidenced by the fact that he sometimes

capitalized words that were actually lower-case in the original papers from which he is supposedly quoting – for example (in Pearson's words):

> *Edgeworth replaces Galton's "Index of Co-relation" and Weldon's "Galton's Function" by the term "coefficient of correlation."*

> (Pearson, 1920, p. 42; underlining added; those underlined
> words were not capitalized in the original papers by
> Galton and Weldon that Pearson referenced).

It seems valid for Pearson to have attributed the term "coefficient of correlation" to Edgeworth, but it seems invalid to trace its origin to Edgeworth's 1892 paper titled "Correlated Averages", as Pearson did on page 42 of his "Notes on the History…". The first appearance of those words in Edgeworth's paper was:

> *…to find the coefficient of correlation between the stature and left cubit….*

> *(Edgeworth, 1892, p. 191)*

Edgeworth introduced those words without explicitly stating that they represented a new term, without highlighting them in quotation marks; nor were they underlined, capitalized, or italicized in the original. In contrast, Edgeworth did use quotation marks to highlight his paper's first use of the term "correlation" (p. 190). Although he did use "coefficient of correlation" (or its plural, all without quotation marks) eight more times in the paper, he also used the phrase "the Galtonian coefficient" (p. 195) and many times spoke of other coefficients such as the *"coefficient of $x_1 x_2$"* (from page 194).

It was actually the next year, 1893, in which Edgeworth revealed his nomenclature preference:

> *Now if there were given a certain constant which may be <u>called</u> the coefficient of correlation (Mr. Galton's r…).*

> *(Edgeworth, 1893, p. 674; underlining added)*

Thus, it might have been more accurate for Pearson to have referenced Edgeworth's 1893 paper, not his 1892 one, as the origin of the *formal* term "coefficient of correlation". In defense of Pearson, historian S. M. Stigler agrees with him that the 1892 paper is the correct origin for the term "coefficient of correlation" (Stigler, 1986, p. 319n).

Correlation Coefficient

In 1896, Karl Pearson published what was to become the most widely used formula for calculating the correlation coefficient (Pearson, 1896a, p. 265). Regarding the *r* calculated by it, Bowley advised: "To distinguish it from other measurements [,] it is sometimes called the sum-product coefficient of correlation" (Bowley, 1920, p. 355). Some authors called it the "coefficient of product-moment correlation" (e.g., Kendall, 1943, p. 330). However, many authors used names that seem to have been coined solely to honor Pearson; for example:

- *Pearsonian Correlation Coefficient* (Magee, 1912, p. 178).
- *Pearsonian Coefficient of Correlation* (Persons, 1914b, p. 348).
- *Bravais-Pearson correlation r* (Garnett, 1919, p. 96).
- *Pearson coefficient of correlation* (Thurstone, 1925, p. 206).
- *product-moment (Pearson) coefficient of correlation* (Symonds, 1926, p. 458).
- *Pearson's r* (Healey, 1984, p. 266).
- *Pearson product-moment correlation coefficient* (Bohrnstedt and Knoke, 1988, p. 271).

It is interesting that Yule at least once used three names for r in the same paper, i.e., Galton's function, coefficient of correlation, and correlation coefficient (Yule, 1895, pp. 604–605), and curiously many years later once felt the need to hyphenate, i.e., "correlation-coefficient" (Yule, 1921, p. 498).

As the decades passed, the Pearsonian Correlation Coefficient came to be known simply as the correlation coefficient; no other nomenclature for r is in general use today. However, one mid-20th-century hyphen-loving historian of mathematics called it the "correlation co-efficient" (Kline, 1953, p. 356).

Tables, Plots, and Charts

Since at least the late 19th century, the term "Correlation Table" has been used in reference to paired data set arrangements that facilitated understanding of trends or causes of inter-relations. In 1911, it was claimed that Francis Galton *invented* the correlation table in 1886 (Gray, 1911); however, it is more accurately stated that Galton may have been the first to use such a table (minus the name) to study generational data (e.g., heights of sons vs. heights of parents) and to study biometric data (e.g., length of arms vs. legs in the same individuals) – those uses will be discussed here in upcoming chapters. Interestingly, C. F. Pigdin later equated the single word "correlation" with the term "correlation tables" (Pidgin, 1889, p. 478).

Data in a correlation table might be text (e.g., names of towns) paired with text (e.g., primary exports) or text paired with numbers (e.g., population). Mathematical analysis could be performed on such data sets if they were all numerical (e.g., age vs. height vs. weight of children). When there were only two sets of numerical data, a particularly useful arrangement is what will here be called a *correlation matrix table*. Table 1.1 is an example of such a table.

TABLE 1.1

Correlation Matrix Table of Height vs. Arm Length

Height (in Inches) of the Adult Human Male	Length (in Inches) of the Lower Left Arm in the Adult Human Male (i.e., Left Cubit)						Total Number of Cases Observed
	16.5–17.0	17.0–17.5	17.5–18.0	18.0–18.5	18.5–19.0	19.0–19.5	
	Interval Midpoints						
Interval Midpoints	16.75	17.25	17.75	18.25	18.75	19.25	
71 and above	0	0	1	3	4	15	23
70	0	0	1	5	13	11	30
69	1	1	2	25	15	6	50
68	1	3	7	14	7	4	36
67	1	7	15	28	8	2	61
66	1	7	18	15	6	0	47
65	4	10	12	8	2	0	36
64	5	11	2	3	0	0	21
Total number of cases observed	13	39	58	101	55	38	304

Source: Derived from Table II in Galton (1888c, p. 138).

In such a table, arbitrarily assigned intervals of one data set (e.g., 16.5–17.0 and 17.0–17.5) are listed across the top of the columns (i.e., the column headers), and arbitrarily assigned intervals or midpoints of the other set are listed in the left-most cells of the rows (i.e., the row headers). In the body of the table, the intersection of a row and column is called a *cell*, in which there is some representation of a count, such as a numeral or its equivalent number of tick marks or dot marks. Such a table has been referred to by many names, such as:

- *Correlation table* (Yule, 1895, p. 609; Davenport, 1899, p. 32).
- *Frequency table* (Elderton, 1906, p. 107).
- *Scatter diagram* (Gavett, 1925, p. 213).
- *Two-way frequency table* (Burgess, 1927, p. 200).
- *Correlation chart* (Walker, 1929, p. 101; she used the word "graph" to indicate a line or dot plot that was created based upon a correlation chart).

It can be confusing to compare textbooks, even ones published in the same year. For example, Gavett (1925, p. 213) used the term "scatter diagram" for a correlation matrix table, whereas Crum and Patton (1925, p. 223) and Day (1925, p. 182) reserved that term for an *X,Y* graph (i.e., a plot) and used other terms for a table.

It is interesting that a 1939 dictionary of statistics gave two *opposing* definitions for "correlation table":

> *A two-dimensional table…. Also called… scatter diagram. Sometimes used in*
> *contradistinction to scatter diagram.*
>
> *(Kurtz and Edgerton, 1939, p. 40)*

The word "contradistinction" in that quotation is defined in regular dictionaries as a difference between things that can be clarified by comparison (Dictionary: Contradistinction).

The term "correlation table" has also been used in reference to a bi-variate table wherein each cell in the matrix contains the value of the correlation coefficient derived from the data sets indicated in the corresponding row and column headers (e.g., Garnett, 1919, p. 100). Such a table has also been called a "correlation matrix" (e.g., Chiang, 2003, p. 338).

When graphically or formulaically determining a correlation coefficient based on data in a correlation matrix table, Francis Galton initiated the practice of using the terms "Subject" and "Relative" in reference to *how* the data were analyzed (e.g., Galton, 1888c, p. 140, and Davenport, 1899, p. 30). Karl Pearson recommended using the word "type" for Subject and "array" for Relative (Pearson, 1896a, p. 260), although he later suggested calling the Relative the "Co-relative" (Pearson, 1930a, p. 51).

In effect, the mathematics involved in calculating Subject and Relative values derived from correlation matrix tables was a time-saving short-cut, as shown by an example from Galton's December 1888 paper (Galton, 1888c): instead of performing transformative calculations on 281 X,Y points and then plotting the resulting values, he calculated and plotted a total of only 13 transformed Subject and Relative values. The mathematics of calculating Subject and Relative values can seem confusing, but the following example (using data from Table 1.1) helps to clarify the math:

- To calculate Stature=Subject and Cubit=Relative for the last row:

  Subject $= 64$, Relative $= 17.3 = [5(16.75) + 11(17.25) + 2(17.75) + 3(18.25)] / (5 + 11 + 2 + 3)$

- To calculate Cubit=Subject and Stature=Relative for the last column:

  Subject $= 19.25$, Relative $= 69.9 = [15(71) + 11(70) + 6(69) + 4(68) + 2(67)] / (15 + 11 + 6 + 4 + 2)$

Notice that interval midpoints were used in those calculations (e.g., 19.25 is midway between the interval boundaries of 19.0 and 19.5 that are given in the header of the last column). Typically, such a midpoint value was used in place of an actual median mean when "deviations from the mean" were plotted or used in formulas; other authors preferred to use the arithmetic mean when calculating deviations. Notice also that "71" was used in place of the table's first-row interval-midpoint non-numeric value of "71 and above"; that method was used by some authors (e.g., Edgeworth, 1892, p. 191). Other authors did not make such substitutions; instead, they would simply not use any row or column that had non-numeric header values. For example, Galton's paper that contained Table 1.1's data also contained a line-plot figure based upon his full version of that table (Galton, 1888c, p. 138; he called his table "Table II", which is reproduced here as Table 4.1). For inclusion in that figure, he did not calculate and plot values derived from his table's two rows and two columns with non-numeric headers; instead, he referred to them as "flanking" and said that "None of the entries lying within the flanking lines [rows] and columns of Table II were used" – more about that quote, in a later chapter.

In the late 19th and early 20th centuries, creating tables from statistical data was considered a science, as evidenced by the title of a 1915 paper in the *Quarterly Publications of the American Statistical Society*: "Theory of Statistical Tabulation" (Watkins, 1915). As explained in that paper, not all tables "show correlation" (p. 745). Watkins explained that to show correlation the table's data must have a "correlational arrangement" in which "The correlational use… supposes the captions [on columns and rows] as well as the stub-items [are] arranged according to the degree of some quality, and thus it involves cross-classification" (p. 746). In other words, the headers ["captions"] of the rows and columns were numerical values arranged in numerical order; the body of the table contained the counts of individuals that exhibited the corresponding header values. An example of such a correlational table is one listing age vs. height for children in New York City; such a table might be subdivided by boroughs (i.e., by "stub-items"); an example of a non-correlational table is one listing New York City boroughs horizontally vs. residents' most common country-of-origin vertically.

Prior to the invention and widespread adoption of the correlation coefficient, correlation plots of two or more correlated variables were commonly graphed versus another variable (e.g., time, in which case the plot would be called a "time series" or "time index series"); the amount of correlation would therefore be indicated by how well the plot lines of the variables appeared to track each other. Figure 1.1 is a generic example of such a "correlation chart", which shows a strong correlation between the two variables plotted as separate wavy lines vs. a third variable whose values are arranged in order of magnitude on the horizontal axis.

When the data from a single plotted line from such a time series is plotted against itself, with the goal of identifying the periodicity of the "harmonic" pattern of the "various types of oscillatory series", the result has been called a "correlogram" (e.g., Kendall, 1946, pp. 404, 406). The relevant method involved calculating a correlation coefficient between original values and values farther down the time line; how far down the line was called "k" in Kendall's book. The k-value is the "lag" between one number and the value

FIGURE 1.1
Generic correlation chart that shows two variables that correlate well with each other vs. a third variable.

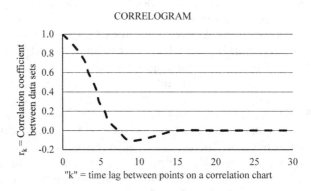

FIGURE 1.2
This correlogram shows strong autocorrelation at low values of k (the "lag" between one number and the value in its own series to which it is to be compared) but virtually no autocorrelation at high k values. (Adapted from Kendall (1946), p. 405.)

to which it is to be compared; if the original values are correlated with values that are k-values further down the line, then the data are said to be autocorrelated (i.e., correlated with their self). The correlation coefficient for a lag of 0 (i.e., $k=0$) is of course 1.00. On the X-axis, the correlogram listed other lag values (i.e., $k=1$, to $k=$ the maximum lag value to be examined); on the Y-axis, the correlogram listed the corresponding correlation coefficients (r_k) from −1 to +1. The correlogram in Figure 1.2 exhibits autocorrelation when the lag is small (1–5) but exhibits little if any autocorrelation at higher lag values (i.e., at higher values of k, the correlation coefficients have values close to zero).

Even noted historians tend to interpret the past through the lens of later years. For example, one historian found a "correlation table" in Francis Galton's unpublished research from 1875 and saw a "correlation chart" in one of his published papers from 1885 (Walker, 1929, pp. 103–104).

Another historian saw a "correlation table" in one of Galton's 1877 publications (Pearson, 1930a, pp. 7). Such descriptions are technically correct, but the historians were not as careful as they could have been to stress that Galton himself did not call them nor even recognize them as examples of mathematical correlation; instead, he recognized and called them examples of reversion (a.k.a. regression), which only years later he discovered to be mathematically linked to his new concept of correlation.

Formulas

There are literally dozens of formulas that have been developed for calculating the linear correlation coefficient, starting it seems with Weldon (1892, p. 3); they have been cataloged by other authors (e.g., Symonds, 1926; Walker, 1929, p. 92ff; Rogers and Nicewander, 1988). To fully discuss their history would be a monumental task; however, a brief discussion here is needed.

Typically, the author of an introductory statistics textbook chooses to discuss only one of those dozens of formulas; the choice seems to be based upon the aspect of correlation theory that the author considers most important. For example, a common approach is historical, that is, to first introduce the regression coefficient and then the correlation coefficient; the author explains that correlation can be viewed as being based upon and historically derived from regression. Textbooks that take that view typically highlight this next formula (where $b=$ the regression coefficient, and Sx and Sy refer to the sample standard deviations of the X and Y values, respectively):

$$r = b\left(Sx/Sy\right)$$

Many other textbooks introduce correlation before regression, explaining that regression can be viewed as being based upon correlation. Textbooks that take that view typically highlight this re-arrangement of the just given formula:

$$b = r\left(Sy/Sx\right)$$

A few authors have wanted to highlight the fact that the correlation coefficient cannot validly be viewed as a proportion; that is, they stress that the correlation coefficient is not really a "measure" of anything (more about this, later; also see discussion in Zorich, 2018). They focus instead on the fact that the square of the correlation coefficient (which is called the *coefficient of determination* and which is typically symbolized by R^2) *can* be viewed as a valid proportion (more about this, later). Some of those authors (e.g., Bohrnstedt and Knoke, 1988, pp. 266–272) *first* discuss how to calculate R^2 and *then* provide this next formula to introduce the correlation coefficient:

$$r = \sqrt{R^2}$$

It is surprising that Galton never discovered a formula for the correlation coefficient; however, he could have provided one on December 20, 1888, if he had only followed his own advice. A decade earlier, he'd published an essay titled "The Geometric Mean in Vital and Social Statistics", in which he recommended that "It is, therefore, an object not only of theoretical but of practical use, to thoroughly investigate a Law of Error, based on the geometric mean…" (Galton, 1879, p. 367). If he had calculated the geometric mean of his regression coefficients, he would have found that it equaled his new coefficient. That is,

$$r = \sqrt{(b_1 b_2)}$$

where (in effect) b_1 is the slope of a linear regression plot of Y on X, and b_2 is the slope of X on Y. On page 136 of Galton's December 1888 paper, he gave $b_1 = 2.5$ and $b_2 = 0.26$ and then graphically determined that the co-relation coefficient equaled 0.8. If he'd followed the advice in his 1879 paper, he would have discovered that $\sqrt{(2.5 \times 0.26)}$ also equaled 0.8. It is unlikely that he'd forgotten about his essay on geometric means, as evidenced by the following:

- Earlier in 1888, he'd written a book titled *Natural Inheritance*, in which he applied the geometric mean to human-stature data (on pages 95, 118, 119, and 195).
- A complete copy of his 1879 geometric mean essay was included as "Appendix E" in that book.
- He'd reviewed, corrected, and sent to his publisher the final printer's proofs of that book just "a few months before [December 1888]" (Galton, 1890a, p. 419).

Was this a case of laziness, disinterest, haste, or distraction? I have found no records that explain this curious historical factoid.

Textbook authors who are inclined more mathematically than historically tend to use Karl Pearson's 1896 formula to introduce correlation and its coefficient. In its most detailed version, Pearson's formula has so many characters and mathematical operators that it can, on a single line, stretch across an entire page in a textbook; thankfully there are many simpler versions, including this one provided by Pearson in his original 1896 paper:

$$r = S(xy)/(n\sigma_1\sigma_2)$$

where S=sum (i.e., Σ in modern terminology), x and y are the deviations from the means of the X and Y values, respectively, n is the number of X,Y data pairs, and σ_1 and σ_2 are the standard deviations (using n in the denominators, not $n-1$) of the X and Y values respectively.

By the end of the 20th century, the linear correlation coefficient for virtually any size paired-data set could be calculated within seconds using electronic computers. However, at the beginning of that century, there were no statistical software programs, no electronic spreadsheets, no electronic computers, no electronic calculators, and mechanical adding machines were just starting to become available. In 1912, statistician W. I. King advised that an "adding machine" was essential when analyzing a large data set; the ones he recommended were available "from $325 up" (King, 1912, p. 233), which is about $10,287 in the year 2023 dollars (calculated using https://www.officialdata.org/us/inflation). At that price, it is no wonder that many published papers provided simplified formulas, methods, and/or forms for calculating or approximating correlation coefficients (e.g., Whipple, 1907; Holzinger, 1923), including one interestingly titled "A Use for Trigonometric Tables in Correlation" (Working, 1921).

Classes of Correlation

This book introduces new terms for various uses of correlation that can be found in pre- and post-1888 literature. The new terms are Observational Correlation, Relational Correlation, and Co-Relational Correlation, each of which is defined here next. They are not meant to replace other descriptors of correlation but rather to organize correlation into what I call *classes* in order to simplify historical discussion.

Observational Correlation

This class of correlation involves pairs of observations or terms that are either naturally or obviously associated. As discussed here in the next chapter, a numeric value based upon counts can be used as a measure of the closeness of some of those associations; but Observational Correlations are otherwise *non-numeric*. Examples include:

- People with brown eyes typically have dark-colored hair.
- Colorful flowers are typically insect-pollinated.
- The words singular and plural.

Relational Correlation

This class of correlation is based on paired sets of *numeric* data, also known as an *X,Y* data set; such data may be individual counts or measurements or may be values taken from a correlation matrix table. To be considered a Relational Correlation, the interest or focus of an individual analysis or plot is the randomly distributed *Y*-data only; the *X*-data are either fixed (e.g., by dividing the observed range of school-children ages into evenly spaced intervals each represented by a median, midpoint, or arithmetic average) or are otherwise known in advance (e.g., 0, 2, 4, or 6 kg of fertilizer per vegetable plot).

W. S. Jevons called the *X*-axis values the "variable", and the *Y*-axis values the "variant" (Jevons, 1874a, vol. 2, pp. 51, 118). More commonly today the *X*-axis is called the "independent variable" and the *Y*-axis the "dependent variable". Example data sets include studies of the age of school children (*X*) vs. their height (*Y*); amount of fertilizer applied (*X*) vs. subsequent crop yield (*Y*); rope diameter (*X*) vs. rope strength (*Y*). In each of those Relational Correlations, the magnitude of *X* can be viewed as possibly "causing" the magnitude of *Y*, but not vice-versa; the outcome of the study and therefore the main focus of the analysis are the *Y* values, not the *X* ones.

Relational correlations can either be...

- Reported in a descriptive manner rather than mathematical.
- Shown in two columns of a table.
- Shown graphically as a plot of *Y* vs. *X*.

In some publications, the units used on one or both of the axes of such plots are semi-quantitative, e.g., small, medium, and large. However, if the units on both axes are numeric, then Relational Correlation can be quantitated by the slope of a hand-drawn or least-squares *linear regression* line. The numeric value of that slope is today called the *regression coefficient*. A formula can be used to calculate that slope, without the need to plot anything. When analyzing this class of correlation, virtually all researchers consider the regression coefficient to be an important statistic.

If the units of measure of both the *X* and *Y* data are specially transformed ("*transmuted*" per Francis Galton, as we shall see in a future chapter), and if those transformed data are plotted rather than the original data, then the slope of the resulting linear regression line is today called the *correlation coefficient*. In virtually all the literature that I've researched, that special transformation is achieved in one of two ways:

- By dividing each individual or group-average *X* value by S_x (the standard deviation or probable error of the individual *X* values), and dividing each individual or group-average *Y* value by S_y (the standard deviation or probable error of the individual *Y* values).

- By subtracting each individual or group-average (or median) X value from the average (or median) of all the X values and then dividing each of those deviations by Sx; and by subtracting each individual or group-average (or median) Y value from the average (or median) of all the Y values, and then dividing each of those deviations by Sy.

A formula can be used to calculate such a slope, without the need to plot anything. When analyzing this class of correlation, not all researchers consider the correlation coefficient to be an important statistic; their dissenting view is that the correlation coefficient provides no additional information beyond that provided by other common regression analysis statistics (more about this later).

Co-Relational Correlation

Co-Relational Correlation is identical to Relational Correlation in every way except that the interest or focus is on *both* the X and Y data. Example data sets include arm length vs. leg length in humans and number of petals vs. number of stamens in flowers. When plotting Co-Relational Correlation data, it cannot be objectively determined which variable should be plotted as X and which as Y (i.e., the concept of dependent and independent variables does not apply). Therefore, the importance of the regression coefficient is greatly diminished because a *different* value is obtained for it depending on which variable is plotted as X and which is Y. On the other hand, the importance of the correlation coefficient is greatly enhanced because the *same* value is obtained for it, no matter which way the data are plotted (more about this later).

As mentioned in the discussion of correlation matrix tables earlier in this chapter, when calculating X,Y plotting points, either variable can be calculated as the "Subject" and either can be calculated as the "Relative". Therefore when plotted, either can be the Subject and the other the Relative, or *both* lines can be plotted (e.g., X as Subject vs. Y as Relative, and Y as Subject vs. X as Relative). Later in this book, there will be much more discussion about such 2-line plots.

As we shall see in subsequent chapters, Francis Galton in December 1888 started to use the word "relation" to describe Co-Relational Correlation prior to the data being transmuted and to use the words "co-relation" or "correlation" after the data were transmuted.

Oxford English Dictionary

In the late 19th and early 20th centuries, the *magnum opus* of dictionaries that provided histories of English word usage was titled *A New English Dictionary on Historical Principles* (referred to here as *NED*); it was edited by James A. H.

Murray and was published one volume at a time, from 1884 to 1928. The volume containing "co-relation" and "correlation" was published in 1893, and the one containing "relation" was published in 1914. In 1933, the dictionary was re-published as a complete set and re-titled the *Oxford English Dictionary*; the texts of the definitions in that complete set were the same as in *NED* and therefore will *not* be referred to here as *OED*. The second edition of the *Oxford English Dictionary* (referred to here as *OED*) was edited by J. A. Simpson and W. S. C. Weiner and published in 1989; the "Introduction" to the second edition includes a brief history of NED and OED and states that the second edition "contains the whole text, unaltered in all essential details, of the twelve-volume first edition [*and*]... the complete text of the four-volume *Supplement to the Oxford English Dictionary*, published between 1972 and 1986; this superseded the previous *Supplement*, which was issued in 1933 as a companion to the main work." As of 2021, that second edition was still the most current *print* edition.

Relation in NED

In *NED*, the word "relation" was given two-thirds of a page (Murray, 1914, p. 398), which is four times more than was given to the word "correlation" (Murray, 1893, p. 1016). Surprisingly, no mention was given to the fact that prior to 1888, Galton, Darwin, Lyell, Spencer, Jevons, and other authors preferentially used the word relation when describing what I have here called Relational or Co-Relational Correlation. *NED* did provide one definition of relation that at first glance seems to be an Observational Correlation:

> *That feature or attribute of things which is involved in considering them in comparison or contrast with each other; the particular way in which one thing is thought of in connexion with another; any connexion, correspondence, or association, which can be conceived as naturally existing between things.*

However, among the several example historical references given immediately after that definition, the following Co-Relational Correlation can be found:

> *The nucleus increases on its part, always preserving the same relation with the entire crystal.*

Co-Relation in NED

In *NED*, co-relation was defined using only five words:

> *Joint or mutual relation; Correlation.*

(Murray, 1893, p. 991)

The entire section in *NED* that was devoted to co-relation, co-relative, etc., comprised less than 100 words, even counting the words in the historical-use quotations. The quotations given there concerned biology, international politics, grammar, and art. The quotation concerning biology was from a mid-19th-century encyclopedia on anatomy and physiology; *NED* gave that quote as:

> *A necessary co-relation between the result of the function, and the aliment.*

NED did not mention any mathematical meaning for co-relation, let alone the statistical meaning that Galton gave it in his 1888 paper; that oversight may be due to the fact that…

- Within a couple weeks of December 1888, Galton himself changed the spelling of co-relation to correlation (more about this, later); and
- It was rare for any post-December-1888 scientist to use the word co-relation as Galton had done in his December 1888 paper (more about this, later).

Correlate in NED

In *NED*, the definitions of correlate were not mathematical (Murray, 1893, p. 1016). For example:

- *Each of two things so related that the one necessarily implies or is complementary to the other.*
- *More generally: Each of two related things; either of the terms of a relation, viewed in reference to the other.*
- *To place in or bring into correlation; to establish or indicate the proper relation between (spec. geological formations, etc.).*
- *To have correlation, to be intimately or regularly connected or related (with, rarely to); spec. in Biol. of structures or characteristics in animals and plants.*

Correlation in NED and OED

In *NED*, correlation was defined differently in different fields of study; those fields included biology, geometry, physics, philosophy, geology, paleontology, and religion (Murray, 1893, p. 1016). Surprisingly, statistics was not one of those fields, an omission that Karl Pearson mentioned in his biography of Galton:

> *…in 1893 when the volume containing the letter C of the Oxford English Dictionary was issued, the Galtonian or biometric sense of "correlation" was not given.*

> (Pearson, 1930a, p. 50n; notice that Pearson misnamed the dictionary — in 1893, it was still called *A New English Dictionary on Historical Principles*)

In Section 1 of *NED* (in what *OED* would call Section "1*a*"), the meaning of correlation was given as:

> *The condition of being correlated; mutual relation of two or more things (imply-ing intimate or necessary connexion).*

All of the literature references given for that definition pre-date 1888.

In *NED's* Section "b" (i.e., in effect, 1b), example quotations were given from the 17th century for this meaning:

> *Relationship (of persons). Obs.*

where "Obs." means obsolete. This meaning was highlighted by a 20th-century biographer of Galton's who was trying to guess why Galton in 1888 initially chose the word "co-relation" rather than "correlation" (more about this, later).

In the previously mentioned 1933 *Supplement*, correlation was given a math-ematical meaning in newly added Section 1c, which defined correlation as:

> *In statistics, the relation of two or more variable quantities.*

> (Murray, 1933, p. 238)

It is humorous that the earliest historical quotation that the *Supplement* pro-vided for that definition was from an 1899 book by Richmond Mayo-Smith, an American "Professor of Political Economy and Social Science" (Mayo-Smith, 1899, cover page) – in other words, none of Galton's many mathematical uses of the word correlation in his 1888, 1889, and 1890 papers were men-tioned (they will be discussed here, in a future chapter). As a further afront to Galton, the meaning Mayo-Smith gave for the word correlation was: "to compare different phenomena with each other, in order to establish rela-tions of co-existence or of sequence" (p. 10). A relation of "co-existence" is an Observational Correlation, and his relation of "sequence" was in reference to the temporal lag between numerical phenomena, an example of which he provided in his very next sentence: "Thus we can compare the fluctua-tions in price of a commodity with its supply in order to discover the relation of demand and supply to price." Neither of those two meanings came close to Galton's December 1888 correlation-coefficient-related biometric mean-ing; not surprising, Mayo-Smith's book had no mention of the correlation coefficient.

A half-century later, in the 1989 second edition of *OED*, that 1933 definition in Section 1c was modified and expanded to this...

> *In Statistics, an interdependence of two or more variable quantities such that a change in the value of one is associated with a change in the value or the expectation of the others; also, the value of this as represented by a correlation coefficient. So [sic] correlation coefficient or coefficient of correlation: a number between −1 and 1 calculated so as to represent the linear interdependence of two variables or two sets of data; spec. the product-moment coefficient.*

> *(Simpson and Weiner, 1989, vol. III, p. 964)*

The first historical reference that *OED* gave in that expanded section 1c was (finally!) from Galton's December 1888 paper; the reference included only part of a sentence; the full sentence is given here:

> *The statures of kinsmen are co-related variables; thus, the stature of the father is correlated to that of the adult son, and the stature of the adult son to that of the father; the stature of the uncle to that of the adult nephew, and the stature of the adult nephew to that of the uncle, and so on; but the index of co-relation which is what I there called "regression," is different in the different cases.*

> *(Galton, 1888c, p. 143)*

It is humorous that that reference quotation for the word correlation did *not* contain the word correlation, but rather only the words co-related, correlated, and co-relation.

The second historical reference that *OED* gave in section 1c was from an 1896 paper by Karl Pearson; *OED*'s version of what Pearson wrote is:

> *Let r_0 be the coefficient of correlation between parent and offspring. We conclude that there is a sensible correlation (circa 0.18) between fertility and height in the mothers of daughters.*

Surprisingly, *OED* misquoted Pearson, who had included the word "therefore" between "We" and "conclude" (Pearson, 1896b, p. 303).

In Section 3, which deals with biology, the definition was:

> *Mutual relation of association between different structures, characteristics, etc. in an animal or plant; 'the normal coincidence of one phenomenon, character, etc., with another' (Darwin, Orig. Species, Gloss.).*

That describes Observational Correlation.

The remaining *NED* definitions for correlation dealt with physics, geometry, and generic uses; they related to neither Relational nor Co-Relational Correlation.

Chapter Summary

The word "correlation" has had more than a two-century history: The first half was almost exclusively non-mathematical, and the second half was primarily mathematical. The pivotal moment in that history was December 20, 1888, when Francis Galton presented a paper that introduced a coefficient for mathematical correlation. The formal name for that coefficient, as well as its symbol and formula, have long histories of their own. This book focuses on three classes of correlation, which will be referred to here as Observational Correlation, Relational Correlation, and Co-Relational Correlation.

2

Pre-1888 Authors, Other Than Francis Galton

Introduction

This chapter begins in France but spends most of its time in England and the USA; along the way, it corrects or clarifies some important aspects of the historical record. Quotations provided here have been sourced from a very wide range of intellectuals and scientists; over the span of a century, their uses of the words correlation and relation were remarkably similar.

Prior to 1888, the word correlation was not being used in a mathematical sense as we do today. Indirect evidence for such a conclusion is a book published in 1865 by **Isaac Todhunter** (1820–1884), who was at that time an English mathematician and a member of the Royal Society; he eventually authored a number of popular mathematics books (MAA.org-Todhunter). His book's full title was: *A History of the Mathematical Theory of Probability from the Time of Pascal to that of Laplace.* It included quotes from and discussions of relevant mathematicians who wrote between the late 15th and early 19th centuries. In all of its more than 600 pages, the word correlation was not used even once; where today's readers might expect to see the word correlation, we instead see the word relation or the term "linear relation" (e.g., on page 170).

Among the 19th-century authors discussed in this chapter are: Georges Cuvier (paleontologist), Charles Lyell (geologist), August Bravais (polymath), William Grove (physicist), Charles Darwin (naturalist), Thomas Huxley (biologist), Herbert Spencer (philosopher), Richard Owen (zoologist), Charles Bray (sociologist), Joseph Le Conte (botanist), William Jevons (economist), Richard Littledale (reverend), Henry Bowditch (physiologist), and Alexander Bell (inventor). Also discussed is an 1859 encyclopedia on anatomy and physiology that included articles from more than 100 scientists. As we shall see, the non-mathematical meanings given to the words relation, co-relation, and correlation did not vary much from author to author nor decade to decade, except in a few incidental cases that foreshadowed the change that would occur on December 20, 1888, when Francis Galton introduced a completely new set of meanings.

DOI: 10.1201/9781003527893-2

1812

Georges Cuvier (1769–1832), a French paleontologist, is credited with being the first to use fossil evidence to clearly show that some species have gone extinct and that new ones have taken their place; "Another concept introduced by Cuvier was that of correlation of parts" (de Beer, 1965, p. 8). Such correlation was detailed in 1812 in his *Recherches sur les ossemens fossiles de quadrupèdes* (*Research on the Fossil Bones of Quadrupeds*), sections of which (including the discussion of correlation) were then published separately in his *Discours sur les révolutions de la surface du globe* (*Discourse on the Revolutions of the Surface of the Globe*). In those writings, Cuvier's use of the French word corrélation did not convey to readers that he intended any new meaning for it, as evidenced by the fact that not all English translations included the English word correlation. For example, the original 1812 version of *Recherches* and the 1825 French edition of *Discours* read "c'était celui de la corrélation des formes dans les êtres organisés" (pages 58 and 95, respectively from Cuvier 1812 and 1825a; underlining added), an American translation published in 1831 read "it was that of the natural relation of forms in organized beings" (Cuvier, 1831, p. 58. NOTE: The first word in the just-given quote from the 1825 edition is "c'était"; but the word in the original 1812 edition is c'étoit, which I conclude is a printer's error misspelling of c'était.). However, a more faithful translation of the entire sentence in that 1825 edition is...

> *Fortunately, comparative anatomy possessed a principle which, well developed, was able to make all the trouble vanish: it is the principle of the* correlation *of structures in organic beings, by means of which each sort of creature could in a pinch be recognized by each fragment of each of its parts.*
>
> *(Cuvier, 1825b, p. 41; underlining added)*

Cuvier was there describing Observational Correlation. Readers may be tempted to infer that he was discussing the correlated growth of structures in an individual from infancy to adulthood (e.g. as a child matures and its legs grow longer, it is typically observed that its arms also lengthen). But Cuvier was instead talking about comparative anatomy, as his book's very next sentences reveal:

> *The entirety of an organic being forms a coordinated whole, a unique and closed system, in which the parts mutually correspond and work together in the same specific action through a reciprocal relationship. None of these parts can change without the others changing as well. Consequently, each of them, taken separately, points to and reveals all the others. Thus, as I have said elsewhere, if the intestines of an animal are organized in such a way as to digest only meat and meat that is fresh, it is necessary also that its jaws be constructed to devour its prey, its claws to seize and tear it apart, its teeth to cut and chew it, the entire*

system of its organs of motion to rush and catch the prey, its sense organs to perceive it from far away (p. 41).

Cuvier was also thinking in terms of Co-Relational Correlations; for example, later in that same book, he discussed how knowledge of "proportions" based upon the "laws of organic economy" can be used to construct an "equation" that can help to "reconstruct the complete animal" from a single bone:

> *The bones of the shoulder will have to have a certain firmness in the animals which use their front limbs for seizing prey, and from this will result once more particular structures for them. The interplay of all these parts will demand certain <u>proportions</u> in all the muscles, and the patterns of the muscles thus <u>proportioned</u> will again determine more particularly the structures of the bones.... In a word, the structure of the tooth entails the structure of the condyle, of the shoulder blade, of the nails, in just the same way as the <u>equation</u> of a curve controls all its characteristics. Moreover, by taking each separate characteristic as the basis of a particular <u>equation</u>, we can find both the ordinary <u>equation</u> and all the other properties whatsoever, even the claws, shoulder blade, condyle, femur, and all the other bones each taken separately, reciprocally indicating or being indicated by the tooth. <u>Starting with each of them, the person who possesses rationally the laws of the organic economy could reconstruct the complete animal</u>* (p. 42; underlining added).

Even though he discussed such co-relational correlations, he did not use the word correlation when doing so. When he used the word correlation, he did so not in regard to such numerical relationships but rather *only* in regard to the presence or absence of structures (e.g. claws). Therefore, Cuvier used the word correlation only in the sense of Observational Correlation.

1832

Charles Lyell (1797–1875) was a famous English geologist best known for popularizing the initially controversial theory of uniformitarianism, which stated that geologic processes of past times are essentially the same as those of present day (Britannica.com-Uniformitarianism). He also wrote on biological topics; for example, the entire second volume of the (1832) first edition of his *Principles of Geology* focused on what today would be called paleontology and evolution. Charles Darwin – who read that volume while voyaging on the H. M. S. Beagle (Darwin, 1876, p. 77n) – dedicated to Lyell the second edition of his *Journal of Researches* regarding that voyage "as an acknowledgment that the chief part of whatever scientific merit this journal and other works of the author may possess, has been derived from studying the well-known and admirable *Principles of Geology*" (Darwin, 1845, p. 7; Darwin's *Journal* was subsequently renamed *The Voyage of the Beagle*).

Lyell's *Principles of Geology* was published in three volumes, in 1830, 1832, and 1833, respectively, at just the right time to be both a scientific and commercial success:

> *From the 1830's onward, and increasingly in the forties and fifties, geology was the most popular of the sciences* [in Britain]... *It was observed that the geology sections of the British Association were always by far the best attended.... At a meeting in Newcastle in 1838 over a thousand people sat through the regular meetings presided over by Lyell...*
>
> (Himmelfarb, 1968, p. 233)

Lyell was given much of the credit for that popularity: "There is no branch of natural history that has made of late years such an advance in general estimation as geology.... The works of Mr. Lyell, especially, have largely contributed to this end..." (Anonymous, 1839, p. 102).

Given that Lyell's main geologic thesis was what we today might call the correlation of causes and effects, past and present, we would expect that the word "correlation" would appear frequently in *Principles of Geology*. However, that is not the case. For example, in the entire three volumes of Lyell's first edition of that work, there is no use of the word correlation or any of its related parts of speech. On the other hand, in his 12th and final edition (1875) he did use those words a few times, but only in the sense of Observational Correlation, and all but one use being on the pages he devoted to explaining Charles Darwin's *Origin of Species* concept of "correlation of growth" (which will be discussed later in this chapter).

Lyell did describe a Relational Correlation in his writings, but he chose not to use the word correlation; for example:

> ...*the average size of the blocks* [of stone moved by glaciers]...*lessens sensibly in proportion as we recede from the principle points of departure.*
>
> (Lyell, 1873, p. 408; underlining added).

In other words, the farther a stone has been pushed by a glacier, the smaller it becomes. Another example was:

> M. Perrey.... *thinks he has detected a* relation *between the frequency of earthquakes and our winter and summer solstices, the greatest number of shocks occurring in perihelion when the sun is nearest, and the least number in aphelion on when it is farthest from the earth.*
>
> (Lyell, 1875, vol. 2, p. 233; underlining added)

Lyell's infrequent use of the word correlation in his writings contrasts greatly with the writings of geologists later in the 19th century. The word correlation had by then become popular short-hand to mean "of the same age", in the sense that a rock formation in France is said to be correlated to one in England if it can be shown that they had both been formed during the same geologic time period. The word correlation in that sense appeared in the title of many geologic books and papers; for example: *Correlation Papers: Cambrian* (Walcott, 1891), and "Principles and Methods of Geologic Correlation by Means of Fossil Plants" (Ward, 1891).

1846

Auguste Bravais (1811–1863), a French physicist best known for his work in crystallography (Britannica.com-Bravais), is widely credited with being the first person to use the word correlation in a mathematical sense. That first use was in his paper titled "Analyse mathématique sur les probabilités des erreurs de situation d'un point" ("Mathematical analysis on the probability of errors of the location of a point"). In it, we find this sentence:

> *La coexistence des mêmes variables m, n, p.. dans les équations simultanées en x et y, amène une corrélation telle, que les modules h_x, h_y, cessent de représenter la possibilité des valeurs simultanées de (x,y) sous le vrai point de vue de la question.*
>
> (*Bravais, 1846, p. 263*)

which translates into the following sentence (which is difficult to understand without the rest of Bravais's paper, which is not provided here):

> *The coexistence of the same variables m, n, p.. in the simultaneous equations in x and y results in a correlation such that the functions h_x, h_y, no longer represent the possibility of simultaneous values of (x,y) from the perspective of the true point of view of the question.*

It is important to note that in the 78 pages of text in that paper, Bravais used the noun correlation only once, and the verb or other forms not at all. His use of correlation in that sentence appears to have been incidental rather than intentional – that is, he did not intend to assign a new meaning to the word. An indication that such a conclusion was shared by his contemporaries can be found in the text of a memoir of his life's work, presented in his honor at the French Academy of Science, a couple years after his death (de Beaumont, 1865): In none of its 24 pages was there any mention of the word or concept of correlation.

In the decade following Galton's December 1888 paper, some authors gave credit to Bravais for the discovery of mathematical correlation (e.g. Pearson, 1896a, p. 261; Yule, 1897a, p. 481). However, in subsequent decades, the general consensus became the reverse, at least among historians of statistical method in the *English*-speaking world (including Yule and Pearson, who retracted their original statements); for example:

- *Bravais... discussed the theory of error for points in space, regarding the errors as either independent or correlated, from the standpoint of the normal law of errors. He did not, however, use a single symbol for a correlation coefficient.... Sir Francis Galton was the first to devise the practical statistical method, and it is to him that we owe the conception of a numerical measure of the intensity of correlation...* (Yule, 1909, p. 722).
- *Bravais has no claim, whatever, to supplant Francis Galton as the discoverer of the correlation calculus* (Pearson, 1920, p. 28).
- *The Bravais treatment leads nowhere so far as correlation theory is concerned.... Galton alone seems deserving of being called the father of correlation* (Kelley, 1923, pp. 152–153).
- *Bravais recognized the existence of a relationship, a "correlation," between his principal variables, but gave it merely passing notice.... Bravais had no single term equivalent to our coefficient of correlation* (Walker, 1929, p. 97).
- *Laplace (1811), Plana (1813), Gauss (1823) and <u>Bravais</u> (1846) all derived normal correlations as the joint distribution of linear forms in independently distributed normal variables but <u>did not define a coefficient of correlation</u>* (H. O. Lancaster, in Kendall and Plackett, 1977, p. 293; underlining added).
- *His [Bravais'] passing reference to the 'corrélation' of x and y was not followed by any attempt to study or measure this 'corrélation'.... No notion of regression is to be found in it, nor that of correlation beyond the basic sense of non-independence* (MacKenzie, 1981, pp. 70, 233).
- *Bravais's (1846) investigations of spatial laws of error were advances within the Laplace-Gauss tradition but contained no hint of an idea of correlation or regression* (Stigler, 1986, p. 353n).

William Robert Grove (1811–1896) was a physicist who was well known and well respected for his 1846 ground-breaking book titled *On the Correlation of Physical Forces*, which went through six editions spanning four decades. In the British journal *Nature*, a reviewer of the last edition lauded the first edition as having been "one of the documents which serve for the construction of the history of science... It has certainly exercised a very considerable effect in moulding the mass of what is called scientific opinion..." (Maxwell, 1874, p. 303). That reviewer summarized his interpretation of the book:

> *The design of the book is to show that of the various forms of energy existing in nature, any one may be transformed into any other, the one form appearing as the other disappears. This is what is meant in the essay [i.e., Grove's book] by the "correlation of the physical forces," and the whole essay is an exposition of this fact, each of the physical forces in turn being taken as the starting-point, and employed as the source of all the others."*

> *(Maxwell, 1874, p. 303)*

Grove's book was well known to other writers who themselves were major influencers of scientific opinion:

- Karl Marx and Friedrich Engels reportedly were both profoundly influenced by the philosophical view provided by Grove in his *Correlation* book (Morus, 2017, p. 63).

- Herbert Spencer prominently mentioned Grove's book in his *First Principles*, in his chapter on "The Correlation and Equivalence of Forces" (1862, p. 263).

- W. S. Jevons repeatedly mentioned Grove, usually in reference to Grove's *Correlation* book (Jevons, 1874a, vol. 1, p. 397; vol. 2, pp. 143, 257, 267, 268).

- Francis Galton spoke of Grove's "masterly book on the 'Correlation of Physical Forces'" (Galton, 1908b, p. 219).

Grove provided a lengthy explanation of what he meant by "Correlation"; the wording was substantially the same in all editions. The following quotation is taken from the sixth and final edition (the edition owned by Francis Galton):

> The term Correlation, which I selected as the title of my Lectures in 1843, strictly interpreted, means a necessary mutual or reciprocal dependence of two ideas, inseparable even in mental conception: thus, the idea of height cannot exist without involving the idea of its correlate, depth; the idea of parent cannot exist without involving the idea of offspring. The word itself had not been previously used [by writers on physics]; and although, as I have said, I object to the introduction of new terms without strong reason, there are a vast variety of physical relations which cannot certainly be so well expressed by any other term. The extent to which it has been since used has, I think, justified me. Its use has, in my judgment, been carried too far in applying it to subjects quite beyond its fair meaning. There are many facts, one of which cannot take place without involving the other; one arm of a lever cannot be depressed without the other being elevated – the finger cannot press the table without the table pressing the finger. A body cannot be heated without another being cooled, or some other force being exhausted in an equivalent ratio to the production of heat; a body cannot be positively electrified without some other body being negatively electrified, &c. To such cases the term correlation may be usefully applied, but hardly to adaptations of structure, &c....
>
> The sense I have attached to the word Correlation, in treating of physical phenomena, will, I think, be evident, from the previous parts of this Essay, to be that of a necessary reciprocal production; in other words, that any force capable of producing another may, in its turn, be produced by it – nay, more, can be itself resisted by the force it produces, in proportion to the energy of such production, as action is ever accompanied and resisted by reaction: thus, the action of an electro-magnetic machine is reacted upon by the magneto-electricity developed by its action.

(Grove, 1874a, pp. 165–167)

Later in the book, he provided a nuance to his definition, wherein he equated *correlation of forces* with *conservation of energy*:

> *It would be out of place here, and treating of matters too familiar to the bulk of my audience, to trace how, by the labours of Oersted, Seebeck, Faraday, Talbot, Daguerre, and others, materials have been provided for the generalisation now known as the correlation of forces or conservation of energy...*
>
> (Grove, 1874a, p. 196)

According to at least one historian, Grove's use of the word correlation in relation to the mutual convertibility of energy became commonplace:

> *The language of correlation* [as used by Grove] *remained pervasive throughout the second half of the* [19th] *century, to the dismay of some proponents of the new theory of the conservation of energy...*
>
> (Morus, 2017, p. 63)

Grove's sixth edition contained more than 30 instances of the word correlation and related parts of speech (exclusive of titles found at the top of pages); in contrast, the word co-relation did not appear anywhere, in any of the editions, except once in the sixth edition. That appearance was in a sentence where "correlated" in the earlier editions was replaced by "co-related". That change was not the result of an end-of-line hyphenation; instead, the word appeared in the middle of a printed line. The one long sentence in which "co-related" occurred was this:

> *His experiments show that when a current of positive electricity traverses a portion of the muscle of a living animal in the same direction as that in which the nerves ramify – i.e. a direction from the brain to the extremities – a muscular contraction is produced in the limb experimented on, showing that the nerve of motion is affected; while, if the current, as it is termed, be made to traverse the muscle in the reverse direction, or towards the nervous centres, the animal utters cries, and exhibits all the indications of suffering pain, scarcely any muscular movement being produced, showing that in this case the nerves of sensation are affected by the electric current; some definite polar condition therefore exists, or is induced in the nerves, to which electricity is <u>co-related</u>, and probably this polar condition constitutes or conveys nervous agency.*
>
> (Grove, 1874a, p. 156; underlining added)

That same sentence appeared in all editions two through five, but with the word correlated instead of co-related. Compared to earlier editions, later editions incorporated other changes, such as minor changes in spelling (e.g. the use of modern English's "show" in place of old English's "shew"), in punctuation (e.g. a semi-colon replacing a comma, or vice-versa), and in wording (e.g. "some definite polar condition therefore exists" replacing "therefore that some definite polar condition exists" (Grove, 1850, 1855, 1862, 1867, 1874a, pp. 89–90, 190–191, 233, 233, and 156, respectively)). However, in the preface to the fifth edition (which he also included in the sixth edition), he explicitly rejected the idea of globally substituting "co-relation" for "correlation" (Grove, 1867, p. v, and 1874a, p. vii). He gave no explanation nor highlighting (e.g. italics) in the sixth edition for the surprising use of the hyphenated word; therefore, that one use of co-related in the sixth edition seems to have been unintentional.

Like Darwin and other contemporaries, when Grove discussed comparisons of quantitative values, he typically used the word "relation", as illustrated by the following examples from the sixth edition:

- *When, however, we examine substances of very different physical characters, we find that their specific heats have no relation to their density or rate of expansion by heat...* (p. 40)
- *The most trustworthy general relation which has been ascertained is, that the magnetic attraction is as the square of the electric force...* (p. 94)
- *Faraday proved that it bore a direct equivalent relation: that is... the amount of oxygen which united with the zinc in each cell of the battery was exactly equal to the amount evolved at the one platinum terminal...* (pp. 142–143)

As we shall see, Grove may have played a key part in Galton's choice of words in December of 1888.

1859

As previously mentioned summarily, *NED*'s source for its "co-relation" quotation relating to biology was *The Cyclopaedia of Anatomy and Physiology*, which in its more than 5000 pages and several volumes included papers from more than 100 authors. It also contained many instances of the word "correlation", which were always in the sense of an Observational Correlation regarding structure or function. Similarly, two of its three instances of the word co-relation involved Observational Correlations (shown here next):

- *The materials of the blood being supplied by the digestive apparatus, we might judge, all things else being equal, of the perfection of the blood by the perfection of this apparatus. But there is likewise a necessary co-relation between the result of the function, and the aliment; for instance, when the apparatus shall be found nearly alike in any two cases, the difference of food necessarily influencing the qualities of the blood, the comparison must be established, every other circumstance being equal, according to the higher or lower nutritive qualities of the food* (Todd, 1859, vol. II p. 652, underlining added).

- *With regard to the arguments adduced by Mr. Simon from Comparative Anatomy, to the effect that a thymus has some essential connection with pulmonary organs of respiration, I would remark that though it is certainly of weight, yet it cannot be regarded as proving absolutely that the two organs are <u>co-related</u> in function* (vol. IV, p. 1100, underlining added).

The third instance of co-relation occurred in a paper on "Animal Heat"; its author discussed changes caused by the act of respiration, namely alteration of the blood, alteration of the air, and production of heat. He first described the "co-relation" between the amount of blood aerated and the extent to which the air has changed, and then described the "relation" between the quantity of altered air and the amount of heat produced. Thus, apparently, he thought the words co-relation and relation were interchangeable; on the other hand, the word correlation does not appear anywhere in his 36-page paper:

> *Since it is necessary that the venous blood should pass through the lungs in order to become arterial from contact with the air of the atmosphere, it is obvious that it cannot undergo any change in its constitution without the air at the same time suffering the change. That the air is altered by the respiratory act is well known to all, and as there is a necessary <u>co-relation</u> between the blood aerated during respiration and the air which it alters, the amount of alteration undergone by the one may be estimated from the change suffered by the other. The quantity of air altered by respiration, all other things being equal, ought to be found in <u>relation</u> with the production of heat* (vol. II, p. 652; underlining added).

Because the blood can be viewed as changing the air (i.e. by absorption or desorption of gases), this can be viewed as a Relational Correlation. But what can be said of altered air vs. production of heat? In this example, neither is clearly causing the other; as such, that is a Co-Relational Correlation.

In regard to the word "relation", the encyclopedia's many authors typically used it in descriptions of a physical or functional association between two or more organs. It was also used to refer to Co-Relational Correlations; for example (all from Todd, 1859; all underlining added):

- *In some animals, the size of the middle sacral artery is scarcely inferior to that of the aorta itself, as in the cetacea and fishes. In all animals furnished with tails, the size of this artery bears a constant <u>relation</u> to the size of that member* (vol. I, p. 197).
- *The intensity of the electrical power seems to bear no <u>relation</u> to the size of the fish, at least after it has attained mature age; small fish are almost always actively electrical* (vol. II, p. 83).
- *The length of the rima glottidis bears no <u>relation</u> to the stature of the individual* (vol. III, p. 112).

Charles Darwin (1809–1882) is possibly the most famous natural scientist of all time, primarily for his having convinced the 19th-century scientific community that natural mechanisms of organic evolution led to both the origin and extinction of species.

Darwin was fond of the word "correlation", using it many times in his 1859 first edition of *Origin of Species* (the full title of which was *On the Origin of Species by Means of Natural Selection, or the Preservation of Favoured Races in the Struggle for Life*); but it wasn't until the fifth and sixth editions that he included a formal definition of the word as he meant it:

> *Correlation.– The normal coincidence of one phenomenon, character, &c., with another.*
>
> (*Darwin, 1872, 6th ed., glossary, p.411*)

As previously mentioned, *NED*'s definition of "correlation" as related to biology was taken in part from Darwin's. Additionally, *NED* provides two incomplete sentences from *Origin*, complete versions of which are shown here:

> *In the next chapter I shall discuss the complex and little known laws of variation and of correlation of growth…. Some instances of correlation are quite whimsical: thus cats with blue eyes are invariably deaf; colour and constitutional peculiarities go together, of which many remarkable cases could be given amongst animals and plants.*
>
> (*Darwin, 1859, pp. 5, 11–12*)

The "whimsical" instance just mentioned is an example of Observational Correlation. The "next chapter" Darwin referred to is chapter five (titled "Laws of Variation"), in which there is a sub-section titled "Correlation of Growth", which is defined there as:

> *I mean by this expression that the whole organisation is so tied together during its growth and development, that when slight variations in any one part occur, and are accumulated through natural selection, other parts become modified. This is a very important subject, most imperfectly understood. The most obvious case is, that modifications accumulated solely for the good of the young or larva, will, it may safely be concluded, affect the structure of the adult; in the same manner as any malconformation affecting the early embryo, seriously affects the whole organisation of the adult. The several parts of the body which are homologous, and which, at an early embryonic period, are alike, seem liable to vary in an allied manner: we see this in the right and left sides of the body varying in the same manner; in the front and hind legs, and even in the jaws and limbs, varying together, for the lower jaw is believed to be homologous with the limbs. These tendencies, I do not doubt, may be mastered more or less completely by natural selection: thus a family of stags once existed with an antler only on one side; and if this had been of any great use to the breed it might probably have been rendered permanent by natural selection (p. 143).*

In Darwin's definition, and in both quotations just provided, Darwin was describing Observational Correlation.

It is interesting that in *Origin*'s fifth edition (1869), Darwin changed some of the first-edition quotes just given. For example: "laws of variation and of correlation of growth" became simply "laws of variation", and the sub-section titled "Correlation of Growth" was re-titled "Correlated Variation" (Darwin, 1869, pp. 20, 146).

In his 1868 book titled *The Variation of Animals and Plants under Domestication*, Darwin almost always used the word correlation in regard to Observational Correlation. However, a few times he used it in regard to Co-Relational Correlation in a seemingly unconscious manner without the slightest indication that he meant to assign a new, mathematical meaning to the word; examples include:

- *These modifications* [increased number of toes; increased height of foot bones]... *in the feet of dogs.... are interesting from being correlated with the size of the body, for they occur much more frequently with mastiffs and other large breeds than with small dogs* (vol. 1, p. 35).
- *... in carriers, runts, and barbs* [3 different breeds of pigeons] *the singular reflexion of the upper margin of the middle part of the lower jaw... is not strictly correlated with the width or divergence... of the pre-maxillary bones, but with the breadth of the horny and soft parts of the upper mandible* (vol. 1, p. 169).
- *... in all the breeds of the pigeon the length of the beak and the size of the feet are correlated* (vol. 2, p. 323).
- *When an organ, such as the beak, increases or decreases in length, adjoining or correlated parts... tend to vary in the same manner* (vol. 2, pp. 353–354; the fourth word in this quote in the printed original is "sueh", which is obviously a printer's error for the word "such").

In his 1871 book *The Descent of Man*, there is one use of the phrase "close correlation", which a reader might misinterpret as a conceptual pre-cursor to Francis Galton's measure of "the closeness of the co-relation". However, Darwin was comparing colors, not numeric data; and therefore it was an Observational Correlation. Darwin's entire sentence was:

> *In the first place, it may be observed that the colors* [sic] *of caterpillars do not stand in any close correlation with those of the mature insect.*

> *(1871, vol. 1, p. 402)*

In contrast to the just-mentioned *few* instances of using the word correlation in a mathematical sense, Darwin *many* times used the word "relation" to express an approximation to our present day mathematical meaning of correlation for example:

- *The eggs of differently sized breeds naturally differ much in size; but, apparently, not always in strict relation to the size of the hen* (Darwin, 1868, vol. 1, p. 248; this is a Relational Correlation).
- *With the breeds of sheep the number of hairs within a given space and the number of excretory pores stand in some relation to each other* (Darwin, 1871, vol. 1, p. 239; this is a Co-Relational Correlation).

However, Darwin was not consistent: He sometimes used relation where he meant correlation (as he defined it in his glossary); examples (all from *Variation*, 1868) include:

- *When parts stand in such close relation to each other as the fleshy covering of the fruit... and the seed, when one part is modified, so generally is the other...* (vol. 2, p. 218).
- *In some cases a relation apparently exists between certain characters and certain conditions, so that if the latter be changed the character is lost...* (vol. 2, p. 290).
- *All the parts of the organisation are to a certain extent connected or correlated together; but the connexion may be so slight that it hardly exists.... Even in the higher animals various parts are not at all closely related; for one part may be wholly suppressed or rendered monstrous without any other part of the body being affected* (vol. 2, p. 319; it is interesting that in this single paragraph, he used both *correlated* and *related* with exactly the same meaning).

Thomas Henry Huxley (1825–1895) was an English biologist and anthropologist who is famous for his 1860 debate with Bishop Wilberforce in defense of the theories, principles, and conclusions in Darwin's *Origin of Species*, which had been published the previous year (Britannica.com-Huxley). At that time, Huxley was considered by some to be "Britain's leading Zoologist" (Gallant, 1972, p. 151).

At the turn of the century, Huxley's *Scientific Memoirs* were published in four volumes totaling more than 2500 pages; they contained an almost exhaustive collection of his papers that had been previously published in scientific journals. The papers in those four volumes were arranged in chronological order, starting with one from 1845. In all four volumes, whenever he used the word correlation (which he did many times) it was in reference to Observational Correlation. On the other hand, whenever he discussed Relational or Co-Relational Correlation, he used the word relation; there are no such uses of relation in volume one, one example each in volumes two and three, and several examples in volume four. The following list contains examples of such uses (the date that precedes each quote is the year in which the relevant paper was originally published):

- 1869: *He [Kant] accounts for the relation of the masses and the densities of the planets to their distances from the sun...* (Huxley, 1901, vol. 3, p. 416).

- 1880:....*the system of measurement hitherto usually adopted gives the absolute sizes of the teeth and their dimensions relatively to one another, but affords no clue to their proportions in relation to the size of the skull, or to the increase or diminution of individual teeth* (Huxley, 1902, vol. 4, p. 405).
- 1880: *After a mammalian embryo, for example, has taken on its general mammalian characters, its further progress towards its specific form is effected by the excessive growth of one part in relation to another...* (Huxley, 1902, vol. 4, p. 458).

In his book titled *Evidence as to Man's Place in Nature*, we find another significant example of his using the word relation where we today might use the word correlation:

> *The external surface of the skull varies considerably in size, as do also the zygomatic aperture and the temporal muscle; but they bear no necessary relation to each other, a small muscle often existing with a large cranial surface, and vice versâ.*

> *(Huxley, 1863, p. 41)*

1862

Herbert Spencer (1820–1903) was one of the most famous English philosophers of the Victorian era (Britannica.com-Spencer); his fame was due to his books and articles that applied his own view of evolution to all branches of science, including cosmology, biology, sociology, and psychology. He was so well known that, in J. G. Hibben's 1896 book on *Inductive Logic*, Spencer was referred to only as "Mr. Spencer"; no additional information about him or his publications was provided, and yet Hibben included lengthy passages that paraphrased Spencer's philosophical and scientific views (Hibben, 1896, pp. 257, 284–287).

Throughout Spencer's writings, we find the word correlation and associated parts of speech. For example, in his 1862 book titled *First Principles*, he used *correlative* many times, almost always in the sense of a pair of what he considered opposing forces or complementary views, i.e. as Observational Correlations:

- *It is a doctrine called into question by none, that antimonies of thought as Whole and Part, Equal and Unequal, Singular and Plural, are necessarily conceived as correlatives: the conception of a part is impossible without the conception of a whole; there can be no idea of equality without one of inequality* (p. 89).
- *Religion and Science are therefore necessary correlatives* (p. 107).

In that book, he used *correlation* many times in reference to Relational Correlations. For example:

- *...physical forces stand not simply in qualitative correlations with each other, but also in <u>quantitative correlations</u>. Besides proving that one mode of force may be transformed into another mode, experiments illustrate the truth that from a definite amount of one, definite amounts of others always arise* (p. 264; underlining added).
- *Either mental energies, as well as bodily ones, are <u>quantitatively correlated</u> to certain energies expended in their production, and to certain other energies which they initiate; or else nothing becomes something and something becomes nothing* (p. 284; underlining added; note that the period at the end of the sentence was missing in the original).

It is important to note that, in the same paragraph in which he provided the just-given page-264 quotation, he also used the word relation and connexion for the meaning for which he'd just used the word correlation:

The investigations of Dulong, Petit and Neumann, have proved a <u>relation</u> in amount between the affinities of combining bodies and the heat evolved during their combination. Between chemical action and voltaic electricity, a quantitative <u>connexion</u> has also been established.... The well-determined <u>relations</u> between the quantities of heat generated and water turned into steam, or still better the known expansion produced in steam by each additional degree of heat, may be cited in further evidence. Whence it is no longer doubted that among the several forms which force assumes, the quantitative <u>relations</u> are fixed (p. 264; underlining added).

All of the foregoing quotations are from the first edition of *First Principles*. In the sixth edition (1904a), he provided a detailed 10-page Index; however, it did not include the word "correlation" as a separate topic, but rather only as a descriptor of a topic (e.g. "Food-supply, correlation of vital and physical forces"). Although the text of the sixth edition frequently made use of the word correlation, he removed it from the title of one chapter: The first-edition chapter on "The Correlation and Equivalence of Forces" became the sixth-edition chapter on "The Transformation and Equivalence of Forces".

In his *Principles of Biology*, he typically used *relation* rather than *correlation* when linking numeric or ordinal data-sets. All of the following quotes from that book discussed Relational Correlations (all underlining added):

- *We see that among classes of organisms, and among the parts of each organism, there is a <u>relation</u> between the amount of nitrogenous matter present and the amount of independent activity* (1865, p. 39).
- *Among the Hydrozoa it is common for any portion of the body to reproduce the rest.... Some of the inferior Vertebrata also, as lizards, can develop new limbs or new tails in place of those that have been cut off.... The highest animals, however, thus repair themselves to but*

> *a very small extent. Mammals and birds do it only in the healing of*
> *wounds.... the power of reproducing lost parts is greatest where the*
> *organization is lowest; and almost disappears where the organization*
> *is highest. And though we cannot say that between these extremes*
> *there is a constant <u>inverse relation</u> between reparative power and*
> *degree of organization... we may say that there is some approach to*
> *such a relation (1865, p. 175).*
> - *...the only exception to the <u>relation</u> between decreasing bulk and*
>   *increasing number of eggs, occurs in the cases of the Pheasant and the*
>   *Black-cock... (1867, p. 434).*

In the detailed 9-page Index to his autobiography (1904b), the word "correlation" is not listed. In its almost 1150 pages of text, he used *correlative* and *correlation* only a few times and only for Observational Correlations.

1866

Richard Owen (1804–1892) is credited for being the first scientist to recognize that some of the newly-discovered huge lizard-like fossils were truly different from today's reptiles; in 1842, he classified them into a taxonomic group that he called *Dinosauria* (Britannica.com-Owen). His reputation in paleontology and comparative anatomy earned him his appointment as "Superintendent of the Natural History Departments of the British Museum" (Owen, 1866, title page). He was also an unwavering proponent of Cuvier's theory of correlation (previously discussed) and a very public opponent of Darwin's theory of evolution (Singer, 1990, p. 394). In his college professorial lectures on comparative anatomy, he stressed the "laws of correlation" (Owens, 1843, pp. 9, 333, 336).

His many publications included a famously massive 1866 tome titled *On the Anatomy of Vertebrates*, in which he used the word correlation several times, always in the sense of an Observational Correlation; examples include:

> - *As vertebrates rise in the scale and the adaptive principle predomi-*
>   *nates, the law of <u>correlation</u>, as enunciated by Cuvier, becomes more*
>   *operative.... As we descend in the scale of life from the grade illustra-*
>   *tive of 'Cuvier's Law,' the method of empirical observation becomes*
>   *more and more essential, the tact with which it is applied being, how-*
>   *ever, in the ratio of the discernment of the <u>correlations</u> of structures*
>   *(pp. xxvii, xxx; underlining added).*
> - *Present reptiles form a mere fragmentary remnant of the great and*
>   *varied class of cold-blooded air-breathing vertebrates which prevailed*
>   *in the mesozoic age. More than half of the ordinal groups of the class,*
>   *indicated by osteal and dental characters, have perished; and it is only*
>   *by petrified faeces [sic] or casts of the intestinal canal, by casts of the*

> brain-case, or by <u>correlative</u> deductions from characters of the petrifi-
> able remains, that we are enabled to gain any glimpse of the anatomical
> conditions of the soft parts of such extinct species... (p. xxvii; under-
> lining added).

- ...both the food and the teeth of these Sauria [a type of lizard] indicate a certain amount of mastication, with which the sense of taste is
  <u>correlated</u> (p. 327; underlining added).

He also used the word co-related, but curiously only once in *On the Anatomy of Vertebrates*, in reference to an Observational Correlation:

> *The longus colli* [muscle] *at the fore or upper part of the spinal column... the
> greater extent and developement* [sic] *of which in Ophidians* [snakes] *is indi-
> cated by the number and length of hypapophyses* [spinal projections]... *the <u>co-
> related</u> muscle, having its foremost insertion into the occipital hypapophysis...
> brings down the head in the blow inflicted by the venom-fangs with proportion-
> ate force* (p. 225; underlining added).

Imagine a physicist entering a London bookstore in 1867; he's there to buy a copy of the just-published fifth edition of W. R. Grove's *The Correlation of Physical Forces*. His eye is drawn to another book, whose title is atypically shown not only on the book's spine but also on its front cover. That book catches his eye because the title is *On Force*; with such a title, a physicist is certainly going to take at least a quick look at it. He opens it and is disappointed to discover that the full title is: *On Force, its Mental and Moral Correlates*. He then laughs to himself a bit upon realizing that the whole title (counting the subtitle) is the first he's ever encountered that included a comma, a semicolon, a colon, and a period. The whole title (which included the period at the end) was:

> *On Force, its Mental and Moral Correlates; and on that which is Supposed to
> Underlie all Phenomena: with Speculations on Spiritualism and other Abnormal
> Conditions of Mind.*

The book's author was **Charles Bray** (1811–1884), a prosperous manufacturer of ribbons; his wealth afforded him the opportunity to become a promoter of radical social and political ideas (Britannica.com-Bray).

He published a number of books and large essays throughout his life. In 1866, he published *On Force*, in which he used the word correlation more than 40 times. Although he referenced and quoted from William Grove's book a few times, Bray formulated his own meaning for that word. The following are interesting and representative examples of how the word correlation was used in *On Force* (all page references are to Bray, 1866):

- *The Correlation of Forces shows that in the cycle of forces we can always
  return to the same starting point without a break, and the Persistence
  of Force shows that this is always done without loss; now these truths,*

not stopping short in Physics, but carried, as they ought to be, into the higher field of Mind, furnish, I think, the most probable explanation of "the Phenomena of Modern Spiritualism," at the present time so much puzzling earnest investigators (p. iv; the phrase "Persistence of Force" in that quote was misprinted as "Persistence of Forco" in the original).

- *There is but one thing known to us in the universe; this, Physical Philosophers have called "Force."…. Everything around us results from the mode of action or motion, or correlation of this one force, the different Forms of which we call Phenomena…* (the page on which those words were found was not paginated; it is titled "Argument" and placed just after the book's table of contents).
- *Cause and Effect are mere correlation of Force, produced by organization or the manner in which forces are concentrated and arranged* (p. 4).
- *…as Life is thus the mere correlate of Physical forces, so Mind is the correlate of Vital forces* (pp. 12–13).
- *Pain and Pleasure are transformed force — mental correlates;…* (p. 72).
- *The Brain contains a whole reservoir of correlated force called soul or spirit…* (p. 126).

Bray seems to have coined (but not defined) the word "re-correlation" and may have been the only person to have ever used it:

> *We have discovered the law of gravitation, and we now want a Newton in the department of mind. We want now to know the law, not of gravitation, but of Levitation, by which Brahmins, and Saints, and Mr. Home, and tables float. We want to know the exact conditions under which vital force becomes mental or conscious force, and of its re-correlation into unconsciousness in sleep or under pressure on the brain;…*
>
> *(Bray, 1866, p. 141; underlining added)*

1873

In 1873, **Joseph Le Conte** (1823–1901) was a professor of geology, natural history, and botany at UC Berkeley when he authored a paper titled "Correlation of Vital with Chemical and Physical Forces". That 2-page paper summarized views that (he claimed) he'd expressed in more detail in prior papers.

By "vital" force he meant any force or energy caused by the inner workings of a biological organism:

> *Vital force….is derived from the lower forces of Nature; it is related to other forces much as these are related to each other – it is correlated with chemical and physical forces…. The mutual convertibility of forces into each other is called*

> *correlation of forces; the persistence of the same amount, amid all these protean*
> *forms, is called conservation of force.*

> *(Le Conte, 1873, p. 266)*

His paper was published long after the 1846 first edition of Grove's *On the Correlation of Physical Forces* (previously discussed), but Le Conte claimed that he'd published his own correlation ideas prior to his having *read* anything relevant by Grove. Le Conte then stated:

> *Grove, many years ago, brought out, in a vague manner, the idea that vital force*
> *was correlated with chemical and physical forces.... I do not, therefore, now*
> *claim to have first advanced this idea, but I do claim to have in some measure res-*
> *cued it from vagueness, and given it a clearer and more scientific form* (p. 266).

Le Conte's paper provided examples drawn mainly from the plant kingdom, e.g. seed germination, bud-break, and photosynthesis. In his discussion of photosynthesis, he said:

> *Light falling on living green leaves is destroyed or consumed in doing the work*
> *of decomposition; disappears as light, to reappear as nascent chemical energy;*
> *and this in its turn disappears in forming organic matter, to reappear as the vital*
> *force of the organic matter thus formed* (p. 267).

The point he wanted to make is that not only is there no energy lost when chemical forces and inorganic substances are transformed into vital forces and biochemical substances, but also there is no energy lost when one vital force or biochemical substance is changed into another – all such forces and substances are "correlated" in that they exhibit conservation of energy. Le Conte felt the need to promote that idea:

> *The correlation of physical forces with each other and with chemical force is now*
> *universally acknowledged and somewhat clearly conceived. The <u>correlation of</u>*
> *<u>vital force</u> with these is not universally acknowledged, and, where acknowledged,*
> *is only imperfectly conceived* (p. 266; underlining added).

In the same year that the just-discussed Le Conte paper was published, a paper titled "The Conservation and Correlation of Vital Force" appeared in the journal *The American Naturalist*. It was authored by **Joseph Trimble Rothrock** (1839–1922), a professor in botany, human anatomy, and physiology at what is now called Pennsylvania State University (PSU.edu-Rothrock). Based upon that paper's title, a reader might assume that it would promote ideas similar to those of Le Conte. Although Rothrock did address biological forces and structures, his approach was less that of Le Conte and more that of Herbert Spencer (previously discussed).

Rothrock's paper focused primarily on what he at first called the "principle of compensation" but later called a "law" (Rothrock, 1873, pp. 332, 333), which he explained by saying that...

> ...all organic things, plants or animals, have a certain proportionate amount of developing force, actual or predestined, and that this synergy is under the direction of inherited tendencies; which being at times misdirected, one organ or set of organs may take on excessive growth. Should this occur, there will be a corresponding atrophy in some other organ or set of organs (p. 333).

It is surprising that even though his paper's title contained the word correlation, the paper itself included only one use of that word, namely:

> I quote the following at second hand from Meckel.... "A girl had on each extremity a superfluous digit, and one hand of her sister wanted four, being the number of digits which her sister had in excess, reckoning the four extremities together." These are a few out of the immense mass of similar illustrations I might bring forward in support of my belief in an absolute law at the bottom of these correlations of structure... (p. 339).

That is an extreme example of his law; a more typical example is this:

> The typical anther of [a flower]... is possessed of two cells. Sometimes, however; there is but one, which may often be explained by the partition wall being obliterated, and so causing the confluence of these usually separate cells. In Salvia (sage), however, there is but one cell where two might certainly have been expected. One has gone, entirely, or at most a mere knob of cellular tissue may remain to suggest the missing cell. Interposed between the perfect and the imperfect cells is a connective, unduly elongated, which from its very length and association with the separated halves of the anther serves to explain the want of development in the one. In other words the connective is vigorous and lusty at the expense of the impoverished cell (p. 334).

In a mathematical sense, he was describing what we might call a zero-sum game of negative correlation; that is, above average growth in one organ or tissue is correlated (compensated) by below average growth in another. From a Spencerian point of view, the two growths are inherently correlated, like the words up and down, left and right; that view is highlighted in Rothrock's paper's summary points, the first two of which were:

> 1st. That organs anatomically or physiologically related tend to compensate among themselves for any aberration of structure or function.
> 2nd. That an organ over-developed in one direction will be under-developed in some other... (pp. 339–340).

As we shall see in our discussion of A. G. Bell, later in this chapter, data from a study of humans could be viewed as contradicting Rothrock's "1st." point.

1874

William Stanley Jevons (1835–1882) was "one of the... notable figures in econometrics... an example of a many-sided genius.... he may be said to have put statistics into economics once and forever" (Davis and Nelson, 1937, pp. 342, 347). Given that Jevons died in 1882, that claim by Davis and Nelson needs to viewed from a pre-1888 perspective, when the word statistics denoted tables, charts, and index numbers, and not correlation coefficients, regression analysis, etc. In support of that view, it is significant that Jevons's accomplishments were not discussed in Stigler's *The History of Statistics: The Measurement of Uncertainty before 1900* (Stigler, 1986).

Jevons's interests were many, as evidenced in 1874 by his authoring an approximately 900-page book on *The Principles of Science: A Treatise on Logic and Scientific Method*; its 1877 second edition went through many reprints through at least 1920. Francis Galton owned a copy of the first edition (Stigler, 1986, p. 298; Jevons, 1874b). Jevons promoted the terms "variable" and "variant" for what is plotted respectively on the X and Y axes of graphs of measurements.

In both editions of *Principles*, Jevons use the word correlation many times. One section of the chapter on "Classification" is entitled "Correlation of Properties", in which he defined his meaning solely as Observational Correlation:

> *Things are correlated (con, relata) when they are so related or bound to each other that where one is the other is, and where one is not the other is not. Throughout this work we have then been dealing with correlations. In geometry the occurrence of three equal angles in a triangle is correlated with the existence of three equal sides; in physics gravity is correlated with inertia; in botany exogenous growth is correlated with the possession of two cotyledons, or the production of flowers with that of spiral vessels.*
>
> *(Jevons, 1874a, vol. 2, p. 354).*

In the second edition, that first edition text was augmented by this sentence:

> *Wherever a proposition of the form $A = B$ is true [,] there correlation exists.*
>
> *(Jevons, 1883, p. 681).*

Additionally, in that second edition, these un-italicized first edition words were now highlighted with italics:

> *where one is the other is, and where one is not the other is not* (p. 681).

A detailed search of both editions reveals that he always used the word correlation in reference to Observational Correlation. On the other hand, when describing a Relational or Co-Relational Correlation, he used words such as "connexion" (i.e. connection) and "correspondence"; but most frequently he preferred the word "relation". For example (all from Jevons, 1874a, vol. 2; all underlining added):

- *We are said to experiment when we bring substances together under various conditions of temperature, pressure, electric disturbance, molecular attraction, &c., and then record the changes observed.... Perfect and exhaustive experimentation would, in short, consist in examining natural phenomena in all their possible combinations and registering all <u>relations</u> between conditions and results which are found capable of existence* (pp. 22, 24).
- *The whole question of the <u>relation</u> of quantities thus resolves itself into one of probability. When we can only rudely [i.e. crudely] measure a quantitative result, we can assign but slight importance to any <u>correspondence</u>* (p. 107).
- *Even when we have no means of accurately measuring the variable quantities [,] we may yet be convinced of their <u>connexion</u>.... The facility with which we can time after time observe the increase or decrease of one quantity with another sufficiently shows the <u>connexion</u>, although we may be unable to assign any precise law of <u>relation</u>* (p. 110).
- *Let us now proceed to consider the modes in which from numerical results we can establish the actual <u>relation</u> between the quantity of the cause and that of the effect* (p. 113).

1876

The Rev. **Richard Frederick Littledale** (1833–1890) is provided here as an example of the way the word correlation was used on religious topics prior to 1888. Littledale's use was closer to philosopher Herbert Spencer's (previously discussed) than to anyone else's, as shown in the following example:

We are... reasonable beings praying to a reasonable God, and believing in the correlation of moral forces.... [Therefore there] must not only be a correlation of physical forces and a correlation of moral forces, but the physical and moral forces must be also mutually correlated....

(Littledale, 1876, pp. 69, 78–79)

Correlation continued to appear in the title of philosophical and religious works for many years. For example: *The Pauline Theology: A Study of the Origin and Correlation of the Doctrinal Teachings of the Apostle Paul* (Stevens, 1911).

1877

Henry Pickering Bowditch (1840–1911), was an M.D. and professor of human physiology at Harvard's medical school (Harvard.edu-Bowditch) when he authored a more than 120-page document titled "The Growth of Children". It was officially part of the "Eighth Annual Report of the State Board of Health of Massachusetts", but the "State Printer" of Massachusetts also published it as a stand-alone book for informal dissemination (*"With the compliments of the Writer"* was printed on its front cover).

Based primarily upon that report, H. M. Walker described Bowditch as one of only six "Writers before Galton" who were important enough for her to mention in the chapter on correlation in her *Studies in the History of Statistical Method*. She described that chapter as a "somewhat abbreviated and schematic account of the development of the theory of correlation and its applications". Among the five other authors were Laplace, Gauss, and Bravais (Walker, 1929, pp. 92f). However, it seems that other historians such as K. Pearson and S. M. Stigler did not hold Bowditch in such high regard. Pearson did not mention Bowditch in his "Notes on the History of Correlation" (Pearson, 1920), nor in his mammoth biography of Galton, in which there is much discussion of correlation history (Pearson, 1914, 1924, 1930a, 1930b). Stigler did not include Bowditch in his 2-volume *American Contributions to Mathematical Statistics in the Nineteenth Century* (Stigler, 1980), and did not mention Bowditch in his *The History of Statistics: The Measurement of Uncertainty before 1900* (Stigler, 1986).

In Walker's discussion of Bowditch's report, she said that he...

> ...*was not far from the idea of correlation and not far from the idea of regression.... Bowditch had six large charts which we would now call correlation charts* [Walker here used the word charts for correlation matrix tables]. *In three of them the height in inches is plotted against the age in years, and in the remaining three the weight in pounds is plotted against the age in years. No lines are shown on these charts, but there is a separate series of remarkable graphs in which a line is drawn through the points which represent the average height (or weight) for each year of age. These are nothing short of curvilinear regression lines for height on age and weight on age. In four plates, the regression of weight on height is also shown.*

> *(Walker, 1929, p. 101)*

All of Bowditch's tables and graphs paired either age with height, age with weight, or height with weight (Figure 2.1 is an example); as such, they are Relational Correlations (not Co-Relational Correlations) because height and weight are dependent upon age, and weight is dependent upon height, in the sense mentioned in Chapter 1's definition of Relational Correlation.

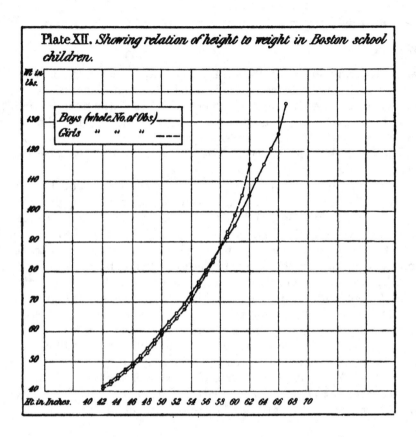

FIGURE 2.1

This is a plot of Y=weight vs. X=height for boy and girls in Boston schools. It shows that boys and girls tend to weigh about the same until they near five feet tall; at greater heights, girls tend to weigh more than boys. (From the 1894 reprint of Bowditch (1877).)

The following quote from his "Summary of Results" helps to clarify that he did *not* intend to determine how well variables were correlated (in a Galtonian sense) but rather his goal was to calculate rates of growth of the *average* child of a given age who lived, studied, or worked under given conditions:

> The growth of children takes place in such a way that until the age of eleven or twelve years, boys are both taller and heavier than girls of the same age.... Children of American-born parents are, in this community, taller and heavier than children of foreign-born parents.... The <u>relation</u> of weight to height in grow-ing children is such that at heights below 58 inches, boys are heavier than girls in proportion to their stature. At heights above 58 inches the reverse is the case.
>
> (Bowditch, 1877, p. 35; underlining added)

None of his data involved generational measurements (e.g. father vs. son), and therefore none of his analyses involve Regression (what Galton originally called Reversion) in either of the two Galtonian senses (i.e. either a "tendency of that ideal mean type to depart from the parent type, 'reverting' towards what may be roughly and perhaps fairly described as the average ancestral type" (Galton, 1877a, p. 291), or a tendency for the offspring of individuals to regress toward "mediocrity", i.e. the present-day population average (Galton, 1885a, p. 2)). The only generational differences on which Bowditch reported involved children; for example, he concluded that their height and weight were positively effected by improved living conditions. He discussed how the living condition of emigrants to the US typically improved, in comparison to the living condition that they'd had in Europe; he then speculated that "whatever tendency residence in America may have to increase the size of growing children, will, in the cases, be intensified by transmission through several generations" (Bowditch, 1877, p. 26). In a related discussion, he addressed an anti-immigrant sentiment regarding the negative affects that certain races were suspected of having on the established races in the USA; he dismissed that sentiment by saying "It will thus be seen that the theory of the gradual physical degeneration of the Anglo-Saxon race in America derives no support from this investigation" (Bowditch, 1877, p. 32).

Walker seems to have misinterpreted Bowditch's frustration, when she said that…

> *The discussion on page 93* [of Bowditch's report] *indicates that Bowditch felt an <u>urgent need</u> of some method of <u>measuring the strength of the relationship</u> between his variables* [,] *and that he was dissatisfied with the method he adopted, which he said was defective.*
>
> (Walker, 1929, p. 101; underlining added; Walker's "page 93" reference is to an 1894 reprint of the original 1877 paper; her reference to that page in that reprint corresponds to page 29 in the 1877 version quoted here below)

As shown in the next quotation, what Walker called an "urgent need" had nothing to do with Bowditch's "method of measuring" but rather had to do with the fact that the data were arranged such that a calculation "with absolute accuracy" of that measure would have required a laborious "retabulation" of tens of thousands of observations. He therefore decided on a less costly approach that involved "interpolation". The full measure of what Bowditch actually said in his 1877 paper was this:

[from the main body of text]

> *The data collected in this investigation afford the means for ascertaining the <u>relation</u> of height to weight in growing children of both sexes and of various races. <u>This relation is for each age most simply expressed by the quotient of the weight in pounds divided by the height in inches.</u> Series of quotients thus obtained are given in Table No. 3, in the columns headed "pounds per inch." Since, however, these quotients increase with the increasing height, it is manifestly impossible to use them for ascertaining the relative stoutness of children who at a given age*

differ from each other in stature. To do this with absolute accuracy, it would be necessary to determine for each age, and in each set of observations, the average weight corresponding to each height. Since, however, the direct determination of this value would necessitate a complete retabulation of all the observations, it has been thought best to adopt an indirect and somewhat less accurate method of getting at the result. This method consists in arranging the heights and weights corresponding to each age, opposite to each other in parallel columns, and then determining by interpolation the weights corresponding to each even inch of height (p. 29, underlining added).

[this following text is from the footnote referenced by a mark at the end of that main body text]

This method [i.e. the one described at the end of that text] *is defective, first, because it does not take into account the possible influence of age upon the ratio of a given height to its corresponding weight; and secondly, because it rests upon the assumption that the average weight for a given age is the same as the average weight of all individuals, without regard to age, whose height is equal to the average height for that age. This assumption clearly involves a trifling error, for, since the weights of growing children increase approximately as the 2.7 powers of the heights, it is evident that at any given age the weight of those children who are above the average height will tend to raise the average weight for that age more than the weights of the children below the average height will tend to lower it, supposing the observations to be uniformly distributed on both sides of the average according to the binomial curve of Quetelet; consequently the average weight for a given age will be somewhat greater than the average weight of all the individuals, regardless of age, whose height is equal to the average height for that age. Notwithstanding these defects, the method has been adopted, first, because it is believed that the errors involved are so small as to be of no practical importance; and secondly, because relative rather than absolute values were sought, and a comparison between several sets of observations is not prevented by a small constant error running through them all* (pp. 29–30n; underlining added).

Walker's "urgent need" conclusion, together with her claim that Bowditch "was not far from the idea of correlation and not far from the idea of regression" gives the impression that he would have made extensive use of Galton's index of regression and index of co-relation, if he'd known about them. Bowditch did have a long history of communication with Galton (per Walker, 1929, p.101; and at https://Galton.org there are references to the many extant letters that had been exchanged between Galton and Bowditch from the 1870s to the 1890s). However, in Bowditch's 1891 revision of his 1877 report, in which the original data were re-analyzed, he did *not* make use of either Galton's 1877 index of reversion nor his 1888 index of co-relation; and neither the 1877 nor 1891 reports contained even a single instance of the words co-relation, correlation, reversion or regression. That might seem surprising, given the fact that the 1891 report was titled "The Growth of Children, Studied by Galton's Method of Percentile Grades", and on its first page it referenced Galton's book on *Natural Inheritance*, which had been published two years prior.

To end this discussion of Bowditch, I propose that the foregoing evidence supports these next two statements, which contradict those of Walker:

- Bowditch in 1877 was nowhere near the ideas of correlation and regression.

- Bowditch was *not* one of the six most important writers before Galton in the history of the development of the theory of correlation and its applications.

1885

In 1885, three years before Francis Galton invented his own numerical measure of the degree of correlation, he was (in a crude sense) beaten to it by **Alexander Graham Bell** (1847–1922), who is best known as the inventor of the telephone. I have not determined if Bell was the first to invent this particular measure of correlation, but only that his is the first explicit example that I've found of it.

Bell's paper comprised a bit more than two pages and was titled "Is There a Correlation between Defects of the Senses?"; he started it by saying:

> *People sometimes assume that a defect of any important sense is balanced to the individual by the increased perception of the remaining senses. For instance: it is often thought that deaf persons have better eyesight than those who hear, and that blind persons have better hearing than those who see. The returns of the tenth census of the United States (1880) concerning the defective classes show clearly the fallacy of such a belief. They indicate that the deaf are much more liable to blindness than the hearing, and the blind more liable to deafness than the seeing.*

> *(Bell, 1885, p. 127)*

Some of Bell's evidence is given here in Figures 2.2 and 2.3, which could be viewed as contradicting Rothrock's view of compensating organic correlation (previously discussed).

Bell found evidence of other correlations:

> *The* [data in the census summary] *tables seem to indicate that in the case of deafness, blindness, idiocy, and insanity, some correlation exists* [idiocy and insanity being two different degrees of mental retardation]; *for persons having one of those defects appear more liable to the others than persons normally constituted, and doubly defective persons appear to be more liable to be otherwise*

TABLE I.

Analysis of the defective classes as returned in the tenth census of the United States (1880).

Singly defective.		
Deaf and dumb [1]	30,995	
Blind	46,721	
Idiotic	73,370	
Insane	91,133	
Total singly defective	. . . .	242,219
Doubly defective.		
Blind deaf-mutes	246	
Idiotic deaf-mutes	2,122	
Insane deaf-mutes	268	
Blind idiots	1,186	
Insane blind	528	
Total doubly defective	. . . .	4,350
Trebly defective.		
Blind idiotic deaf-mutes	217	
Blind insane deaf-mutes	30	
Total trebly defective	. . . .	247
Total defective population . . .	. . . .	246,816

[1] The 'deaf and dumb' have no other natural defect save that of deafness. They are simply persons who are deaf from childhood, and many of them are only 'hard of hearing.' They have no defect of the vocal organs to prevent them from speaking. A child who cannot hear our language with sufficient distinctness to imitate it remains dumb until specially instructed in the use of his vocal organs. In the above table, the 'deaf and dumb' are therefore classified with those having a single defect.

FIGURE 2.2
The 1880 USA Census data in this table shows that human defective traits tend to be acquired correlatively (i.e. having one defective trait increases the probability of having others). (From Bell (1885, p. 127).)

TABLE II.

Percentage of the population of the United States who are defective.

	Totals.	Percentage.
Deaf and dumb	33,878	0.0675
Blind	48,928	0.0975
Idiotic	76,895	0.1533
Insane	91,959	0.1833
Defective population	246,816	0.4921
Population not defective	49,908,967	99.5079
Total population	50,155,783	100.0000

FIGURE 2.3
The 1880 USA Census data in this table shows the percentage of the population that exhibited the listed defective classes. (From Bell (1885, p. 127).)

defective than persons having a single defect…. Although the proportion of the insane who are deaf or blind is abnormally large, the evidences of a correlation between insanity and the other defects noted above are not well marked; in regard to deafness, blindness, and idiocy, a marked <u>correlation</u> appears to exist.

1. *Deaf-mutes. — There are fourteen and a half times as many blind persons among the deaf and dumb in proportion to the population as there are in the community at large, and forty-six times as many idiotic.*
2. *Blind. — There are fourteen times as many deaf-mutes among the blind in proportion to the population as there are in the community at large, and nineteen times as many idiots.*
3. *Idiotic. — There are forty-three times as many deaf-mutes among the idiotic in proportion to the population as there are in the community at large, and eighteen times as many blind.*

The apparent correlation between deafness, blindness, and idiocy, may possibly indicate that in a certain proportion of cases these defects arise from a <u>common cause</u>, perhaps arrested development of the nervous system.

(Bell, 1885, p. 128; underlining added)

Bell and Galton both speculated that apparent correlation between characteristics might be due to a common cause. Both men provided a numerical measure of the strength of correlation; Galton's number was a ratio derived from measurements, whereas Bell's was a ratio derived from counts. Galton's number was a measure of Co-Relational Correlation; Bell's was a measure of Observational Correlation, which in future decades would be called association or contingency (e.g. Kurtz and Edgerton, 1939, pp. 27–28). Much more about Galton's measure is provided here in later chapters.

Chapter Summary

Prior to 1888, the word correlation was used by scientists in many fields, but almost exclusively in reference to Observational Correlation. Bravais did use it one time in one sentence in one paper, in reference to a mathematical value that is conceptually linked to Galton's correlation coefficient. Darwin used it a few times instead of the word relation in reference to Co-Relational Correlations, but it seems he did so without intending to add any mathematical meaning to the word. Spencer spoke of energies that are "quantitatively correlated" with each other. Bell used correlation in reference to ratios of counts. Otherwise, the scientific community consistently used the word relation (rather than correlation) in reference to Relational or Co-Relational Correlations. The hyphenated word co-relation was rarely used; when used, it was almost always in reference to Observational Correlation.

3

Francis Galton, before December 1888

Introduction

Francis Galton (1822–1911) was the first cousin to Charles Darwin, with whom he visited and corresponded throughout his life; as of 2021, an entire section of www.Galton.org was dedicated to showcasing copies of the extant correspondence between them. As detailed in any of several biographies (e.g., Pearson, 1914, 1924, 1930a, 1930b; Forrest, 1974; Keynes, 1993; Johnson and Kotz, 1997; Bulmer, 2003), Galton gained fame as a young man for his travels, explorations, and researches; by 1860, he'd been awarded the Royal Geographical Society's Founder's Gold Medal and had been elected to the Royal Society. In subsequent decades, he developed important methods and made important discoveries in various fields, among which were meteorology, heredity, mathematical statistics, finger prints, and social science. For some years, he was president of the Geography and Anthropology sections of the British Association for the Advancement of Science. He was knighted in 1909. Upon his death in 1911, the bulk of his still-large fortune was willed to the "University of London" (Forrest, 1974, p. 288); that may have been what led at least one of his admirers to posthumously incorrectly refer to him as *Professor* Francis Galton (King, 1912, p. 219). As of 2020, the "Complete Works" section of www.Galton.org held about 600 documents, almost all of which are his technical papers and books.

Some of Galton's writing is plagued by sentences that are difficult to understand, even on second or third reading. Galton was aware of the need for clarity in technical writing, given that he published an entire paper on that topic (Galton, 1908a). However, his writing style is reminiscent of the early 19th-century English novelist Jane Austen, in the sense that the lengthy conversations in her stories are notoriously difficult to follow – who is saying what to whom? But she was not concerned about that; she confided in a letter to her sister in 1813 that "...a 'said he' or a 'said she' would sometimes make the Dialogue more immediately clear – but 'I do not write for such dull Elves'" (La Faye, 2011, p. 210). Similarly, some of Galton's sentences could benefit from a few more words.

Galton was aware of the shortcomings of his writing style, as he expressed in 1887:

DOI: 10.1201/9781003527893-3

It is a serious drawback to me in writing, and still more in explaining myself, that I do not so easily think in words as otherwise. It often happens that after being hard at work, and having arrived at results that are perfectly clear and satisfactory to myself, when I try to express them in language [,] I feel that I must begin by putting myself upon quite another intellectual plane. I have to translate my thoughts into a language that does not run very evenly with them. I therefore waste a vast deal of time in seeking for appropriate words and phrases; and am conscious, when required to speak on a sudden, of being often very obscure through mere verbal maladroitness, and not through want of clearness of perception. This is one of the small annoyances of my life. I may add that often while engaged in thinking out something [,] I catch an accompaniment of nonsense words, just as the notes of a song might accompany thought. Also, that after I have made a mental step, the appropriate word frequently follows as an echo; as a rule, it does not accompany it.

(Galton, 1887b, p. 29)

In scientific writing, a writer's emotion does not typically display itself, and so it was usually with Galton. However, he made an exception on page 58 of his 1889 book *Natural Inheritance*. There, he called the "Probable Error" a "cumbrous, slip-shod, and misleading phrase" that is "illegitimate", subject to "misinterpretation", and "absurd when applied to...Stature...". He stated that he preferred instead the terms "Mid-Error" or "Probable Deviation" but tended not to use them because the term Probable Error was "too firmly established for me to uproot it". He proclaimed that "I shall however always write the word Probable... in the form of 'Prob.'; thus 'Prob. Error,' as a continual protest...". During his final review and revision of the printer's proofs of *Natural Inheritance* (a review that was mentioned in Galton, 1890a, p. 419), his protest of Probable Error may have been a factor in his not correcting the many type-setting errors regarding it – in more than half of the instances of that term, there is a colon where the period should be, thus "Prob: Error" rather than "Prob. Error" (e.g. p. 68). Galton's first biographer, Karl Pearson, attributed the mistakes that he found in *Natural Inheritance* to "the haste with which it was prepared" (Pearson, 1930a, p. 57).

Another example of emotion occurred late in 1889 when Galton wrote a paper that described the history of his discovery of the correlation coefficient; in it, he expressed a type of professional fear, which caused him to work overly hastily. The exact quote is: "Fearing that this idea [of mathematical correlation]... would strike many others as soon as 'Natural Inheritance' was published [planned for early 1889],... I made all haste to prepare a paper...." (Galton, 1890a, p. 421; underlining added), a paper that included many haste-caused errors, as we will see here in the next chapter.

A third example of emotion can be found in his 1892 book titled *Finger Prints:*

It is <u>hateful</u> to blunder in calculations of adverse chances, by overlooking cor-relations between variables, and to falsely assume them to be independent, with the result that inflated estimates are made which require to be proportionately reduced.

(Galton, 1892a, p. 109; underlining added)

When reading biographies of Galton that were written in the 20th century, it is important to realize that authors then faced difficulties that authors now do not. Such difficulties include the fact that 20th-century research was a manual task; each of the literally hundreds of references listed in the back of such books had to be actually read or at least skimmed by the authors or their staff. Additionally, there was only one major collection of sources, namely the Galton Library at University College London. In contrast, in the 21st century, the "Complete Works" of Francis Galton are freely available as electronic files to anyone in the world, at www.galton.org.

As of 2020, virtually all of those "Complete Works" were documents authored by Galton (the few exceptions were, for example, reviews of his books). During 2020, I spent a few hours a day for more than a month either reading or electronically searching all those documents for any words con-taining "co-", "relat", "correl", "rever", or "regres". An electronic search was conducted on a file only after it was converted to "Searchable PDF" format using "PDF Converter Professional 8.1 by Nuance Communications"; an elec-tronic search for "relat" (for example) would return all instances of relation, co-relation, and correlation.

Galton's first major biographer, Karl Pearson, stated that Galton's first use of the word "correlation" was in 1874, in *English Men of Science*: "This is the first occasion on which I have noticed Galton using this word…" (Pearson, 1924, p. 150, n1). However, that book did not contain even one use of that word; but it did contain a single use of the word "correlated" (p. 98). Apparently Pearson had also not carefully read Galton's earlier book, *Hereditary Genius* (1869), in which the word correlation does occur (more about this, later). Such mistakes by Pearson may be in keeping with errors in his late-life lec-tures on the history of statistics; in the Preface to the published collection of those lectures, edited by his son Egon, a couple of Karl Pearson's colleagues are reported to have "thought K. P. [Karl Pearson] had not studied all of his sources adequately and that some of what he had written involved errors of fact…" (Pearson, 1978, p. xvi). Another example of Pearson's mistakes is in his 1930 volume of Galton's biography, in which he described Galton as having discovered the coefficient of correlation in 1889, rather than 1888 (Pearson, 1930a, p. 5; however, he stated the date correctly on p. 50). Galton biographer N. W. Gillham speaks more kindly of Pearson: After discovering an error in Pearson's biography of Galton, Gillham stated that "This slip by Pearson is surprising, for his immense biography of Francis Galton is nothing if not meticulous" (Gillham, 2001, p. 258).

Galton's second major biographer, D. W. Forrest, stated that the "first occasion on which Galton uses the term [correlated]" was in *Hereditary Genius*, in the chapter on "Divines", in a paragraph that Forrest summarized as "The two characteristics of strong morality and instability are in no way correlated..." (Forrest, 1974, p. 96). The use to which Forrest referred is either on page 263 and 282 of Galton's book (page 263 points to page 282); however, Forrest had overlooked pages 51, 84, 130, and 227, which contained two other instances of the word correlated and one each of the words correlatives and correlation (none of which uses applied to either strong morality or instability).

A collection of concise biographies by various authors was published in 1997 under the title *Leading Personalities in Statistical Sciences from the Seventeenth Century to the Present*. In it, Cedric A. B. Smith claimed that Galton "...first used the word correlation in his book *Hereditary Genius* in a general mathematical sense... but not with precise definition" (Johnson and Kotz, 1997, pp. 109, 110 ref#1). However, that claim is invalid because all uses of the word correlation in *Hereditary Genius* were Observational Correlations; that is, none of them were mathematical in any sense (they will be discussed in more detail, later in this chapter).

N. W. Gillham and M. Bulmer each wrote book-length biographies of Galton (published in 2001 and 2003, respectively), but Galton's first use of the word correlation was not addressed in either of them.

No historian seems to have taken notice in print of the times that Galton used the word "correlation" in papers published prior to *Hereditary Genius*. Those times will be discussed in this chapter, followed by a discussion of his use of the words "relation" and "co-relation" prior to 1888. Special attention will be given to his books *English Men of Science* and *Natural Inheritance*; it will be shown that neither of them contain any mention or discussion of mathematical correlation, despite claims to the contrary by some historians. Two of his papers from early 1888 are highlighted for being presages of his late 1888 paper that introduced mathematical correlation to the world.

Galton's Pre-December 1888 Use of the Word "Correlation"

Galton did not always use the word correlation where today we would expect it. Like Darwin and others, Galton used different terms; for example, the word "correspond":

- *It is notorious that the movements of the barometric column <u>correspond</u> in some [degree] to the changes of the weather, and especially to those of the wind's velocity...* (Galton, 1870a, p. 31; underlining added. NOTE: The word "degree" in that quotation is a guess, as the copy available on Galton.org shows a single illegible word of approximately six letters in length, and a much revised version that appeared in *Nature* (Galton, 1870b, p. 501) included in its first paragraph the phrase "the degree to which they correspond").
- *...the movement of the* [instrument's] *templet* [sic] *in X <u>corresponds</u> to the ordinate of the wet-bulb trace, just as that of the templet in Y <u>corresponds</u> to the ordinate of the dry-bulb trace, both ordinates having identical abscissae* (Galton, 1872a, pp. 24, 26; underlining added).

The following are examples of how Galton used the word correlation prior to 1888; the un-italicized text following most of the quotations are my comments.

1863

It is hardly possible to conceive masses of air rotating in a retrograde sense in close proximity... without an intermediate area of direct rotation, which would, to use a mechanical simile, be in gear with both of them, and make the movements of the entire system <u>correlative</u> and harmonious.

(Galton, 1863, p. 386; underlining added)

An area of clockwise rotation typically exists between two "close" areas of counter-clockwise rotation; they act on each other in a "correlative" manner. As such, this is an Observational Correlation.

1866

In a recent report... by Mr. Farrer, Capt. Evens and the author [Francis Galton], *they had pointed out many objections to the existing methods of computing ocean statistics. The object of the present paper was to draw attention to yet another objection.... The objection the author made was, that the observations* [of wind direction] *by a sailing ship are more numerous* [within a defined geographical quadrant] *in respect to antagonistic winds or calms than in respect to favourable weather.... It must further be observed, that the error pointed out not only affects the winds, but it affects all the meteorological elements so far as they are <u>correlated</u> with the winds; the temperature and dampness are especially affected by it.*

(Galton, 1866, p. 17; underlining added)

The word "correlated" was used there in reference to sets of numeric data (i.e. wind direction vs. temperature, and wind direction vs. humidity level); a plot of wind direction (*X*) vs. temperature or humidity (*Y*) could be constructed. In this example, variation in humidity and temperature are caused by (are dependent upon) wind direction; and therefore at first glance this seems to be a Relational Correlation. However, because wind direction is artificially numerical (i.e. degrees on a compass), not truly numerical (i.e. east, west, north, south), this use is closer to being an Observational Correlation than a Relational one.

1867

> *A traveller* [sic] *does excellently, who takes latitudes by meridian altitudes, once in the twenty-four hours.... in preparing them, he should bear the following well-known maxims in mind: – Let all careful observations be in doubles. If they be for latitudes, observe a star N. and a star S.; the errors of your instruments will then affect the results in opposite directions, and the mean of the results will destroy the error. So, if for time, observe in doubles, viz. a star E. and a star W. Also, if for lunars, let your sets be in doubles – one set of distances to a star E. of moon, and one to a star W. of moon. Whenever you begin on lunars, give three hours at least to them, and bring away a reliable series; you will be thus possessed of a certainty to work upon, instead of the miserably unsatisfactory results obtained from a single set of lunars taken here and another set there, scattered all over the country, and impossible to* <u>*correlate*</u>*."*

> *(Galton, 1867, pp. 32–33; underlining added)*

The latitudes measured "all over the country" should be the same when taken east/west of each other, but should differ when taken north/south of each other. If north/south distances were plotted as *X* and measurements of latitude were plotted as *Y*, we might call this a Relational Correlation. But that is not Galton's focus here; if it were, he would not have said that such results would be "impossible to correlate". The data that he hopes to "correlate" are the east/west readings, which he expects to be the same; as such, this is an Observational Correlation.

1869

The following are *all* uses of the word correlation etc. that are found in Galton's book titled *Hereditary Genius.* They are all Observational Correlations:

- *It is a consequence of this system of notation, that F. and B. and S. are always printed in capitals, and that their correlatives for mother, sister, and daughter are always expressed in small italicized type, as f., b., and s* (p. 51).
- *Ability must be based on a triple footing, every leg of which has to be firmly planted. In order that a man should inherit ability in the concrete, he must inherit three qualities that are separate and independent of one another: he must inherit capacity, zeal, and vigor; for unless these three, or, at the very least two of them, are combined, he cannot hope to make a figure in the world. The probability against inheriting a combination of three qualities not correlated together, is necessarily in a triplicate proportion greater than it is against inheriting any one of them* (p. 84).
- *Secondly, it appears, that the wives of the Divines were usually women of great piety; now it will be shown a little further on, that there is a frequent correlation between an unusually devout disposition and a weak constitution* (pp. 263–264).
- *The Puritan's character is joyless and morose; he is most happy, or, to speak less paradoxically, most at peace with himself when sad. It is a mental condition correlated with the well-known Puritan features, black straight hair, hollowed cheeks, and sallow complexion* (p. 281).
- *These views will, I think, explain the apparent anomaly why the children of extremely pious parents occasionally turn out very badly. The parents are naturally gifted with high moral characters combined with instability of disposition, but these peculiarities are in no way correlated. It must, therefore, often happen that the child will inherit the one and not the other* (p. 282).
- *The last general remark I have to make is, that features and mental abilities do not seem to be correlated. The son may resemble his parent in being an able man, but it does not therefore follow that he will also resemble him in features* (p. 333).
- *It is curious to remark how unimportant to modern civilization has become the once famous and thoroughbred looking Norman. The type of his features, which is, probably, in some degree correlated with his peculiar form of adventurous disposition, is no longer characteristic of our rulers, and is rarely found among celebrities of the present day…* (p. 348).
- *It is easy to form a general idea of the conditions of stable equilibrium in the organic world, where one element is so correlated with another that there must be an enormous number of unstable combinations for each that is capable of maintaining itself unchanged, generation after generation* (p. 370).

This next quote deserves highlighting, for interest-sake only:

To be a great artist, requires a rare and, so to speak, <u>unnatural correlation</u> of qualities. A poet, besides his genius, must have the severity and stedfast [sic] earnestness of those whose dispositions afford few temptations to pleasure, and he must, at the same time, have the utmost delight in the exercise of his senses and affections (p. 227; underlining added).

That is possibly the only use, ever, by anyone, of the inexplicable term "unnatural correlation". It is an Observational Correlation.

In this next set of quotes, the use of the words "degree" and "how far" at first glance seem to indicate a Relational Correlation:

> *I propose in a future chapter, after I have discussed the several groups of eminent men, to examine the degree in which transcendent genius may be correlated with sterility... (p. 130).*

However, that "future chapter" included this next quotation:

> *I regret I am unable to solve the simple question whether, and how far, men and women who are prodigies of genius, are infertile.... There are many difficulties in the way of discovering whether genius is, or is not, correlated with infertility (p. 330).*

That is an Observational Correlation because he is not discussing degrees of intelligence vs. gradations of sterility but rather top-level genius vs. total sterility; that is, his words "degree" and "how far" were used to indicate that not *every* transcendent genius is sterile.

1872

> *The configuration of every land, its soil, its vegetable covering, its rivers, its climate, its animal and human inhabitants, act and re-act upon one another. It is the highest problem of Geography to analyse their correlations, and to sift the casual from the essential.*

> *(Galton, 1872b, p. 344)*

Here, he described Observational Correlations; for example, the correlation between forests and humidity:

> *...the results of the* [political] *elections at one place* [i.e. one city] *may or may not influence those at another (on the principle of correlation).*

> *(Galton, 1872c, p. 395)*

The source and meaning of the term "principle of correlation" is unknown and unclear, even after reading the entire article; however, this quote clearly involves an Observational Correlation:

> *I must guard myself against the objection, that though structure is largely correlated, I have treated it too much as consisting of separate elements. To this I*

answer, first, that in describing how the embryonic were derived from the struc-
tureless elements, I expressly left room for a small degree of correlation; secondly,
that in the development of the adult elements from the embryonic, there is a per-
fectly open field for natural selection, which is the agency by which correlation is
mainly established; and thirdly, that correlation affects groups of elements, and
rather than the complete person, as is proved by the frequent occurrence of small
groups of persistent peculiarities, which do not affect the rest of the organism, so
far as we know, in any way whatever.

(Galton, 1872c, p. 396)

These are classic Observational Correlations involving biological structures,
as the term was used by Cuvier and Darwin, despite his use of the term
"degree of correlation".

1874

In Galton's book titled *English Men of Science* (hereafter referred to as EMS),
there is only one use of the word correlated and no instances of any other
related parts of speech. That one use occurs in a chapter titled "Qualities",
in which there are four sub-sections titled in the following order: "Energy",
"Energy Much Above the Average", "Cases of Energy Below the Average",
and "Size of Head".

In the subsection on "Energy", Galton said:

It will be seen that the leading scientific men are generally endowed with great
energy; many of the most successful among them have laboured as earnest ama-
teurs in extra-professional hours, working far into the night.

(Galton, 1874c, p. 75)

The entire subsection on "Size of Head" is given here next:

I may mention that energy appears to be <u>correlated</u> *with smallness of head, a fact*
which is well illustrated here, although the average circumference of head among
the scientific men is great. Energy is also, as we have seen, strongly marked
among them; but it is much more strongly marked among those who have small
heads. I have ninety-nine returns [of questionnaires that included head-size
data], *many of which I have verified myself, using the hat maker's whalebone*
hoop, and measuring inside the hats. It appears that the average circumference
of an English gentleman's head is 22¼ to 22½ inches. Now, I have only thirteen
cases under 22 inches, but eight cases of 24 inches or upwards. The general sci-
entific position of the small-headed (who are mostly slender, but not necessarily
short) and large-headed men seems equally good; but the fact is conspicuous
that, out of the thirteen of the former, there are only two or three who have not

remarkable energy; and out of the eight of the latter there is only one who has. A combination of great energy and great intellectual capacity is the most effective of all conditions; but, like the combination of swiftness and strength in muscular powers, it is very rare (pp. 98–99; underlining added).

In 1924, Karl Pearson analyzed the data on which Galton had based that *EMS* Size-of-Head subsection:

Galton concludes that "although the average circumference of head among the scientific men is great," there is a "correlation" between smallness of head and energy.... A modern statistician would be not quite happy in asserting without further investigation [beyond that given in EMS] *that the owners of large heads were the less energetic. But actually... Galton's conclusion* [regarding owners of small heads] *is reasonably justified.*

(Pearson, 1924, pp. 149–150)

That conclusion by Pearson is based upon his performing something like a Chi-square test of significance on Galton's data, a test that yielded a p-value of 0.02 ("the odds... are about 50 to 1", in Pearson's words). For that test, he created the equivalent of a contingency table with three levels of head size vs. only two levels of energy, those two energy-levels being "Energetic" and non-Energetic. Thus, Pearson correctly treated Galton's energy-levels as binomial attributes rather than as variates. (NOTE: The reader of Pearson's analysis needs to subtract the "Energetic" number from the total "Number" for each level of head size, in order to find the non-Energetic number. Pearson used the term "Energetic men uniformly distributed" for what we today would call the "Expected" when calculating a Chi-squared result.)

In 1975, V. L. Hilts claimed, in his "A Guide to Francis Galton's *English Men of Science*", that in the *EMS* Size-of-Head sub-section "Galton argues that energy and head size are inversely related" and that Galton "hints at the idea of correlation.... [although] the hint is not very strong" (Hilts, 1975, p. 24). Such a claim presupposes that Galton is describing Relational or Co-Relational Correlation. If he were, then this would have been the second and not the first mention, the first being in a five-page paper published earlier that same year; in that paper, Galton prominently mentioned that "energy appears to be correlated with smallness of head" (Galton, 1874a, p. 229); interestingly, a one-page summary of it that appeared in *Nature* did not mention anything about head-size (Galton, 1874b).

In his "Guide", Hilts describes what he found in Galton's original note-book that contained *EMS*-related analyses. He describes his find as a "two-dimensional diagram... in which Galton compares the distribution of head size... with the distribution of stature" (p. 24). It is important to note that he did *not* mention finding a plot, diagram, correlation table, or even speculation about an inverse mathematical relationship between head-sizes and energy levels. In other words, Hilts did not find evidence to support an inference that the *EMS* remark about energy level and head size being "correlated" was intended by Galton to be interpreted mathematically.

In 1989, R. E. Fancher published his research into what he called "An Unpublished Step in the Invention of Correlation"; his research included re-evaluating the just-discussed information that Hilts had provided (Fancher, 1989, p. 447). On that page, in a subsection titled "Galton's Steps to Correlation", Fancher discussed Galton's "'scatter plot' of the relationship between head size and height [i.e. stature]" and concluded that Galton "seems not to have closely analyzed this plot mathematically". Fancher made no mention of Galton's remark (nor Hilts's claim) about head size vs. energy level, as if Fancher did not consider it to be one of Galton's steps to correlation.

In summary, it seems reasonable to conclude that Galton in EMS did *not* propose an inverse scale of head-size *vs.* energy level. Instead, he discussed an interesting observation regarding the presence or absence of a particular type of energy that he called variously "energy much above the average", "great energy", "energy...much more strongly marked", or "remarkable energy". Galton's phrase "energy appears to be correlated with smallness of head" refers to Observational Correlation not Relational or Co-Relational Correlation, because his meaning is that high energy is correlated *only* with smallness of head; and therefore, contrary to the claims by Hilts, EMS contains neither an example of inverse correlation nor even a "hint of the idea of correlation".

1877

> We should thus learn how far the more obvious physical characteristics may be
> correlated with certain mental ones.…
>
> (Galton, 1877c, p. 6)

This is an Observational Correlation.

1880

> These diagrams are really helpful because their shape is correlated with the sub-
> ject they portray.
>
> (Galton, 1880a, p. 318)

This is an Observational Correlation:

> …artists might be found whose habit was to visualise numerals… in bold and
> beautiful curves. In the instances I am about to give… there is more tendency to

geometric precision… I should be most curious to learn… [if those instances] are generally correlated with a true eye to straightness, squareness, and symmetry.

(Galton, 1880b, p. 255)

This is an Observational Correlation.

1884

The colour of the hair of animals is often found to be intimately correlated with their power or incapacity to thrive under certain conditions….

(Galton, 1884a, p. 7)

This is an Observational Correlation.

1887

M. Topinard informs me that as the original tints of Broca have already changed colour, he is engaged in preparing a new and much smaller series of only five or six tints, for hair-color [sic] to serve as a fresh departure. These will of course be correlated with Broca's numbers.

(Galton, 1887a, p. 146)

This is an Observational Correlation.

Galton's Pre-December 1888 Use of the Word "Relation"

It is important to recognize that in the years prior to 1888, Galton preferred to use the word "relation" when describing the connection between data sets that could be plotted as X vs. Y. For example:

- *…the angles 30°, 45°, 60°, and 90°…are by far the most useful ones in taking rough measurements of heights and distances, because of the simple relations between the sides of right-angled triangles whose other angles are 30°, 45°, &c* (Galton, 1856, p. 244).
- *The proportion, for instance, of the still-births published in the Record newspaper and in the Times was found by me, on an examination of a particular period, to bear an identical relation to the total number of deaths* (Galton, 1872d, p. 130).

- *Tall men are often thin, and short ones are fat, and the curious fact seems thoroughly verified that the general relation between height and weight is strictly as the squares* (Galton, 1874d, p. 343).
- *The faculty of sense discrimination has in many respects been the subject of most elaborate experiments, chiefly in regard to the relation between the amounts of stimuli, as measured by objective standards… and the corresponding amount of the evoked sensations, measured by subjective standards…* (Galton, 1881a, p. 336).

Galton's 1883 book titled *Inquiries into Human Faculty and its Development* contained no use of the words correlated or correlation, but instead used the words *related* and *relation*:

- *The vital functions are so closely related that an inferiority in the production of healthy children very probably implies a loss of vigor generally, one sign of which is a diminution of stature* (p. 21).
- *I cannot discover any closer relation between high visualizing power and the intellectual faculties than between verbal memory and those same faculties* (p. 111).

In those uses, Galton was not juxtaposing attributes such as blond hair and blue eyes, in which case he would have used the words *correlated* or *correlation* (i.e. for an Observational Correlation). Instead, he was making connections between characteristics that, in his mind, are semi-quantitative. He used the words "inferiority" in the first example, "high" in the second, and "proportion" and "total number" in the third; in the second, he also spoke of "intellectual faculties", which he believed could be measured – starting with the research he conducted for his previous two books, Galton's ideas formed the basis of what other researchers would develop into the mathematical concept of "IQ" (Gillham, 2001, pp. 215–230). He was very close there to describing a Relational Correlation involving semi-quantitative data.

In the next few years, he continued his habit of using the word "relation" rather than "correlation" for comparisons of such types of data; for example:

- *There are many simple and interesting relations to which I am still unable to assign numerical values…* (Galton, 1885a, p. 7).
- *I was surprised to find that there is no close relation between* [hand] *strength of squeeze and breathing capacity* (Galton, 1885b, p. 25).
- *I propose to express by formulae the relation that subsists between the statures of specified men and those of their kinsmen…* (Galton, 1886a, p. 42).

Galton's Pre-December-1888 Use of the Word "Co-Relation"

During my search of the approximately 400 pre-1888 books and papers available in the "Complete Works" section of www.Galton.org, not a single instance was found of Galton having used the words co-relation or co-relate (Zorich, 2020a).

January 1888

When Galton was president of the Anthropological Institute of Great Britain and Ireland, he gave a "President's Address" lecture at the start of each calendar year. This year's speech focused on his recommendation for establishing an anthropometric laboratory "at which children and adults of both sexes could at small cost have their faculties measured by the best methods known to science, and a record kept for their future use" (Galton, 1888a, p. 346). In that address, he spoke of...

> *...the length of the middle finger, which is correlated with the length of the foot....*
>
> *(Galton, 1888a, p. 349)*

Thus, he had begun to use "correlate" instead of "relate" in reference to X, Y paired data sets. However, he had not switched completely, as shown later in his lecture:

> *Another, and a very important question, is as to the <u>degree</u> in which the several bodily proportions that are measured may be looked upon as <u>independent variables</u>. The stature is <u>related</u> with the length of the foot, and with that of forearm, and we should expect a still closer <u>relation</u> to exist between any two of these taken together and the third. We have <u>yet</u> to learn the proportion between the number of elements measured, and their value for purposes of identification* (p. 354; underlining added).

That last sentence revealed his frustration at not "yet" being able to assign a numerical value to the "degree" to which variables are "independent" (or, conversely, he could have said "the degree of their interdependency").

May 1888

In the Spring of 1888, he gave another lecture at the Anthropological Institute; this one focused in detail on a topic that he'd discussed in less detail in his January address, namely the use of anthropometry for identification of criminals. In this lecture, he explained that doubling the number of features measured does not double the amount of useful information, because such measurements are correlated (although he did not use that word):

> *The bodily measurements are so dependent on one another that we cannot afford to neglect small distinctions in an attempt to make an effective classification. Thus long feet and long middle-fingers <u>usually go together</u>. We therefore want to know whether the long feet in some particular person are <u>accompanied by</u> very long, or moderately long, or barely long fingers.... The more numerous*

> *the measures the greater would be their <u>interdependence</u>, and the more unequal would be the distribution of cases among the various possible combinations of large, small, and medium values. No attempt has <u>yet</u> been made to estimate <u>the degree of their interdependence</u>.*

> *(Galton, 1888b, p. 350; underlining added)*

That 15-page paper did not include even one instance of the word correlation. He discussed many of the same data sets as in his January paper; but instead of describing them as being "correlated", he used other terms such as "dependent on one another", "usually go together", "accompanied by", and "interdependence". As seen in the just-given quotation, he ended this discussion, as he did in January, with a sentence indicating his frustration at not "yet" being able to assign a numerical value to the degree of interdependence.

Natural Inheritance

Galton's soon-to-be-famous book *Natural Inheritance* was published in February of 1889 (Gillham, 2001, p. 267). Galton had written it in early-to-mid 1888, and then "<u>After</u> the [printer's] proofs of my book had been finally revised [by me] and had passed out of my hands, it happened that there was a delay of a few months before its actual publication. In the interim I was busily at work upon a new inquiry [that resulted in my discovery of]…. Correlation…. I made all haste to prepare a paper for the Royal Society with the title of 'Correlation'" (Galton, 1890a, pp. 419, 421; underlining added). He presented that now-famous paper to the Royal Society in December of 1888.

Most historians and Galton himself agree about how little of correlation there is in *Natural Inheritance*:

- V. L. Hilts stated that in *Natural Inheritance* Galton "did not note the existence of a single correlation coefficient itself" (Hilts, 1973, p. 228).

- S. M. Stigler correctly pointed out that "the word *correlation* does not appear in *Natural Inheritance*" and that the "idea of correlation was clearly just on the fringe of this work" (Stigler, 1986, p. 297).

- P. M. Porter stated that the "method of correlation was not in *Natural Inheritance*…" (Porter, 1986, p. 299).

- N. W. Gillham concurred: "With the notable exception of correlation, *Natural Inheritance* pulled together in one place much of Galton's work on heredity, anthropometrics, and statistics" (Gillham, 2001, p. 258).

- Galton's own words support those claims: "…in a book by myself, that will be published in a few days, called 'Natural Inheritance'….

All the data to which I shall refer [in today's lecture] will be found in that book also, <u>except such as concern correlation</u>." (Galton, 1889d, p. 297; underlining added).

A search of *Natural Inheritance* finds zero instances of the words correlation, correlated, co-relation, or co-related.

Unjustified Claims Regarding *Natural Inheritance*

Given those analyses and statements, the following claim by Cummins in his introduction to a 1965 reprint of Galton's book *Finger Prints* seems to be unjustified:

> *Natural Inheritance... has special importance as a milestone, since here Galton <u>elaborated</u> on the principal of correlation which marked the origin of the biometric "school".*

> *(Galton, 1892b, p. viii; underlining added)*

C. B. Davenport, in his "A History of the Development of the Quantitative Study of Variation", included a paragraph whose first sentence stated that "During the eighties, Galton, in a remarkable series of papers, developed the quantitative theory of individual variation" (Davenport, 1900, p. 866). Davenport used his *next* two sentences to discuss Galton's contribution to the determination of probable error, regression, and the "law of ancestral inheritance". His *next* five sentences described how "In 1888, Galton made another important step", namely the discovery of the "correlation-index". Davenport's *next* sentence claimed that "The culmination of this epoch-making work of Galton was his [book] 'Natural Inheritance,' 1889, which applied the quantitative methods he had elaborated to the data of human inheritance...". In other words, Davenport incomprehensibly claimed that a book that Galton had finished writing by *mid*-1888 was the "culmination" of work that included Galton's discovery of the correlation coefficient in *late* 1888.

Karl Pearson made similarly unfounded claims:

- *The ideas on heredity and <u>correlation</u> which had been working in Galton's mind during the decade of the 'eighties <u>found final expression in</u> his book entitled <u>Natural Inheritance</u>...* (Pearson, 1930a, p. 57; underlining added).
- *I interpreted that sentence of Galton to mean that there was a category broader than causation, namely correlation, of which causation was only the limit...* (from a speech given by Pearson in 1934, quoted in Porter, 1986, p. 298; Porter indicated that the sentence to which Pearson referred can be found on page three of *Natural Inheritance*).

Alan Treloar, in his 1946 book titled *Correlation Analysis*, wrongly implied that Galton described his discovery of correlation in *Natural Inheritance* (which Treloar mistakenly dated to 1908 rather than 1889):

> *The analysis of properties of the normal bivariate frequency distribution in 1886 by Francis Galton, with subsequent <u>recognition of the correlation coefficient</u> as a general measure of association for rectilinear systems, initiated a new era for quantitative biology in particular and statistical analysis in general. <u>In his classic treatise on "Natural Inheritance" (1908) Galton wrote of his discovery</u> with that inspiring phraseology so characteristic of him: "This part of the inquiry may be said to run along a road on a high that affords wide views in unexpected directions, and from which easy descents be made to totally different goals...."* The concept of correlation, with its replacement of that of causation, ran indeed upon a high level.*
>
> *(Treloar, 1942, p. 1; underlining added)*

That "inquiry" was actually *not* referring to anything about correlation but rather to what Galton mentioned in the sentence (given here next) that immediately preceded the quote that Treloar gave: "It familiarizes us with <u>the measurement of variability, and with curious laws of chance</u> that apply to a vast diversity of social subjects" (Galton, 1889a, p. 3; underlining added).

Similarly humorous and unjustified is Walker's claim in her *Studies in the History of Statistical Method* that *Natural Inheritance* contained a summary of "most" of Galton's research on correlation as well as a presentation of "all" that he contributed to correlation theory:

> *The concepts of regression and correlation did not become generally known until after the appearance of <u>Natural Inheritance</u> in 1889. This <u>sums up most of Galton's work on correlation,</u> and <u>presents all that he contributed to the theory.</u> It would be an excellent thing if every student beginning to study correlation could be urged to read first the introduction to Natural Inheritance and then the section dealing with correlation.*
>
> *(Walker, 1929, p. 106; underlining added)*

What part of the "introduction to *Natural Inheritance*" does Walker urge every student to read? Its approximately 2-page "Introductory" chapter says nothing about correlation of any class; it does have two mathematical statements that relate to two of what Galton calls the three "problems to be dealt with" in the book. Those two statements were:

> - *...the child does not on the average received so much as one-half of his personal qualities from each parent, but something less than half. The question I have to solve, in a reasonable and not merely in a statistical way, is, how much less?* (Galton, 1889a, p. 2)

- *We are all agreed that a brother is* [to his brother] *nearer akin* [sic] *than a nephew, and a nephew* [nearer] *than a cousin, and so on, but how much nearer are they in the precise language of numerical statement?* (p. 2)

In *Natural Inheritance*, Galton addressed those two problems by means of "regression" analysis (not correlation analysis); for example:

- *I call this ratio of 2 to 3 the ratio of "Filial Regression." It is the proportion in which the Son is, on average, less exceptional than his Mid-Parent* (p. 97).
- *We are now able to deal with the distribution of the statures among the Kinsmen in every near degree.... The Regression... is a convenient and correct measure of family likeness* (p. 132).
- *The problem of expressing the relative narrowness of different degrees of kinship... was easily solved. It is merely a question of the amount of the Regression that is appropriate to the different degrees of kinship* (p. 196).

Where in *Natural Inheritance* is Walker's "section dealing with correlation"? The book does not contain any chapter, sub-chapter, or section with that word in its title. I conjecture that she meant either chapter VII, titled "Discussion of the Data of Stature", or chapter IX, titled "The Artistic Faculty", both of which are discussed here next.

Most of the data sets discussed in the chapter on Data of Stature can be interpreted as being either Co-Relational or Relational Correlations; however, none of them were identified as correlations by Galton; instead, he refers to them as "relations". For example:

Description of the Tables of Stature.... Table 11... refers to the relation between the Mid-Parent and his (or should we say its?) Sons and Transmuted Daughters, and it records the Statures of 928 adult offspring of 205 Mid-Parents.... Tables 12 and 13 refer to the relation between Brothers (p. 91).

Galton used the term "Mid-Parent" for his special version of the average height of the mother and father of a given child; and he used "Transmuted Daughters" for his standardized way to increase the recorded heights of all daughters so that on average they are the same as those of all sons. Based upon that page-91 quotation, a plot of stature of Mid-Parent vs. Sons is a Relational Correlation, assuming the Mid-Parent is plotted as X, because parental stature is the primary cause of son stature; in contrast, a plot of brother vs. brother is a Co-Relational Correlation, because it cannot be objectively determined which data set should be plotted as X (i.e., as the independent variable). In his table 11, he provided numerical values that can be converted into an X,Y data set of Mid-Parent height vs. Offspring height. Similarly, his tables 12 and 13 can be converted into X,Y data sets of height of a given son vs. average height of his brothers. Other examples include:

- ...*y* = *the number of individuals who have the stature P ± x.... As the relation between y and x* [is]... *governed by the law of Frequency of Error...* (p. 122).
- *In the case of Stature the relation between the Q of the Co-Fraternity and that of the Population was found to be as 15 to 17* (p. 184).
- *Galton had previously defined "Q" as a value that is "practically the same as the value which mathematicians call the 'Probable Error'"* (p. 53).

The following are examples in *Natural Inheritance* of Galton *not* using the word correlation where, in a post-December-1888 world, he likely would have (all underlining added):

- ...*the values of the Ratios of Regression...from Mid-Parent to Son, and... from Son to Mid-Parent. These and other <u>relations</u> were evidently a subject for mathematical analysis...* (pp. 101–102).
- ...*the various results of my statistics are not casual and disconnected <u>determinations</u>, but strictly <u>interdependent</u>* (p. 103).
- *Organisms are so <u>knit together</u> that change in one direction involves change in many others...* (p. 123).
- *The ratio of filial Regression is found to be so <u>bound up</u> with co-fraternal variability, that when either is given the other can be calculated* (p. 196).

In the chapter on "The Artistic Faculty", Galton's data are summarized in his table 22 (Figure 3.1), wherein he compared the number of artistic parents in a single family (i.e., two, one, or none) to how many artistic children they had.

TABLE 22.

INHERITANCE OF THE ARTISTIC FACULTY.

Parents.	Children.						
	Observed.			Per cents.			
				Observed.		Calculated.	
	Number of Fraternities.	Total children.	Of whom are artistic.	art.	not art.	art.	not art.
Both artistic.............	30	148	95	64	36	60	40
One artistic; one not..	101	520	201	39	61	39	61
Neither artistic...... ...	150	839	173	21	79	17	83
Totals................	281	1507	469	100	100	100	100

FIGURE 3.1

The data in this table shows that more than half of children born to artistic parents are themselves artistic, but that the remaining large fraction of children are not artistic; based upon this analysis, Galton concluded that artistic ability exhibits the same type of generational regression to the mean that stature does (from Galton (1889a, p. 218)).

Because of the paucity of that data, he next performed a *thought* experiment; while doing so, he invented the method that he would later use to discover the correlation coefficient. In a sub-chapter section titled "Regression", his thoughts were these:

> *It is perfectly conceivable that the Artistic Faculty in any person might be somehow measured, and its amount determined, just as we measure Strength.... Let us then suppose the measurement of the Artistic Faculty to be feasible... and that the measures of a large number of persons were* [available].... *Let the graduations of the scale by which the Artistic Faculty is supposed to be measured, be such that <u>the unit of the scale shall equal the</u> Q [probable error]... of the general population* (pp. 158–159; underlining added).

That appears to have been the first time in print that he ever used the Probable Error as a scaling unit on an X,Y plot, real or imaginary (see examples of his earlier uses of the probable error, or probable deviation, in Galton, 1885a, p. 7, and in Galton, 1886c, p. 251). Unfortunately, in *Natural Inheritance* this choice of scale (units of probable error) was not applied to any actual measurement data (e.g. stature) but rather used only in formulas that yielded rough estimates of the number of artistic persons within a sub-population. After he found a "very happy agreement" between observation and estimate, he proclaimed that "We may therefore conclude that the same law of Regression, and all that depends upon it, which governs the inheritance both of Stature and Eye-colour, applies equally to the Artistic Faculty" (pp. 161–162).

Karl Pearson performed a detailed analysis of that artistic faculty data; his conclusion was that "...it is clear that we can<u>not</u> assert that the accordance of percentages between theory and observation given in the... table justifies us in assuming on the present material that the Regression is the same for Artistic Faculty and Stature" (Pearson, 1930a, p. 69; underlining added). Similarly, N. W. Gillham observed that Galton sometimes artificially "smoothed" his data;

> *Smoothing... is a process every scientist is familiar with. However, smoothing can be carried too far. This happens when <u>a scientist tries to make data fit a hypothesis to which they are not properly suited</u>. In such cases the scientist is not attempting to falsify or fudge his results, but is simply too enamored of his own idea. <u>Galton</u> would later be <u>guilty of this</u> when he tried to apply regression to the mean to <u>artistic faculty</u>....*
>
> (Gillham, 2001, p. 254; underlining added)

It is interesting that in at least one early 20th-century textbook, the process of smoothing was encouraged (e.g. Davenport and Rietz, 1907, p. 690, in a section titled "'Smoothing' of Figures").

What are we to conclude about Walker's use of the word "correlation" in the advice mentioned previously? Apparently her concept of correlation there involved neither a correlation coefficient nor a plot of re-scaled *X,Y* data, despite the fact that (as we shall see here in the next chapter) those were the two most important features in Galton's December 1888 paper that introduced mathematical correlation. It could be argued that her advice promoted confusion of the regression coefficient for the correlation coefficient.

Other Interesting Facts about *Natural Inheritance*

Social scientists of the early to mid-19th century (most famously Adolphe Quetelet) considered human deviation from the mean to be a mistake and that "human variability was fundamentally error.... all deviation from [the mean] should be regarded as flawed, the product of error" (Porter, 1986, pp. 7, 104). Such a view mirrored mid-19th century astronomers who developed a concept of contamination of data; they referred to such contamination as "entanglement of observations" (Porter, 1986, p. 273) – their idea was that the raw data from measurements of planetary orbits (for example) fails to graph as a smooth curve because they are contaminated with error. Galton did not agree with such views, at least not in regard to biological data, as evidenced in every chapter of *Natural Inheritance*; many years later, he explained that "these errors or deviations were the very things I wanted to preserve and to know about" (Galton, 1908b, p. 305).

When reading *Natural Inheritance*, it is helpful to know that the mathematical average was a relatively new concept in the 19th century, especially in regard to biometric data:

> ...*the very concept of averaging is a new one* [,] *and before 1650 most people could not observe an average because they did not take averages.*

> *(Hacking, 1975, p. 92)*

Credit for first applying the concept of the average to biometric data is typically given to Quetelet (e.g. by Galton, 1889a, p. 55; and by Stigler, 1986, pp. 161f). In the mid-19th century, Quetelet was an astronomer and mathematician who became interested in descriptive statistics, especially those regarding mankind; he coined the term "average man" and famously promoted what today is called the normal distribution (Hacking, 1990, pp. 105f). Although Galton was a Quetelet disciple when it came to the normal distribution, he did not follow Quetelet in regard to the average; Galton explained in *Natural Inheritance* that:

> *The knowledge of an average value is a meagre piece of information.... So in respect to the distribution of any human quality or faculty, a knowledge of mere averages tells but little; we want to learn how the quality is distributed among*

the various members of the Fraternity or of the Population, and to express what
we know in so compact a form that it can be easily grasped and dealt with…. We
require no more than a fairly just and comprehensive method of expressing the
way in which each measurable quality is distributed among the members of any
group… (pp. 35–36).

Galton's negative attitude toward the mathematical average seems to have
been known even in America, where, in the official journal of the American
Statistical Association, **George K. Holmes** published a rebuttal without
explicitly naming Galton:

An attempt is made in this article to limit, define, and defend the use of the aver-
age, which seems to have fallen into some disrepute among <u>theoretical statisti-</u>
<u>cians</u>…. when accepted for what it truly means, an average has a value beyond
what it is sometimes credited with having, and that, as a concise and comprehen-
sive expression of general significance, there is reason why it may be used, as it
is, for popular understanding, without analysis.

(Holmes, 1891, p. 421, 426; underlining added)

It is interesting that about a year later, Holmes published a short paper that
began with the full text of a letter he'd received from Galton, a letter that
commented on one of Holmes's recently published papers that Galton had
received from him. The text of that short paper referred to Galton as "the
most eminent authority on mathematical measures of distribution" (Holmes,
1892, p. 272). The last paragraph in Galton's letter criticized Holmes's writing
style; in its last sentence, Galton could have found a way to use the word correlation, but curiously he did not:

Permit me to criticize the terms of your query in p. 141, viz., "Is wealth more
widely, evenly, and generously distributed in [Massachusetts than in New
York]*?" Either those three adjectives mean the same thing, or they do not. If they*
do, two of them are superfluous, and in fact, I have assumed them all to mean
evenly. If they do not, your query involves three independent variables, and could
not be answered without explaining how they are to be rendered commensurable.

(Holmes, 1892, p. 271; what Galton called "your query" was Holmes's
paper "Measures of Distribution", 1892, Pub. Am. Stat. Soc., 3: 141–157)

After Karl Pearson first read *Natural Inheritance*, he praised Galton's efforts
but also warned that "there is, in my own opinion, considerable danger in
applying the methods of exact science to problems in descriptive science,
whether they be problems of heredity or of political economy" (Stigler, 1986,
p. 304; quote taken by Stigler from an unpublished but extant copy of a lecture
delivered by Pearson in March 1889). However, upon further consideration,
Pearson reversed himself and became Galton's greatest fan; in Pearson's
biography of Galton, he hyperbolically praised *Natural Inheritance*:

It may be said that this publication created Galton's school; it induced Weldon, Edgeworth and the present biographer to study correlation and in doing so to see its immense importance for many fields of inquiry. It is idle to overlook the haste with which it was prepared and the many slips and positive errors to be found in its pages, but no one who studied it on its appearance and had a receptive and sufficiently trained mathematical mind could deny its great suggestiveness, or be other than grateful for all the new ideas and possible problems which it provided. The methods of Natural Inheritance may be antiquated now, but in the history of science it will be ever memorable as marking a new epoch, and planting the seed from which sprang a new calculus, as powerful as any branch of the old analysis, and valuable in just as many fields of scientific research.

(Pearson, 1930a, pp. 57–58)

Chapter Summary

Prior to 1888, Francis Galton used the word correlation only in reference to Observational Correlation; claims to the contrary by other historians are shown to be in error. Whenever Galton spoke of the mathematical dependency or interdependency of two paired data sets, he typically used the word relation. He never even once used the hyphenated word co-relation. During the Winter, Spring, and Summer of 1888, Galton seemed to be unconsciously preparing his mind for the discovery of the correlation coefficient. Early in that year, he began for the very first time to use the word correlation for not only Observational Correlation but also for Co-Relational Correlation. He expressed frustration at not being able to quantitate those correlations. Most importantly, in mid-1888 he performed a mathematical thought experiment that involved using an analytical method that was similar to the co-relation method that he would announce at the very end of that year.

4

Francis Galton, December 1888

Introduction

This chapter differs from the previous ones and the ones that follow, in that it examines not only the uses of the word correlation but also analyzes other details in Galton's paper that introduced mathematical correlation to the world. The reason for doing so is that this paper is pivotal in the history of the use of the word correlation. Prior to this paper (see Figure 4.1), the word had virtually *no* mathematical meaning; starting with this paper, the word began to have *only* mathematical meaning.

December 20, 1888.

Professor G. G. STOKES, D.C.L., President, in the Chair.

The Presents received were laid on the table, and thanks ordered for them.

The following Papers were read :—

I. " Co-relations and their Measurement, chiefly from Anthropometric Data." By FRANCIS GALTON, F.R.S. Received
December 5, 1888.

" Co-relation or correlation of structure " is a phrase much used in
biology, and not least in that branch of it which refers to heredity, and
the idea is even more frequently present than the phrase ; but I am
not aware of any previous attempt to define it clearly, to trace its
mode of action in detail, or to show how to measure its degree.

FIGURE 4.1
This is a scan of the title and first sentence in Francis Galton's first paper that described his invention/discovery of mathematical co-relation. The date that he presented the paper orally to the Royal Society is shown at the top. (From Galton (1888c, p. 135).)

The major topics covered in this chapter are:

- Historical importance of Galton's December 1888 paper.
- Various erroneous ways that his paper has been referenced.
- How he discovered the correlation coefficient.
- Errors in his paper.

- Why the paper's title starts with "Co-relation" rather than "Correlation".
- False claims in the paper's first sentence.
- An explanation for what Galton meant when he said that the regression coefficients of 0.26 and 2.5 are not "reciprocal".
- Corrected version of the paper's only "figure".
- Analyses of the paper by other historians.

Importance

Historians and biographers have emphasized the importance of Galton's December 1888 paper, Karl Pearson most flamboyantly:

- *Like so much of Galton's work [,] the present paper* [i.e., from December 1888] *reaches results of singular importance by very simple methods; his methods are indeed so simple that we might almost believe they must lead to a fallacy had not Galton deduced thereby the correct answer. It is the old experience that a rude* [i.e., crude] *instrument in the hand of a master craftsman will achieve more than the finest tool wielded by the uninspired journeyman* (Pearson, 1930a, p. 50).
- *Only when we look at what has happened since 1888, do we realise the importance of that short paper on "Co-relations"! Thousands of correlation coefficients are now calculated annually, the memoirs and text-books on psychology abound in them; they form, it may be in a generalised manner, the basis of investigations in medical statistics, in sociology and anthropology. Shortly, Galton's very modest paper of ten pages from which a revolution in our scientific ideas has spread is in its permanent influence, perhaps, the most important of his writings. Formerly the quantitative scientist could only think in terms of causation, now he can think also in terms of correlation. This has not only enormously widened the field to which quantitative and therefore mathematical methods can be applied, but it has at the same time modified our philosophy of science and even of life itself* (Pearson, 1930a, pp. 56–57).
- *It is fair to say that the idea which Galton expressed in his paper of 1888 and the developments which that idea stimulated created a revolution in the methodology of the social sciences* (Hilts, 1973, p. 206).
- *The almost simultaneous appearance of Galton's book Natural Inheritance and his method of correlation in 1889 marks the beginning of the modern period of statistics* (Porter, 1986, p. 296; 1889 is given as the date because Galton's December 1888 paper was not formally published until then).

- [Galton's December 1888 paper] *opened up a new view of the intellectual landscape* (Stigler, 1986, p. 299).
- *Co-relation....* [was] *the greatest single contribution Galton made to science* (Keynes, 1993, p. 116).
- ...*perhaps* [Galton's] *most important contribution was his invention of correlation and his discovery of the phenomenon of regression to the mean* (Stigler, 1999, p. 6).
- *Galton's pathbreaking memoir of 1888 on correlation, together with his greatest scientific book,* Natural Inheritance.... *would form the cornerstone of a new science, biometrics* (Gillham, 2001, p. 258).

Referenced Different Ways

It is surprising how many different ways such an important paper has been referenced. The most commonly accepted and seemingly correct reference is:

> 1888. "Co-relations and their measurement, chiefly from anthropometric data", *Proceedings of the Royal Society of London* 45: 135–145.

For example, that's how it was given in Stigler (1986, p. 383). However, there are many incorrect references to be found. Typically, references are missing the words "of London"; it was many years later that the Royal Society dropped the *of London* from its official title. The following are a sampling of the various other ways Galton's paper has been incorrectly or incompletely referenced:

- Weldon (1890, p. 453): "'Roy. Soc. Proc.,' vol. 45, No. 274, pp. 135 *et seq.*" No year is listed; instead, the publication sequence number is given ("274").
- Weldon (1893, p. 325): "'Roy. Soc. Proc.,' vol. 40, p. 63". That volume contains Galton's 1886 paper titled "Family Likeness in Stature", and that page begins an appendix written by J. D. H. Dixon (which focused on Galton's *regression* measurement, not his *correlation* measurement).
- Davenport (1899, p. 41): "Correlations and their Measurement, chiefly from Anthropometric Data. Proc. Roy. Soc. London, XLV, 136–145." Davenport mistakenly used "Correlations" instead of "Co-relations", and started pagination with page 136 instead of 135.
- Elderton (1906, p. 163): "'Correlations and their Measurement.' Proc. Roy. Soc., vol. xlv., pp. 136–145". Elderton mistakenly used "Correlations" instead of "Co-relations", shortened the title (by not providing the words "chiefly from...", and started the pagination with page 136 instead of 135.

- Galton (1908b, p 328): "Correlations... *(Roy. Soc. Proc.)*... 1889". This reference is found in the Appendix to his autobiography, the table of contents of which states that the Appendix contains "Books and Memoirs by the Author"; the important points to note are:
  - o The first word in the title should be hyphenated, i.e., "Co-relations...".
  - o He listed the year as 1889 because that is the year that the Society officially published Volume 45 (with his paper), although the paper itself is dated 1888 in the header of its printed pages.
- Pearson (1920, p. 39): "In a paper read to the Royal Society on December 5, 1888, entitled 'Correlations and their Measurement chiefly from Anthropometric Data'". As mentioned in the introductory sentences to the published paper itself (see Figure 4.1), it was "Received December 5, 1888" and "read" on "December 20, 1888". The paper's actual title was "Co-relations..." not "Correlations...".
- Walker (1929, p. 106n): *"Proceedings of the Royal Society, XLVII (1888)"*. The correct volume is XLV not XLVII.
- Forrest (1974, p. 312): *"Proceedings of the Royal Society, 45 (1888)...* Also in *Nature, 39,* (1889), p. 238". The original 10-page paper was not "also" published in *Nature*; what was published there was a half-page summary of it.
- September 25, 2023 (at www.galton.org/bibnew/all.html): "1888.... *Proceedings of the Royal Society* 45 (December 13): 135–45". The "December 13" is a mistake. This error seems to have been caused by the fact that in the top right corner of page 134 is the date "Dec 13" with "1888" showing up in the top left corner of the facing page 135. The December 13 does indeed apply to all the information on page 134 and to the top half of page 135 (none of which is Galton's paper). In the middle of page 135, we find the date "December 20, 1888" followed immediately by Galton's 10-page paper on "Co-relations...". Across the top of facing pages 136–137 through 144–145 we find "December 20, 1888".
- September 29, 2020 (on the Royal Society's website, https://royalsociety.org/journals): The publication date is given as January 1, 1889. The Society's formal publication of this lecture would have been as part of the completed Volume 45, publication of which would not be possible until after April 25, 1889, from which meeting the last papers of that volume were obtained. As an example of a similar error, that same website on September 29, 2020, listed January 1, 1889, for the publication date of a paper titled "The Enervation of the Renal Blood-vessels", the published pages of which stated that it was "received" on February 1, 1889, and read in front of the Society on February 21, 1889.

How Did Galton Discover the Correlation Coefficient?

What specifically precipitated Galton's discovery of correlation? At least two historians have speculated on that question:

- "Rivalry between Bertillon and Galton spurred the invention of the theory of correlation..." (Hacking, 1990, p. 182). Bertillon (to be discussed in more detail, in a future chapter) had in 1880 proposed a system of criminal identification using measurements of many body parts; that system assumed that body parts are inherited *independently*, which assumption Galton rejected in the late 1880s.

- "The fact that the regression of *y* on *x* differed from that of *x* on *y* worried him persistently until... he saw that the solution was to normalize *x* and *y* in terms of their own variability" (C. A. B. Smith, in Johnson and Kotz, 1997, p. 110).

The discovery of the correlation coefficient is described by Galton in documents that he published in 1889, 1890, and 1908. The description given in the 1908 publication was for many years considered to be correct (e.g., Pearson, 1924, p. 393; Hilts, 1973, p. 228; Forrest, 1974, p. 197; Smith in Johnson and Kotz, 1997, p. 110). More recently, historians have concluded that the 1908 account (written when Galton was more than 85 years old) is fancifully inaccurate, and that the descriptions given in the 1889 and 1890 publications are much closer to the true story (e.g., Porter, 1986, p. 292; Stigler, 1986, p. 298; Stigler, 1989; Gillham, 2001, p. 258). That 1890 paper seems to have escaped the notice of historians and biographers for nearly 100 years, most likely because it was published in an American rather than English journal.

It is interesting to note that the Preface to Johnson and Kotz's 1997 set of biographies referenced the just-cited 1986 books by Stigler and by Porter; those books provided evidence that Galton's 1908 story could not be trusted. Apparently, C. A. B. Smith, who authored Johnson and Kotz's chapter on Galton, discounted that evidence, given that he instead promoted Galton's 1908 story (based upon which, Smith amazingly dated Galton's discovery of correlation to 1877 (Johnson and Kotz, 1997, p. 110).

Relevant text from all three of Galton's descriptions are given here next:

January 22, 1889: President's Address to the Anthropological Institute – in addition to eventually being published in the Institute's own journal (Galton, 1889e), this was published in a modified version in *Nature* (Galton, 1889d). The relevant text of the President's Address is:

> *Correlation is a very wide subject indeed. It exists whenever the variations of two objects are in part due to common causes; but on this occasion I must only speak of those correlations that are of anthropological interests. The particular problem I first had in view was to ascertain the practical limitations of the*

ingenious method of anthropometric identification due to M. A. Bertillon, and now in a habitual use in the criminal administration of France. As the length of the various limbs in the same person are to some degree related together, it was of interest to ascertain the extent to which they admit of being treated as independent. The first results of the inquiry, which is not yet completed, have been to myself a grateful surprise. Not only did it turn out that the expression and the measure of correlation between any two variables are exceedingly simple and definite, but it became evident almost from the first that I had unconsciously explored the very same ground before. No sooner had I begun to tabulate the data than I saw that they ran in just the same form as those that referred to family likeness in stature, which were submitted to you two years ago. A very little reflection made it clear that family likeness was nothing more than a particular case of the wide subject of correlation, and that the whole of the reasoning already bestowed upon the special case family likeness is equally applicable to correlation in its most general aspect.

> (Galton, 1889e, pp. 403–404; underlining added; the
> corresponding text that appeared in *Nature* (Galton,
> 1889d, p. 296) differed insignificantly from this text.)

1890: "Kinship and Correlation":

After the proofs of my book [Natural Inheritance] had been finally revised and had passed out of my hands, it happened that there was a delay of a few months before its actual publication. In the interim I was busily at work upon a new inquiry that had been suggested to me by two concurrent circumstances.... The other circumstance arose out of the interest excited by M. Alphonse Bertillon, who proved that it was feasible to identify old criminals by an anthropometric process.... Then a question naturally arose as to the limits of refinement to which M. Bertillon's system could be carried advantageously.... The sizes of the various parts of the body of the same person are in some degree related together.... These two problems — namely, that of estimating the stature of an unknown man from the length of one of his bones, and that of the relation between the various bodily dimensions of the same person – are clearly identical.... Reflection soon made it clear to me that not only were the two new problems identical in principle with the old one of kinship which I had already solved, but that all three of them were no more than special cases of a much more general problem – namely, that of Correlation.

> (Galton, 1890a, pp. 419–421; underlining added)

1908: *Memories of My Life*:

As these lines are being written, the circumstances under which I first clearly grasped the important generalisation that the laws of Heredity were solely concerned with deviations expressed in statistical units, are vividly recalled to my memory. It was in the grounds of Naworth Castle, where an invitation had been given to ramble freely. A temporary shower drove me to seek refuge in a reddish recess in the rock by the side of the pathway. There the idea flashed across me, and I forgot everything else for a moment in my great delight.

(Galton, 1908b, p. 300)

Written in Haste, with Errors

Galton wrote his December 1888 paper in "all haste" in order to ensure his priority of discovery; he admitted that in "the hurry of preparation", he made a "number of numerical blunders... though none in the theory or formulas" (1890a, p. 421; as we will see, his claim to have made no formulaic blunders is itself a blunder). He never published those errata, although he'd tabulated a few of them (Pearson, 1930a, p. 55n). Some of the errors have been discussed by historians (e.g., Stigler, 1986, p. 319; Pearson, 1930a, pp. 53–57); Stigler conjectured that Galton's "early readers must have been bewildered" by such errors (Stigler, 1986, p. 291); Pearson said simply that "The paper shows... signs of haste in preparation" (Pearson, 1930a, p. 56n). Galton himself called these blunders "sad" (Galton, 1890a, p. 421).

Blunders that I have confirmed or discovered in his December 1888 paper include:

- In Table II, the "Total cases" values of 48, 48, and 34 should have been 38, 47, and 35, respectively, as determined by addition using Excel; these errors were also identified by S. Stigler, on page 319 of his *The History of Statistics*.
- In Table III, the "No. of cases...Stature" value of 48 should have been 47, as determined from the corrected value in Table II.
- In Table IV, the "lengths of head...Calculated" values of 7.69 and 7.65 (at the top of page 141) are obviously wrong because the Calculated values in that column should uniformly decrease in steps of 0.04 (rounded), in line with the uniformly decreasing Heights. However, the 7.69 is instead 0.09 *larger* than the 7.60 above it. The 7.69 and 7.65 should have been 7.56 and 7.52, respectively.
- In Table IV, the page 142 "No. of cases" of Stature value of 49 should be 47, in order for that entire column of numbers to match the corrected "Total cases" column in Table II (mentioned above).
- In Table IV, the "Left cubit" value of 17.1 (on page 142) should have been 18.1 – the numbers in that column progress in steps of 0.3, from 16.9 to 19.0, and therefore the value between 17.8 and 18.4 should have been 18.1, not 17.1.

- In Table V, for Middle finger vs. Cubit, the f value of 0.61 should have been 0.53 – the correct answer is found using the r value (0.85) and the formula $\sqrt{(1-r^2)}$ that was provided in the table; that is, $\sqrt{(1-0.85^2)} = 0.53$.
- In Table V, the "Stature...Head length...As 1 to..." value of 0.38 should have been 0.038, as determined by the fact that the geometric mean of 0.038 and 3.2 (its paired value) equals 0.35 (which is the value of the relevant r that is given in Table V), whereas r equals a nonsensical geometric mean of 1.10 if 0.38 is used.
- In Table V, for "Height of knee... Stature... As 1 to..." value of 1.20 should likely have been 2.10 (i.e., 1.20 is a simple transposition error), as shown by a plot of Table IV's data of "Height of knee" vs. "Mean of corresponding statures. Observed" (using MS Excel, a slope of 2.12 is obtained). Additional evidence is found in the geometric mean calculation of the correlation coefficient: $\sqrt{(0.41 \times 2.10)} = 0.70$, whereas $\sqrt{(0.41 \times 2.10)} = 0.93$, which rounds to the $r = 0.9$ value that is given in Table V.
- On page 143, he incorrectly referenced a paper that he'd published a couple years earlier; his incorrect reference was:

  > This is precisely analogous to what was observed in kinship, as I showed in my paper read before this Society on "Hereditary Stature" ('Roy. Soc. Proc.,' vol. 40, 1886, p. 42)

  The title of the paper on that page of the Society's journal is not "Hereditary Stature" but rather "Family Likeness in Stature".

- At the top of page 144, the value of 0.44 should be 0.47, as determined by calculation using Excel; this error was also mentioned by Karl Pearson on page 55 of volume IIIA of his *The Life, Letters and Labours of Francis Galton*.
- On page 145, in his paper's final paragraph, his formula is incorrect (more about this later).

Additional blunders can be found in his then just-finished book *Natural Inheritance*; some of those errors he privately admitted were "gross", due partly to a misprint that he failed to catch during his final review and partly to his "blind & careless writing out of a formula from memory" (that self-criticism is from a letter found in the "Galton Archives" and given in Stigler, 1986, p. 291). Also in *Natural Inheritance*, Galton quoted from a paper that he referenced as "Journ. Anthropol. Inst. 1885", but then warned that "There is a blunder in the paragraph, p. 23, headed 'Height Sitting and Standing.' The paragraph should be struck out" (Galton, 1889a, p. 43n). Sadly, that note about a blunder is itself a blunder. The page 23 to which he is referring is *not* in that Journal but rather in a stand-alone pamphlet (Galton, 1885b) in which he'd combined two of his previously published papers (namely Galton, 1885c, 1885d).

My purpose in noting such errors is to highlight the fact that Galton could at times be careless and less than exact. Such a propensity is one of the reasons that his December 1888 paper is difficult to understand. In that paper, he introduced a new concept, new term, and new coefficient, as if they were clear in his mind, but in reality he himself had not yet decided on how to name, interpret, or use them, as we shall see in this and the following chapter.

December 20, 1888: Galton Orally Presented His Paper, "Co-Relations and Their Measurement, Chiefly from Anthropometric Data"

Because this paper is pivotal in the history of the word correlation, it will be examined here in what might seem like excessive detail; however, I am using this paper to bridge what came before with what came after. Additionally, taking such a detailed look at a single paper was fun for me. Subsequent papers and books by Galton and others will be given much less scrutiny.

Why Did His Paper's Title Begin with the Word "Co-Relations?"

This was the first time that Galton had ever used the hyphenated word "co-relation" – I base that claim upon my examination of all of the approximately 400 pre-December-20–1888 documents that were available in the "Complete Works" of Francis Galton on the www.Galton.org website, as of June 2020 (Zorich, 2020a). It was interesting to find (on July 21, 2019) a copy of Galton's 1888 paper that had been transcribed into modern font, in which more than half of Galton's uses of "co-relation" were mis-transcribed into the word "cor-relation"; that copy was found at http://galton.org/essays/1880-1889/galton-1888-co-relations-royal-soc/galton_corr.html.

Why did he choose Co-relations rather than Correlations? Nobody knows. Galton never explained nor even discussed his choice, at least not in writing: Not in his autobiography (Galton, 1908b), nor in his published accounts of the discovery of the correlation coefficient (previously discussed), nor in any extant personal letter (e.g., not in Pearson, 1914, 1924, 1930a, 1930b, nor in any of the letters available as of 2021 on Galton.org). That question seems to have been examined in print by only one historian and one biographer (discussed here next). The following five possible answers to that question are interesting to consider.

Answer #1: Avoidance of a Word That Had
a Different Established Meaning

S. M. Stigler conjectured that "The initial spelling of the term *co-relation* seems to have been a conscious attempt on Galton's part to distinguish his term from the word *correlation*, which was already in common use", and that "the brief usage of 'co-relation' was to emphasize the novelty of the concept to which Galton attached to the term" (Stigler, 1986, p. 297, 1989, p. 76).

One argument against such conjectures is that Galton actually was *not* opposed to giving novel meaning to words that were already in common use. For example, in 1877 he published a paper that introduced his "coefficient of reversion" (Galton, 1877a, p. 298). He certainly was at that time aware of how the word reversion had long been used in natural science literature, most especially in *Origin of Species*, which we know he read intently very soon after its 1859 publication (Darwin and Seward, 1903, letter 82). He'd also encountered that word in Darwin's 1868 *The Variation of Animals and Plants under Domestication*, at least parts of which Galton must have read or he couldn't have subsequently lectured on "Pangenesis" (Galton, 1871), the theory of which Darwin had introduced in that book.

In *Origin*, Darwin used the words reversion or revert many times in his explanations for the sometimes appearance of ancestral characteristics such as zebra-like stripes on the shoulders or legs of the ass or horse (Darwin, 1859, pp. 163f). For example (all page references to Darwin, 1859):

> ...*revert to the wild aboriginal stock* (p. 15).
> ...*the well-known principle of reversion to ancestral characteristics* (p. 25).
> ...*characters reappear from the law of reversion* (p. 196).
> ...*reversions to long-lost characters* (p. 473).

In *Variation*, Darwin used the words "reversion" or "revert" many times. Although those uses can be viewed in at least three subtly different ways (Vorzimmer, 1971, p. 35), none of them bears resemblance to Galton's use. For example (all page references to Darwin, 1868, vol. 1):

> ...*the tendency to reversion to a primordial condition* (p. 38).
> ...*revert completely to the character of their parent-stock* (p. 77).
> ...*the well-known principle of "throwing back" or reversion* (p. 201).
> ...*reversion to a primordial and extinct condition of the species* (p. 291).

NED's biological-section literature references for the word reversion began with Darwin's *Origin* and Lyell's *Principles of Geology*; the *NED* definition was "The fact, or action, of reverting or returning to a primitive or ancestral type or condition; an instance of this." Such a meaning was so very strongly entrenched that it survived into 20th century biological texts:

> ..."*reversion,*" *the reappearance of an ancestral combination of characters* [after a single cross of certain varieties or races].

> *(Goodrich, 1919, p. 52)*

Galton himself sometimes used reversion in line with the *NED* definition; for example:

> ...*a rabbit which, at the age of six months, produced young which reverted to ancestral peculiarities.*

> *(Galton, 1871, p. 394)*

But sometimes he used it differently; examples include:

- *Elephants...are peculiarly apt to revert to wildness if they once are allowed to wander and escape to the woods* (Galton, 1865, p. 134).
- *When animals reared in the house are suffered to run about in the companionship of others like themselves, they naturally revert to much of their original wildness* (Galton, 1865, p. 136).
- *...there will be numerous... cases of reversion in the first and... second generation...when the third generation... has been reached, the race will begin to bear offspring of distinctly purer blood than in the first, and after five or six generations, reversion to an inferior type will be rare* (Galton, 1873, pp. 129–130).

In each of those last three examples, Galton used the word reversion to mean *predictable* changes in either (i) an individual, caused by an altered living situation, or (ii) successive generations of a given sample of individuals. In contrast, the meaning given by Darwin and Lyell and *NED* focused predominately on *unpredictable* changes that occur randomly in a population.

In 1877, Galton's "coefficient of reversion" equaled the fraction that progeny, on average, retain of how far their parents, on average, had deviated *from the population average* for a given measurable characteristic. In his words:

> ...*the progeny of all exceptional individuals tends to "revert" towards mediocrity.*

> *(Galton, 1877a, p. 283).*

The fact that he put quotes around this first appearance of the word revert in his paper may indicate that he knew he was using the word differently than was commonly done. By "mediocrity" he meant (for example) "the average height of the race" (p. 283). He considered reversion to be one of the "processes of heredity" (p. 289). He went on to clarify his meaning:

> *The only processes concerned in simple descent that can affect the characteristics of a sample of a population are those of Family Variability and Reversion. It is well to define these words clearly. By family variability is meant the departure of the children of the same or similarly descended families from the ideal mean type*

of all of them. Reversion is the tendency of that ideal mean type to depart from the parent type, "reverting" towards what may be roughly and perhaps fairly described as the average ancestral type.

(Galton, 1877a, p. 291; note that he again used quotation marks)

In cases where the processes of heredity are "not typical", Galton explained that "Reversion might not be directed towards the mean of the race..." (p. 298). However, he said that in *typical* cases, the "typical laws of heredity" are:

We see by them that the ordinary genealogical course of a race consists in a constant outgrowth from its centre, a constant dying away at its margins, and a tendency of the scanty remnants of all exceptional stock to revert to that mediocrity, whence the majority of their ancestors originally sprang.

(Galton, 1877a, p. 298)

In summary: Starting in 1877, Galton used "reversion" in reference to gradual predictable generational changes in *measurements* from the exceptionally large or small toward the population average. In contrast, everyone else had been and was still using the word reversion in reference to unpredictable sudden changes in *appearances* toward an ancient or primordial characteristic. Having made such a drastic, almost 180° change in meaning to the word reversion, it seems that he would have had no hesitation at all in using "correlation" to refer to Co-Relational Correlation, even though everyone else was still using the word in reference to Observational Correlation. Based upon this analysis, I conclude that Galton did not choose "co-relation" in order to avoid using a word that had a different established meaning.

NOTE: It is interesting that in 1896, science-fiction author H. G. Wells also used the word reversion in reference to changes in individuals. His book titled *The Island of Dr. Moreau* was published that year; its chapter XXI is titled "The Reversion of the Beast Folk". The book is about a doctor who surgically transformed individual mammals into humanoid-looking creatures that for many years post-surgery could walk on two legs, converse in broken English, and act civilly. However, after an alteration to their living situation, they began "reverting, and reverting very rapidly" back into inarticulate wild beasts running on all fours (Wells, 1896, pp. 221, 231).

Additional comments related to Answer #1:

Stigler concluded in a 1999 publication that...

Jevons's Principles of Science did make one contribution to statistics that is worth noting; it helped give us [the word] "correlation." The book does not present the concept of correlation.... Rather it seems to have been Galton's source for the word. The word was then in common use, for example, by Darwin in the Origin of Species (1859), and by W. R. Grove in The Correlation of Physical Forces (1865), but it was Jevons's use that we know caught Galton's eye at a crucial time. In his personal copy of Jevons's book, Francis Galton marked the explanation of the term (Jevons, 1874, p. 354) and wrote in the margin: "Nice wd".

(Stigler, 1999, p. 89n)

What Galton wrote in poor penmanship at the bottom of that page 354 in Jevons's book is difficult to decipher, and my interpretation of it is slightly different from that given by Stigler (1986, p. 298) shown here next:

Nice wd. never so with the common meaning (Grove's). Thus where motion is there heat may not be, but when motion is not then heat will [not] be

Stigler used regular font for that entire quote, except that he italicized the original word "not" in "there heat may not be"; and he added the word "not" and its brackets near the end of the quote. This next quote is my own interpretation of what Galton wrote (underlining is in the original by Galton):

Nice wd – never so with the common meaning. (Grove's) thus where motion is there heat may <u>not</u> be, but when motion is not there heat will be

Galton wrote that note after reading Jevons's sentence (given here next), which Galton had asterisked on the page with his note. Neither Stigler's nor my version of Galton's note agrees in meaning with the quote from Jevons, and therefore it is unclear if Galton meant to agree or disagree with him.

Things are correlated (con, relata) when they are so related or bound to each other that where one is the other is, and where one is not the other is not.

(Jevons, 1874a, vol. 2, p. 354)

Notice that Jevons was there describing an Observational Correlation, not a Relational or Co-Relational one. As previously discussed, a detailed search of Jevons's book reveals that he always used the word correlation in reference to Observational Correlation. On the other hand, when describing a Relational or Co-Relational Correlation, he used words such as "connexion" (i.e., connection) and "correspondence"; but, most frequently he preferred the word "relation".

As discussed earlier in this book, there are differences between the formal definitions of correlation given by Darwin, Grove, and Jevons, but their definitions all clearly refer to Observational Correlation. Stigler's conclusion that Jevons was the "source" for the word correlation may possibly be more accurately stated that Jevons's use of the word "correlated" reminded Galton of his having encountered it many times in the writings of Grove and Darwin (and possibly those of Spencer). The most relevant facts are:

- Galton's December 1888 paper was titled "Co-Relations..." not "Correlations...".
- The body of that paper included only one use of the word correlation and one of correlated *vs.* 14 uses of co-relation and seven of co-related.

It seems reasonable to conclude that, in December 1888, Galton never intended to give us the word "correlation", but rather intended to give us the word "co-relation".

Answer #2: Avoidance of a Word That Connoted "Things" Rather Than "Relationships"

N. Gillham, in his biography of Galton, said that "Galton probably used 'co-relate' [sic] rather 'Correlate' in his title on purpose" (Gillham, 2001, page 379, note# 42; however, "co-relate" did *not* appear in Galton's title, but rather "co-relations"). Gillham summarized Stigler's viewpoint just given and then speculated:

> *But there may be another reason why Galton chose "co-relation" rather than "correlation" (see The Oxford Universal Dictionary [Oxford: Clarendon Press, Third Edition, 1955] 394, 399). "Co-relate," a word of fairly recent vintage first used in 1839, meant a joint or mutual relation. In this sense it could be taken as meaning kinship, a subject in which as we have seen, Galton was intensely interested. The older word "correlation," whose usage data from at least 1551, refers to a mutual relationship of two or more things. Thus, Galton may have felt that co-relate expressed his intent to tie his concept to kinship as opposed to "things."*
>
> *(p. 379, note# 42)*

Gillham's dictionary reference is to a "shorter" version of the *Oxford English Dictionary* (Onions et al., 1937, title page), the longer version of which was discussed here in Chapter 1. In regard to the definitions in that longer version, it can be said that:

- The entire text of its definition of co-relation was only five words: *"Joint or mutual relation; correlation"*.
- None of its example quotations of co-relation had anything remotely to do with kinship but instead apply to things (such as art, grammar, and politics).
- Its biological quotation referred to the co-relation of anatomical parts and physiological functions (i.e., things).

- The word correlation (not the word co-relation) was used to mean "*Relationship (of persons)*" as late at the 17th century, but that meaning is now "obsolete".

Gilham's speculation may possibly be true, but the evidence in *OED* seems to contradict it.

Answer #3: Penchant for Hyphenated Words

Prior to 1888, Galton had a long history of using and/or coining odd hyphenate words, especially ones that started with "co-". For example:

- *...we find that in no place is the Moslem so tolerant, of those not his co-religionists, as here* [in Zanzibar] (Galton, 1861, p. 123).
- In *Hereditary Genius* (1869), he used "co-heiress" many times for a daughter who inherits a fortune jointly with a sibling.
- In *English Men of Science* (1874c, p. vi) he spoke of "pre-efficients", by which he meant "all that has gone to the making of"; he didn't take credit for coining this word, but rather in a footnote said that "The word was suggested to me".
- *Let us view in imagination the stream of travellers* [sic] *who leave London simultaneously and go as quickly as they can to their destinations, starting by the postal routes. Some of the travellers will be seen to leave the main lines at each successive halting-place...* (Galton, 1881b, p. 740).
- *A single* [photographic] *plate... exposed to several negatives yields... a composite. Several of these composites may in their turn be exposed to another plate under similar conditions...; the result is called a co-composite. Several of these co-composites may be combined to produce a co-co-composite, and so on* (Galton and Mahomed, 1882, p. 481).
- *...counting the offspring of like mid-parentages as members of the same co-family...* (Galton, 1885a, p. 6).
- *Co-kinsmen... co-fraternals... co-filials...* (Galton, 1886a, p. 56)
- *co-family* (Galton, 1886c, p. 255).
- In *Natural Inheritance* (published in 1889 but written in early to mid-1888), he used the terms "Co-Fraternity", "Co-kinsmen", and "Mid-Co-Fraternity" (Galton, 1889a, pp. 94, 114, 117, respectively).

Galton's appreciation for new terms and phrases continued into later life, as evidenced by his praise of a contemporary's publication that Galton pronounced as having been written "with a remarkable earnestness, [a] wealth of apposite phrases, and happy turns of expression" (Galton, 1894, p. 755).

The single instance of his use of "co-efficient of reversion" in a paper of his that was published in *Nature* (Galton, 1877b, p. 532) should not be counted among these examples because none of the many other uses of "coefficient" in that paper had a hyphen; and the unhyphenated word "coefficient" rather than the hyphenated word co-efficient appeared in the original published version of that paper (Galton, 1877a, p. 298), which *Nature*'s version otherwise duplicated word for word, table for table, and figure for figure. If we consider the "co-efficient" hyphen to be a printing error, it would not be the only one in that paper: Amazingly, *Nature* failed to print the last sentence and formulae (starting with "Suppose...") that are present in the original.

In closing here, it is interesting to note that in 1967, statisticians Snedecor and Cochran (neither of whom were historians) seemed to claim that Galton himself coined the word co-relation: "The data [that we are using in this part of our textbook] are from the article by Galton in 1888 in which the term 'co-relation' was first proposed" (Snedecor and Cochran, 1967, p. 177).

Answer #4: Advice from a Friend

In Galton's personal copy of Jevons's book on *Principles of Science*, there is evidence that he was familiar with the definition of correlation given in W. R. Grove's book *Correlation of Physical Forces* (see Answer#1, above). Galton's personal copy of the sixth edition of Grove's book is now stored at University College London (Grove, 1874b). Galton may have read the advice given in Grove's fifth edition preface, which is included in that sixth edition. The relevant text from that fifth edition preface is:

> *The phrase 'Correlation of Physical Forces,' in the sense in which I have used it, having become recognised by a large number of scientific writers, it would produce confusion were I now to adopt another title. It would, perhaps, have been <u>better</u> if I had in the first instance used the term <u>Co-relation</u>, as the words 'correlate,' 'correlative,' had acquired a peculiar metaphysical sense somewhat differing from that which I attached to the substantive correlation.*

> (Grove, 1867, p. v, taken from Grove, 1874a, p. vii; underlining added)

Unfortunately, the page on which that preface occurs in Galton's copy has no underlining or markings or handwritten notes, unlike the relevant page in his copy of Jevons's book. However, assuming that Galton had read that preface, it may have influenced his choice of words in December 1888; in support of such a claim, I provide the following evidence (from Galton's autobiography) of how close was the friendship between him and Grove (all page references are to Galton, 1908b; all underlining added):

- *Grove...my valued friend in later years* (pp. 41–42).

- *The late Justice Sir William Grove (1811–1896) is one of those to whom I owe most for sympathy in my inquiries, for <u>helpful criticisms</u>, and for long-continued friendship (p. 219).*
- *It was his [Grove's] practice to rent a large house and shooting [sic] during the autumn vacation, and he most hospitably asked my wife and myself to make long visits to him during three autumns (p. 220).*

Galton and Grove were both members of the Athenaeum Club, which was founded in 1824 as a meeting place for distinguished authors, scientists, and artists (Athenaeum-History). There would likely have been many scientific discussions between Galton and Grove while at the Club, given that by 1888 they had both been members of it for more than 30 years and that they both had been "elected by the [Club's] Committee, under Rule II., as being 'of distinguished eminence in Science, Literature, or the Arts, or for Public Services'" (Waugh, 1894, pp. 55, 62, and unpaginated introductory page). They may have taken walks together or frequently visited each other's home, given that in 1888, Galton lived at 42 Rutland Gate (Galton, 1908b, p. 158) and Grove at 115 Harley Street (Lee, 1901, p. 372); their homes were less than two miles apart, most of which distance would have been through Hyde Park; and both homes were also less than two miles from the Athenaeum Club (at 107 Pall Mall).

Pearson, in his 4-volume biography of Galton (Pearson, 1914–1930), provided other evidence of Galton's relationship with Grove. For example:

- Pearson considered Grove to be "Galton's close friend" (vol. 2, p. 282 n1).
- Over a span of many years, Galton and his wife were periodically over-night house-guests of the Grove's (vol. 2, pp. 130, 161, 180; vol. 3B, p. 465).
- Grove helped Galton with his biometric studies by allowing his finger prints to be taken, and his head to be "often measured" (vol. 3A, pp. 216, 248).
- "Both Galtons [Francis and his wife] were much depressed during this year [1896]. Emily Gurney died, and Sir William Grove died…" (vol. 2, p. 280n).
- Galton confided in a letter that "Grove was one of the very kindest friends I ever had" (vol. 3B, p. 531).
- Galton's home had "on the walls the prints of Galton's friends – Darwin, Grove, Hooker, Brodrick, Spencer, Spottiswooode, etc…" (vol. 2, pp. 11–12).

As we saw in an earlier chapter, Grove had previously formed a strong opinion regarding the misuse of the word correlation:

> *Its use* [i.e. the use of the word correlation] *has, in my judgment, been carried*
> *too far in applying it to subjects quite beyond its fair meaning. There are many*
> *facts, one of which cannot take place without involving the other; one arm of a*
> *lever cannot be depressed without the other being elevated – the finger cannot*
> *press the table without the table pressing the finger. A body cannot be heated*
> *without another being cooled, or some other force being exhausted in an equiva-*
> *lent ratio to the production of heat; a body cannot be positively electrified without*
> *some other body being negatively electrified, &c. To such cases the term correla-*
> *tion may be usefully applied, but hardly to adaptations of structure, &c.*
>
> *(Grove, 1874a, p. 166; underlining added)*

I conjecture that prior to finalizing his December 1888 paper, Galton discussed it with his good friend Grove. I speculate that Galton's word choice
was based on Grove's "helpful criticisms" in connection with Grove's fifth-
edition-Preface advice that it would be "better" to use the word "co-relation"
than the word "correlation".

Answer #5: "Co-Relation" was the Best Word for What He Meant

Michael Bulmer in his biography of Galton briefly focused on the word
co-related:

> *In this paper (Galton 1888), he began by saying that "co-relation or correlation*
> *of structure" was a common idea in biology (for example, the length of the arm*
> *was said to be co-related with that of the leg, because a person with a long arm*
> *usually has a long leg, and conversely), but that no attempt had previously been*
> *made to measure its degree.*
>
> *(Bulmer, 2003, p. 193; underlining added)*

Thus it seems that Bulmer agreed with Galton's claim (to be discussed shortly)
that the hyphenated word co-relation had been in common use to describe
Co-Relational Correlations. Unfortunately, neither Galton nor Bulmer provided any evidence for that claim. As previously discussed, co-relation had
not been used in that way by Galton nor by any other author whom I've
found. On the contrary, virtually all pre-1888 natural scientists had instead
been using a different word, namely *relation*.

It is significant that, prior to December 1888, Galton and his contemporaries
used the word correlation almost exclusively in reference to Observational
Correlation (as previously discussed). Like Lyell, Darwin, Spencer, and others, Galton's earlier writings (and his 1888 paper) usually described the dependency of one numeric data set upon another as being a "relation"; therefore,
in late 1888, Galton predictably referred to his measure of *co*-dependency as

a *co*-relation. In support of such a conjecture, it is noteworthy that in 1893, NED elaborated on the various ways that the prefix "co-" had been used in the English language; it stated that the "general sense is 'together', 'in company', 'in common', 'joint,-ly', 'equal,-ly', 'reciprocally', 'mutually'" (Murray, 1893, p.543).

In his December 1888 paper, what difference in meaning did Galton give to the words "relation" and "co-relation"? He told us on the second page of the paper, where he said...

> The relation *between the cubit* [a measure of arm length] *and the stature* [height] *will be shown to be such that for every inch, centimetre, or other unit of absolute length that the cubit deviates from the mean length of cubits, the stature will on the average deviate from the mean length of the statures to the amount of 2.5 units, and in the same direction. Conversely, for each unit of deviation of stature, the average deviation of the cubit will be 0.26 unit. These* relations *are not numerically reciprocal, but the exactness of the* co-relation *becomes established when we have transmuted the inches or other measurement of the cubit and of the stature into units dependent on their respective scales of variability.... The particular unit* [of variability] *that I shall employ is the value of the probable error of any single measure in its own group.*
>
> (Galton, 1888c, p. 136; underlining added)

There he used the word "relation" to describe a plot of *X,Y* data wherein the probable errors of *X* and *Y* have different magnitudes. The word "co-relation" was used only after the data had been converted ("transmuted" he called it) by dividing each *X* and each *Y* by their respective probable error. On another page, he again used that same vocabulary:

> The statures of kinsmen are co-related *variables; thus, the stature of the father is* correlated *to that of the adult son, and the stature of the adult son to that of the father; the stature of the uncle to that of the adult nephew, and the stature of the adult nephew to that of the uncle, and so on; but the index of* co-relation, *which is what I there called "regression," is different in the different cases. In dealing with kinships there is usually no need to reduce the measures to units of Q* [i.e. probable error], *because the Q values are alike in all the kinsmen, being of the same value as that of the population at large. It however happened that the very first case that I analysed was different in this respect. It was the* reciprocal rela-tion *between the statures of what I called the "mid-parent" and the son* (p. 143; underlining added).

That one and only use of the word correlated in the body of the paper is puzzling. I speculate that Galton may have drafted his paper using the word correlation throughout and may have at the last moment changed his mind to co-relation; in implementing that switch, he may have blundered in not changing that one instance of correlated. Such speculation would explain why, in the entire paper (other than in the first sentence in which he mentions the classic term "correlation of structure"), there is only one instance of the un-hyphenated word, vs. 21 instances of the hyphenated. Additionally, he stated that the probable errors of kinsmen are typically "alike", and so there is no need to transmute them, and so their statures can be said to be "co-related" variables as is, without transmutation. However, in a particular set of data involving the mid-parent and son values, the probable errors were not the same; and he therefore says that there is only a "relation" between the data sets rather than a "co-relation".

About a month later, in a lecture given on January 22, 1889, he tried to clarify himself (by this time, is seems that he'd already decided to substitute the word "correlation" for his original word "co-relation" – more about that, later):

> *The various measurements made at the* [Galton Biometric] *laboratory have already afforded data for determining the general form of the <u>relation</u> that connects the measures of the different bodily parts of the same person.' We know in a general way that a long arm or a long foot implies on the whole a tall stature – ex pede Herculem; and conversely that a tall stature implies a long foot. But the question is whether their <u>reciprocal relation</u>, or <u>correlation</u> as it is commonly called, admits of being precisely expressed. Correlation is a very wide subject indeed. It exists whenever the variations of two objects are in part due to common causes.....*

> *(Galton, 1889e, p. 403; underlining added)*

In that lecture, he also discussed regressions that are *not* reciprocal (and are therefore *not* referred to as correlations), which result in different regression values depending on whether they are derived from a plot of X on Y or Y on X. Then he said:

> *The lengths of head-lengths and head-breadths are akin to each other in the same sense as kinsmen are. So it is in the closer <u>relation</u> between the lengths of symmetrical limbs, left arm to right arm, left leg to right leg. The regression would be strictly <u>reciprocal</u> in these cases. When, however, we compare limbs whose variations take place on different scales, the differences of scale have to be allowed for before the regression can assume a <u>reciprocal form</u>. The plan of making the requisite allowance is perfectly simple; it merely consists in dividing each result by the probable error of any one of the observations from which it was deduced.... In some cases the scale of variation in the two correlated members is very different, and this divisor may be very large. Thus the length of the middle finger varies at so very different a rate from that of the stature that 1 inch of the difference of the middle finger length is associated on the average with 8.4 inches of stature. On the other hand, 10 inches of stature is associated on average with 0.6 inch of middle finger length. There is no reciprocity in these numerals; yet, for all that, <u>when the scale of their respective variations is taken into account by using the above-mentioned divisor, the values become strictly reciprocal</u>.*

> *(Galton, 1889e pp. 404–405; underlining added)*

In a large footnote on page 408 of the published version of that lecture (given here next), he clarified some of the text that he'd provided in the concluding paragraph of his December 1888 paper; his clarification is an almost exact quote from the end of the January 3 abstract of his December paper:

> *The general result of the inquiry* [described in the December 20 paper] *was that, when two variables that are severally* [sic] *conformable to the law of frequency of error, are correlated together, the conditions and measure of their closeness of correlation admits of being easily expressed. Let x_1, x_2, x_3, &c., be the deviations in inches, or other absolute measure of the several "relatives" of a large number of "subjects," each of whom has a deviation, y, and let X be the mean of the values of x_1, x_2, x_3, &c. Then (1) $y = rX$, whatever may be of value of y. (2) If the deviations are measured, not an inches or other absolute standard, but in units, each equal to the Q (that is, to the probable error) of their respective systems, then r will be the same, whichever of the two correlated variables is taken for the subject. In other words, the relation between them becomes reciprocal; it is strictly a correlation. (3) r is always less than 1. (4) r (which, in the memoir on hereditary stature, was called the ratio of regression) is a measure of the closeness of correlation. (5) The probable error, or Q, of the distribution of x_1, x_2, x_3, &c., about X, is the same for all values of y, and is equal to $\sqrt{(1-r^2)}$ when the conditions specified in (2) are observed. It should be noted that the use of the Q unit enables the variations of the most diverse qualities to be compared with as much precision as those of the same quality. Thus, variations in lung-capacity which are measured in volume can be compared with those of strength measured by weight lifted, or of swiftness measured in time and distance. It places all variables on a common footing.*

> *(Galton, 1889e, pp. 408n–409n; underlining added;* the only difference between this quote and the one in the January 3 abstract is the absence here of a unimportant sentence that had been placed just in front of the "(5)")

He continued in that paper to use the word relation in reference to untransmuted relationships, and the word correlation in reference to transmuted ones:

> *It has already been shown that the correlation connects the deviations, and has nothing to do with the mean or average values. Now, to express this relation truly, so that it shall be reciprocal, the scale of deviation of the correlated limbs, say, for example, of the cubit and of the stature of adult males, must be reduced to a common standard. We therefore reduce them severally to scales in each of which their own Q is the unit. The Q of the cubit is 0.56 inch, therefore we divide each of its deviations by 0.56. The Q of the stature is 1.75 inch, so we divide each of its deviations by 1.75. When this is done the correlation is perfect. The value of regression is found to be 0.8, whether the cubit be taken as the "subject" and the mean of the corresponding statures as the "relative," or vice versâ.*

> *(Galton, 1889e, p. 416–417; underlining added)*

A year later, he continued to clarify his terminology, in his paper titled "Kinship and Correlation":

- *I will take another of the same kind of examples in order to emphasize* <u>*the difference between relation and correlation, of which no explanation has thus far been attempted*</u> (Galton, 1890a, p. 425; underlining added; this statement (i.e., "no explanation…") is not true, given the attempts he made in his 1889 papers, previously discussed.).

- *There is not* <u>*that reciprocal relation between them which is conveyed by the word correlation*</u>*…. There is* <u>*relation*</u> *between stature and length of finger, but no real correlation. On the other hand, the scale of variation of symmetrical limbs, such as that of the right and left cubit, is so nearly the same that they can justly said to be correlated…. <u>Whenever the resulting variability of the two events is on a similar scale, the relation becomes correlation</u>…. When it is not the same, and when the variations are of the character shortly to be described as quasi-normal, a simple multiplication will be found to suffice… to <u>transform the relation into a correlation</u>. Thus we may speak of the length of the middle finger and that of the stature being correlated together under a recognized understanding that the variations are quasi-normal, and that the multiplication in question shall be made. <u>Henceforth I will use the word correlation subject to these tacit understandings</u>* (p. 426; underlining added).

In summary for Answer #5: Galton observed that when probable errors differ between the X and Y variables, the slope of an X-as-relative + Y-as-subject plot differs from the slope of the corresponding Y-as-relative + X-as-subject plot, and that therefore there are *two* different "relations". However, after X and Y have been transmuted by dividing by their respective probable error, the resulting two relationship plots (i.e., X on Y, and Y on X) surprisingly have the same slope (i.e., the same relation). Therefore, Galton probably chose the word "co-relation" because it was the best term for his transmutation of two "relations" into one.

The remainder of this chapter includes a detailed examination of the most important and/or interesting text in his December 1888 paper.

The Entire First Sentence in Galton's December 1888 Paper was

"Co-relation or correlation of structure" is a phrase much used in biology, and not least in that branch of it which refers to heredity, and the idea is even more frequently present than the phrase; but I am not aware of any previous attempt to define it clearly, to trace its mode of action in detail, or to show how to measure its degree.

(Galton, 1888c, p. 135)

Galton put quotation-marks around the first four words in that sentence, which is a standard indicator that the words have been derived from some other publication or possibly from common usage. Let's now look at his sentence, phrase by phrase.

> *"Co-relation... of structure" is a phrase much used in biology...*

That is not a true statement. "Co-relation of structure" may have *never* been used in *any* field of study. The only biology-related instances of the word "co-relation" found during research for this book were thrice in a mid-19th-century encyclopedia of articles on animal anatomy and physiology (Todd, 1859, previously discussed) and once in Richard Owen's *Anatomy* (previously discussed). In all of that encyclopedia's more than 5000 pages written by more than 100 authors, the phrase "co-relation of structure" was not used at all; the word co-relation was used twice by one author and once by another, all of which uses were in reference to co-relation of function, not co-relation of structure (quotes previously discussed). The use by Owen ("the co-related muscle") is notable for its being the only one in that book's more than 2000 pages, vs. several uses of the word correlation (e.g., "the correlations of structures"; full quotes and references were previously discussed). As earlier mentioned, Galton himself had never used that phrase nor even the word "co-relation" in any of the approximately 400 pre-1888 documents that are available on Galton.org (Zorich, 2020a). His making this unfounded statement seems to be another blunder; or possibly he was referring to himself when he subsequently said that "the idea is even more frequently present than the phrase".

> *"...correlation of structure" is a phrase much used in biology...*

That is a true statement; as previously discussed, that phrase had been widely used for about a century, e.g., by Cuvier, Owen, and Darwin.

> *I am not aware of any previous attempt to define it clearly...*

At first glance, that is not a true statement. Galton had thoroughly read *Origin of Species*, as proved by Galton's 1859 letter to Darwin (letter 82 in Darwin and Seward, 1903). In *Origin*, Darwin had used the term "correlation of growth" for the overarching concept that included Cuvier's correlation of structure. Darwin clearly defined his term in *Origin* and followed it with several pages of explanation. That definition (shown here next) used the phrase "tied together" where, post-1888, the word correlated might have been used:

> *I mean by this expression that the whole organisation is so <u>tied together</u> during its growth and development, that when slight variations in any one part occur, and are accumulated through natural selection, other parts become modified.*

> *(Darwin, 1859, p. 143; underlining added)*

Additionally, Galton was familiar with the definitions for the word correlation provided by Grove and Jevons (as proved by a record documented
by Stigler, 1986, p. 298), although those definitions did not relate to bodily
structures.

Whether or not Galton's "I am not aware..." statement is false depends
upon the meaning Galton intended for "co-relation or correlation of structure". At this point in the paper, the 19th-century reader would assume that
Galton meant the same as Cuvier, Owen, and Darwin, i.e., an Observational
Correlation; however, it was only later in the paper that Galton explained that
he meant (for example) the relationship between length of arm and length of
leg, i.e., a Co-Relational Correlation. It is true that *those* relationships had not
previously been clearly defined *mathematically*. Therefore, it is fair to say that
his statement that "I am not aware..." is at least misleading, given its location
in Galton's paper.

> *I am not aware of any previous attempt... to trace its mode of action in detail...*

That statement is not true. Or possibly Galton had forgotten that in *Origin
of Species*, Darwin had focused exactly on that topic in his 7-page sub-section titled "Correlation of Growth" (Darwin, 1859, pp. 143–150); and in fact,
Darwin's explanations were in much more detail than anything provided by
Galton in his December 1888 paper.

> *I am not aware of any previous attempt... to show how to measure its degree.*

That statement is likely true. The only previous attempt encountered during research for this book was that by Alexander Graham Bell (previously
discussed). Most likely, Galton was not aware of Bell's attempt – it was published in 1885 in an American journal and it focused on *counts* of defects
of the senses rather than *measurements* of body parts. Even if Galton had
been aware of Bell's paper, he would have likely dismissed it because it was
based upon *averages*; and as Galton would explain on the second page of his
Co-relations paper:

> ...it is well to point out that the subject in hand has nothing whatever to do
> with the <u>average</u> proportions between the various limbs... (p. 136; underlining
> added).

In *Natural Inheritance*, Galton had explained his reasoning for such negativity
about averages:

> *The knowledge of an average value is a meagre piece of information.... So in
> respect to the distribution of any human quality or faculty, a knowledge of mere
> averages tells but little....*

> (Galton, 1889a, pp. 35–36)

It is interesting that in Galton's December 1888 paper's first sentence, he used the word degree in reference to the aspect of co-relation that he'd be measuring ("how to measure its <u>degree</u>"); but in the very next paragraph, he instead talked about "how the <u>closeness</u> of co-relation in any particular case admits of being expressed by a simple number" (pp. 135–136; underlining added). In the next paragraph, he changed his mind again, when he spoke of "the <u>nearness</u> with which they [height and arm-length] vary together" (p. 136; underlining added). In the paper's penultimate paragraph, he returned to the word degree when he discussed "how to measure the degree in which one variable may be co-related with the combined effect of n other variables" and "a method by which the degree may be measured" (p. 144). Finally, in the last words of the paper's ultimate paragraph, he switched back to "closeness": "r measures the closeness of correlation" (p. 145). In his only subsequent new publications that focused at least in part on correlation, he continued to vacillate when he said that r is...

- *a measure of the closeness of correlation* (Galton, 1889e, p. 408n).
- *the measure of correlation between any two variables* (Galton, 1889d, p. 296).
- *an admirable measure of the closeness, or weakness, of correlation* (Galton, 1890a, p. 430).

He seems to have never decided on a best word to describe the feature of correlation that his r was measuring. In the 20th century, there continued to be no uniformity among textbook authors, who variously described r as measuring (for example) the amount, extent, degree, intensity, or strength of correlation (see Zorich, 2018, for a lengthier discussion of this point).

The Next Two Sentences in Galton's December 1888 Paper were

Two variable organs are said to be co-related when the variation of the one is accompanied on the average by more or less variation of the other, and in the same direction. Thus the length of the arm is said to be co-related with that of the leg, because a person with a long arm has usually a long leg, and conversely (p. 135).

That meaning was not unique to Galton. For example, as previously discussed, Darwin had used "correlated" with exactly that meaning in at least four instances in *Variations of Plants and Animals*, the most obvious example being this: "in all the breeds of the pigeon the length of the beak and the size of the feet are correlated" (Darwin, 1868, vol. 2, p. 323).

Galton continued his explanation in his next sentence:

> *If the co-relation be close, then a person with a very long arm would usually have*
> *a very long leg; if it be moderately close, then the length of his leg would usually*
> *be only long, not very long; and if there were no co-relation at all then the length*
> *of his leg would on the average be mediocre* (p. 135).

That sentence is not clearly written. What he should have written is this following revision of it that I have created, wherein the most relevant text that I added is underlined:

> If the co-relation be close, then a person with a very long arm would
> usually have a very long leg, <u>and a person with a very long leg would</u>
> <u>usually have a very long arm</u>; if it be moderately close, then a person with
> a very long arm (<u>or leg</u>) would typically have a leg (<u>or arm</u>) that is only
> long, not very long; and if there were no co-relation at all, then a person
> with a very long arm (<u>or leg</u>) would have leg (<u>or arm</u>) length that is on the
> average mediocre.

That added text may have been obvious to Galton, and therefore it did not occur to him to state his thoughts fully. If he hadn't been in a hurry, I think he would have written more clearly, given the clarity with which he explained correlation in papers written at greater leisure during the following year. Such "haste" (as he would call it in one of those papers) is what he had called "on a sudden" in a description of his writing style (previously discussed).

At this point in his December 1888 paper, Galton had slightly confused his audience; however he clarified himself in his next sentences:

> *It is easy to see that co-relation must be the consequence of the variations of the*
> *two organs being partly due to common causes. If they were wholly due to com-*
> *mon causes, the co-relation would be perfect, as is approximately the case with*
> *the symmetrically disposed parts of the body. If they were in no respect due to*
> *common causes, the co-relation would be nil. Between these two extremes are an*
> *endless number of intermediate cases, and it will be shown how the closeness of*
> *co-relation in any particular case admits of being expressed by a simple number*
> (p. 135).

There, he focused on biometrics; as such, of course a long arm does not *cause* a long leg, nor vice-versa. Instead, some unknown biological forces are synchronously at work, tending to make those limbs either both long or both short. That tendency Galton says can be "expressed by a simple number". Next, he said:

> *The fact that the average ratio between the stature and the cubit is as 100 to 37,*
> *or thereabouts, does not give the slightest information about the nearness with*
> *which they vary together* (p. 136).

That sentence helps the reader understand what Galton was trying to do in this paper: He wanted to show how to assign numerical values to the "nearness" with which paired data-sets "vary together". Given his denigration of averages in that sentence, the reader may be surprised to find them prominently mentioned in his next paragraph:

> The <u>relation</u> between the cubit and the stature will be shown to be such that for every inch, centimetre, or other unit of absolute length that the cubit deviates from the mean length of cubits, the stature will on the <u>average</u> deviate from the mean length of the statures to the amount of 2.5 units, and in the same direction. Conversely, for each unit of deviation of stature, the <u>average</u> deviation of the cubit will be 0.26 unit (p. 136; underlining added).

Flanking Data

Those "average" deviations of 2.5 and 0.26 that Galton gave in that last quote were determined by him using values in his Table III, which was constructed based on data in his Table II. The data of other pairings of (e.g., stature vs. knee height) were given in his Table IV. He then used the data from Tables III and IV (and other data not provided in his paper) to create plots (only one of which was provided in the paper) on the basis of which he calculate the correlation coefficients that he showed in his Table V. In his words:

> The values derived from Table II, and from <u>other similar tables</u> [not provided in this paper], are entered in Table III.... Six other tables are now given in summary form [in Table IV].... <u>From Table IV</u> [and Table III] <u>the deductions in Table V can be made</u>; but they may be made directly from tables <u>of the form</u> of Table III, whence Table IV was itself derived (pp. 140, 142; underlining added).

Modified versions of Galton's Tables II and III are provided here as Tables 4.1 and 4.2, respectively; his Tables IV and V are discussed later in this chapter. In Table 4.1, the core set of data is highlighted in black background with white font; the data outside that core set is what Galton in his paper called the "flanking" data. In Table 4.2, the numbers in white font on a black background are the "transmuted" non-flanking data that he plotted in his one and only "figure" (which will be discussed in detail, later in this chapter).

There are four "flanking" rows and columns in his Table II; he labeled them as "71 and above", "Below 64", "19.5 and above" and "Under 16.5". In the last sentence in a paragraph that focused solely on how "Tables were then constructed... like Tables II and III" (p. 137), he stated that...

TABLE 4.1

Table II was the Only Correlation Matrix Table Provided in Galton's December 1888 Paper

Galton's Table II, Modified for Clarity. "Total Cases" 38, 47, 35 Are Corrections to Galton's Incorrect Totals of 48, 48, 34

Stature in Inches, as Interval Midpoints	Length of Left Cubit in Inches, 348 Adult Males								Total Cases
	Under 16.5	16.5– 17.0	17.0– 17.5	17.5– 18.0	18.0– 18.5	18.5– 19.0	19.0– 19.5	19.5 & above	
				Interval Midpoints					
	N/A	16.75	17.25	17.75	18.25	18.75	19.25	N/A	
71 and above....	0	0	0	1	3	4	15	7	30
70.................	0	0	0	1	5	13	11	0	30
69.................	0	1	1	2	25	15	6	0	50
68.................	0	1	3	7	14	7	4	2	38
67.................	0	1	7	15	28	8	2	0	61
66.................	0	1	7	18	15	6	0	0	47
65.................	0	4	10	12	8	2	0	0	36
64.................	0	5	11	2	3	0	0	0	21
Below 64......	9	12	10	3	1	0	0	0	35
Totals...........	9	25	49	61	102	55	38	9	348

Source: Adapted from Galton (1888c).

> *None of the entries lying within the flanking lines* [rows] *and columns of Table II were used* (p. 138).

What did Galton mean by "None... were used"? Did he mean it literally? For example, for men whose average (i.e., "mean") cubit length is 18.25, did he calculate the average stature for only the men whose heights were from 64 to 70 (*excluding* those whose height were "71 and above" and "Below 64")? That is, did he use *only* the data with white font on a black background in my Table 4.1? The answer to that question is historically important because "[Table II contains the] data from which Galton in 1888 found the *first* published correlation coefficient" (Stigler, 1986, p. 319, table 9.1; italics added).

One way to answer that question involves calculating averages based only on my version of Table II's black-cell entries (i.e., excluding all flanking entries) and comparing those averages to the corresponding ones in Galton's Table III. For brevity's sake, I've focused here only on the stature averages calculated for men whose cubit lengths are at the extreme "interval midpoint" values of 16.75 and 19.25 (headed by "16.5–17.0" and "19.0–19.5", respectively). In Table II, the 16.75 column has flanking-row statures that are all "Below 64" and are therefore all *smaller* than those for the black-cell entries. Similarly, the 19.25 column has flanking-row statures that are all "71 and above" and are therefore all *larger* than those for the black-cell entries. If Galton had included

TABLE 4.2

Galton Derived This Table from His Table II

Galton's Table III, Modified for Clarity.
The Corrections Shown in His Table II (i.e., Table 4.1) Are Shown Here Also.

Stature: Median = 67.2 inches; Probable Error = 1.75 inch
Left Cubit: Median = 18.05 inches; Probable Error = 0.56 inch

		Stature = Subject			Left Cubit = Relative		
			Deviation from Median, in Units of...				Deviation from Median, in Units of...
Number of Non-flanking Cases (Total = 284)	Chosen Interval Midpoints in Inches	Inches	Probable Errors	Calculated Mean of Corresponding Left Cubits	Inches	Probable Errors
30	70	2.8	1.60	18.8	0.8	1.42
50	69	1.8	1.03	18.3	0.3	0.53
38	68	0.8	0.46	18.2	0.2	0.36
61	67	−0.2	−0.11	18.1	0.1	0.18
47	66	−1.2	−0.69	17.8	−0.2	−0.36
36	65	−2.2	−1.25	17.7	−0.3	−0.53
21	64	−3.2	−1.83	17.2	−0.8	−1.46

		Left Cubit = Subject			Stature = Relative		
			Deviation from Median, in Units of...				Deviation from Median, in Units of...
Number of Non-flanking Cases (Total = 330)	Chosen Interval Midpoints in Inches	Inches	Probable Errors	Calculated Mean of Corresponding Statures	Inches	Probable Errors
38	19.25	1.2	2.14	70.3	3.1	1.8
55	18.75	0.7	1.25	68.7	1.5	0.9
102	18.25	0.2	0.36	67.4	0.2	0.1
61	17.75	−0.3	−0.53	66.3	−0.9	−0.5
49	17.25	−0.8	−1.42	65.0	−2.2	−1.3
25	16.75	−1.3	−2.31	63.7	−3.5	−2.0

Source: Adapted from Table III in Galton (1888c).

in his stature averages the values for the flanking entries, then those averages should differ predictably from the averages calculated using only the values for the black-cell entries. Using standard frequency-table mathematics, it is a simple matter to calculate the stature averages for *only* the black-cell entries in columns 16.75 and 19.25 (i.e., without the flanking rows); the resulting stature averages are 65.4 and 69.1, respectively. The corresponding stature averages given by Galton in the lower part of his Table III are 63.7 and 70.3. Thus, Table II's black-cell average for the 16.75 column is larger than that in Table III, and Table II's black-cell average for the 19.25 column is smaller than that in Table III. Such differences are exactly what would be expected if Galton had indeed used *all* the data rather having used "None of the entries lying within the flanking lines and columns".

Based upon the results of that investigation, we can conclude that Table III's averages *do* included Table II's flanking entries. By extension, that statement applies to the averages of all other data-pairings shown in Table IV and to all other tables not in his paper but which he used to calculate the correlation coefficients shown in his Table V. This is an example of a Galton verbal blunder: Instead of "None of the entries lying within the flanking lines and columns of Table II were used.", Galton should have written: "None of the entries lying within the flanking lines and columns of Table II are listed in Table III, although they were used to calculate the averages shown in that table."

However, Galton did not *plot* the flanking data in his figure (discussed later in this chapter); that is, he plotted only the data with white font on a black background in my version of his Table II (he plotted it after the data had been transmuted, as in his Table III). His sentence about not using the flanking data was misplaced in his paper; that is, instead of placing it (as he did) at the end of a paragraph in which he discussed creation of his Tables II and III, he should have placed it two paragraphs later where he explained how he created his figure.

Reciprocal

After he stated that the values of those two converse relations between cubit and stature were 2.5 and 0.26, he immediately informed the reader that...

> These <u>relations</u> [2.5 and 0.26] *are not <u>numerically reciprocal</u>*... (p. 136; underlining added).

At this point in his lecture or published paper, what meaning would the listener or reader assume for "not numerically reciprocal"? It seems that there are only two possibilities, both of which will be discussed here next. The fact that both meanings might seem self-evident may explain why Galton did not clearly explain himself in December of 1888; but surprisingly he did feel the need to explain himself in the following month (which is discussed later in this section).

First, he might have meant that the two numbers are (in my words) "not mathematical inverses of each other"; that is, the reciprocal of 2.5 (i.e., 1/2.5) is 0.40 not 0.26, and the reciprocal of 0.26 is 3.8 not 2.5. If that was his meaning for "reciprocal" at this point in his paper, then his purpose might have been to emphasize the data-source for his numbers, as explained here next.

Using the data from only the *lower* part of Galton's Table III (see Table 4.3 here), the least squares linear regression (LSLR) slope of Stature on Cubit is

TABLE 4.3

The Deviations from the Median (in Inches) of Subject and Relative
Values that Galton Calculated for Stature and Cubit

Data from LOWER Part of Galton's Table III	
Left Cubit as Subject, Deviation from Median, in Units of Inches	Stature as Relative, Deviation from Median, in Units of Inches
1.2	3.1
0.7	1.5
0.2	0.2
−0.3	−0.9
−0.8	−2.2
−1.3	−3.5
Data from UPPER Part of Galton's Table III	
Left Cubit as Subject, Deviation from Median, in Units of Inches	Stature as Relative, Deviation from Median, in Units of Inches
2.8	0.8
1.8	0.3
0.8	0.2
−0.2	0.1
−1.2	−0.2
−2.2	−0.3
−3.2	−0.8

Source: Derived from Table III in Galton (1888c).

found to be 2.58, which is essentially the same as Galton's value of 2.5 that
he derived from a hand-drawn plot. Using that *same* lower-part data set, the
LSLR slope of Cubit on Stature is 0.39; there is indeed numerical reciprocity
between 2.58 and 0.39 – that is, 1/2.58 = 0.39, and 1/0.39 = 2.58. Such close
numerical reciprocity between Y on X and X on Y slopes calculated with the
same data set is seen when the data points plot as nearly a straight line (as
do Galton's data points from the lower part of Table III). However, Galton did
not derive his 2.5 and 0.26 from the *same* data set: He used the *lower* part of
Table III to obtain his 2.5, but he used the *upper* part of Table III to obtain his
0.26. Both the upper and lower parts of that table were created with the same
raw data, but the upper part used Stature as Subject (i.e., subjectively chosen
intervals) and Left Cubit as Relative (i.e., the average of all cubit values whose
Subject measurements were in a given interval), whereas the lower part used
Left Cubit as Subject and Stature as Relative. Therefore, one possible reason
for stating that 2.5 and 0.26 are not numerically reciprocal might have been
his desire to emphasize that those two values were not derived from a single
paired data set but rather from the two related but different sections of his
Table III.

In the sentences immediately following his "not numerically reciprocal"
statement, he provided evidence for a second possible meaning for it, i.e.,

(in my words) "2.5 and 0.26 are not the same number – they are not equal". He began by saying:

> These <u>relations</u> are not numerically reciprocal, but the exactness of the <u>co-relation</u> becomes established when we have transmuted the inches or other measurement of the cubit and of the stature into units dependent on their respective scales of variability. We thus cause a long cubit and an equally long stature, as compared to the general run of cubits and statures, to be designated by an identical scale-value. The particular unit that I shall employ is the value of the probable error of any single measure in its own group. In that of the cubit, the probable error is 0.56 inch = 1.42 cm.; in the stature it is 1.75 inch = 4.44 cm. Therefore the measured lengths of the cubit in inches will be transmuted into terms of a new scale in which each unit = 0.56 inch, and the measured lengths of the stature will be transmuted into terms of another new scale in which each unit is 1.75 inch. After this has been done, we shall find the deviation of the cubit as compared to the mean of the corresponding deviations of the stature, to be as 1 to 0.8. Conversely, the deviation of the stature as compared to the mean of the corresponding deviations of the cubit will also be as 1 to 0.8. Thus the existence of the <u>co-relation</u> is established, and its measure is found to be 0.8 (p. 136; underlining added).

Farther on in the paper, he spoke of the "<u>reciprocal</u> relation between the statures of what I called the 'mid-parent' and the son" (p. 143, underlining added; this was the only other time in the entire paper that the word reciprocal was used). Galton's words there do not at first appear to have a numerical connotation; but later in the same paragraph, he calculated the slope of an X,Y plot of transmuted mid-parent on transmuted son and the slope of transmuted son on transmuted mid-parent, and then concluded that the two values (which he gave as 0.47 and 0.44) are "practically the same". As Pearson would later point out (Pearson, 1930a, p. 55), Galton made a calculation error that when corrected would have allowed him to have said "identically the same", because *both* slopes in fact equaled 0.47. Notice that both of those values are actually correlation coefficients, because they are the slopes of X,Y plots in units of probable error. Given that result, it seems that Galton there used the word reciprocal to mean "practically the same" or almost equal.

Galton's future publications provided additional evidence for the just-discussed second meaning. For example, in the summary section of an abstract of his December 20, 1888, paper, an abstract that was published on January 3, 1889, he wrote:

> If the deviations are measured, not an inches or other absolute standard, but in units, each equal to the Q (that is, to the probable error) of their respective systems, then <u>r will be the same</u>, whichever of the two correlated variables is taken for the subject. <u>In other words, the relation between them becomes reciprocal;</u> it is strictly a correlation.

> (Galton, 1889b, p. 238; underlining added)

The word reciprocal also appeared in a paper that Galton presented orally a few weeks later, on January 22, 1889. In a subsection titled "Correlation", he first discussed the "reciprocal relation" between human stature and foot size (Galton, 1889e, p. 403); and then he discussed similar relations:

> *Now the relation of head-length to head-breadth, whose variations are on much the same scale, or speaking in technical language, whose probable errors are the same, is identical in character to the relation between kinsmen. There is regression in both cases, though its value differs. The lengths of head-lengths and head-breadths are akin to each other in the same sense as kinsmen are. So it is in the closer relation between the lengths of symmetrical limbs, left arm to right arm, left leg to right leg. <u>The regression would be strictly reciprocal</u> in these cases. When, however, we compare limbs whose variations take place on different scales, the differences of scale have to be allowed for before the regression can assume <u>a reciprocal form</u>. The plan of making the requisite allowance is perfectly simple; it merely consists in dividing each result by the probable error of any one of the observations from which it was deduced. Unfortunately the method cannot be briefly explained except by using these technical terms. In some cases the scale of variation in the two correlated members is very different, and this divisor may be very large. Thus the length of the middle finger varies at so very different a rate from that of the stature that 1 inch of difference of middle finger length is associated on the average with 8.4 inches of stature. On the other hand, 10 inches of stature is associated on the average with 0.6 inch of middle finger length. <u>There is no reciprocity in these numerals</u>; yet, for all that, when the scale of their respective variations is taken into account by using the above-mentioned divisor, <u>the values become strictly reciprocal</u>.*

> (Galton, 1889e, pp. 404–405; underlining added)

Later in that paper, he included virtually all of the end-of-paper summary from his January 3 abstract (Galton, 1889e, pp. 408n–409n). That summary mentioned that…

> *…r will be <u>the same</u>, whichever of the two correlated variables is taken for the subject. <u>In other words</u>, the relation between them becomes <u>reciprocal</u>…* (underlining added).

The full text of that summary was provided in a tiny-font footnote that spanned the bottoms of two pages (I wonder how many scientists read such long footnotes). Thankfully, many pages later in the main body of the paper, he explained that reciprocal means equal (although this final explanation was not nearly as clear as the one in the footnote, at least in regards to the word reciprocal):

> *Lastly, <u>as regards the correlation</u> of lengths of the different limbs. [sic] It has already been shown that the correlation connects the deviations, and has nothing to do with the mean or average values. Now, to express this relation truly, so that it shall be <u>reciprocal</u>, the scale of deviation of the correlated limbs, say,*

for example, of the cubit and of the stature of adult males, must be reduced to a
common standard. We therefore reduce them severally to scales in each of which
their own Q is the unit. The Q of the cubit is 0.56 inch, therefore we divide each
of its deviations by 0.56. The Q of the stature is 1.75 inch, so we divide each of
its deviations by 1.75. When this is done the correlation is perfect. The value of
regression [i.e. the correlation coefficient] is found to be 0.8 [i.e. the same],
whether the cubit be taken as the "subject" and the mean of the corresponding
statures as the "relative," or vice versâ.

(Galton 1889e, pp. 416–417; underlining added)

Why might he have chosen the word "reciprocal" instead of the word
"equal"? I speculate that he may have been following the example set by his
good friend Grove, who had defined correlation using the word reciprocal.
Grove had said:

The term Correlation, which I selected as the title of my Lectures in 1843, strictly
interpreted, means a necessary mutual or reciprocal dependence of two ideas,
inseparable even in mental conception.... The sense I have attached to the word
Correlation, in treating of physical phenomena, will, I think, be evident, from
the previous parts of this Essay, to be that of a necessary reciprocal production;
in other words, that any force capable of producing another may, in its turn,
be produced by it – nay, more, can be itself resisted by the force it produces, in
proportion to the energy of such production, as action is ever accompanied and
resisted by reaction: thus, the action of an electro-magnetic machine is reacted
upon the magneto-electricity developed by its action.... It would be out of place
here, and treating of matters too familiar to the bulk of my audience, to trace
how... materials have been provided for the generalisation now known as the
correlation of forces or conservation of energy...

(Grove, 1874a, pp. 165, 167, 196; underlining added)

Grove used the word reciprocal in the sense of "equal"; that is, when energy
is transformed from one form to another, there always remains an equal (i.e.,
the same) amount of energy ("conservation of energy").

Intraclass Coefficient

It is important to realize that Galton's December 1888 co-relation coefficient
was not the same as our present-day correlation coefficient because he analyzed tabular data (from a correlation matrix table), not individual raw
data points. As Stigler has emphasized, Galton "interpreted the correlation

coefficient both as a regression coefficient and as what we would now term as an intraclass correlation coefficient" (Stigler, 1999, p. 182). A relevant definition is:

> *Intra-class correlation* [is] *A measure of correlation within members of certain natural groups or "families".*
>
> (Kendall and Buckland, 1960, p. 145)

Today's reader might slight Galton for not analyzing and plotting the individual data points, but that would be an unfair criticism for a 19th century scientist handling what was then *big data*. The graphical method that Galton invented required him to transmute each value that would be plotted, i.e., to divide each value by its corresponding probable error. He had 348 cubit measurements and 348 stature measurements; he segregated them into a total of 13 groups, resulting in only $2 \times 13 = 26$ divisions by probable errors (see Galton's Table III). If he had plotted individual values, he (or an assistant) would have had $2 \times 348 = 746$ divisions to perform *by hand!* He had a similarly large number of values for each of his other variables that were mentioned in his paper (i.e., head length, head breadth, finger length, and knee height), which if transmuted one by one would have required many hundreds more divisions. Working in haste to a self-imposed deadline, he chose to perform about 200 divisions (using groups) instead of many times that (using individual values).

Galton could have used a mechanical calculator to ease his computation burden, given the fact that ones that could do division had become commercially available decades prior to 1888 (Britannica.com-Arithmometer). Did he own or have access to one? The following evidence suggests that he did not:

- In 1889, he did not blame data-entry-errors into such equipment for the many "numerical blunders" that he published in his December 1888 paper (Galton, 1890a, p. 421).
- In 1907, Pearson recommended that Galton employ "Miss Elderton" for "rapid and correct calculations" (Pearson, 1930a, p. 328).
- In Galton's 1908 autobiography, such a machine is not mentioned; nor, in Karl Pearson's 4-volume biography of Galton, was it mentioned as having been used by him.
- In 1924, when mechanical devices for performing arithmetic were commonly available, Pearson remarked that "mechanical calculators" were now being stored "in the Galton Laboratory...[in] the quaint stunted wardrobe from Galton's [home] dressing-room... a use which would have delighted his heart!" (Pearson, 1924, p. 12n).

As Yule summarized: "The method used by Galton to determine his coefficient was a <u>graphical approximation</u>" (Yule, 1910, p. 540; underlining added).

His Figure

Galton next provided some details on how he had collected his data and had arranged them into tables. Then he described how he created his paper's only "figure"; strangely, he did not label its axes, but rather provided that information only in the text of his paper. What is also strange is that the range of the axes was such that two of the plotted points (the uppermost and lowermost "x" points) actually fell outside the range of his plot's vertical axis. Additionally strange from a modern perspective is that the co-relation coefficient equals the slope of the line as seen from the Y-axis, not the X-axis.

Most strangely, he combined both measurements (cubit and stature) on the same axis. At first that might seem like heresy, but it fit his purpose, which was to show that for every 1.0 probable-error-unit change in the Subject (cubit or stature), there was only a 0.8 unit change in the Relative (stature or cubit). "This decimal fraction is consequently the measure of the closeness of the co-relation" (Galton, 1888c, p. 140).

The data that Galton intended to plot in his "figure" are identified here in Table 4.2 with white font on a black background. The paired values starting with 1.60, 1.42 (in the upper part of the Table) and ending with −1.83, −1.46 were plotted by Galton using the symbol "o" (which he called "circles"); they represent the correlation matrix table *row* data, i.e., Stature as Subject and Cubit as Relative. The remaining pairs (in the lower part of that table) were plotted using the symbol "x" (which he called "crosses"); they represent the correlation matrix table *column* data, i.e., Cubit as Subject and Stature as Relative. His figure has been recreated and modified here as Figure 4.2.

In a subsequent chapter of this book, we will discuss formulas for the correlation coefficient that were invented by Weldon, Edgeworth, and Davenport/Bullard. It will be helpful during such discussions to recall that Galton's December 1888 figure used one mark (circle or cross) for each non-flanking row or column in his correlation matrix table. There were a total of seven such rows and six such columns, and therefore seven circles and six crosses appeared in his figure.

There is at least one blunder in Galton's published figure. In the fourth row of Table 4.2's transmuted values can be found −0.11 and +0.18 as a data pair. However, Galton plotted that +0.18 value as −0.18. The corrected value is plotted in the version of the figure shown here. The figure's slope, using the corrected value and viewed from the Y axis, is 0.79 (per MS Excel's linear regression functions).

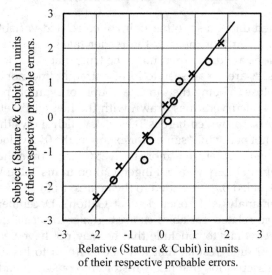

STATURE & CUBIT DEVIATIONS
"Observed" Data from Galton's Table III

Galton's December 1888 "Figure", modified & clarified:
  · Additional words added to main title.
  · Range of axes expanded from +/− 2 to +/− 3.
  · Titles and units of measure added to axes.
  · "o" = Stature as Subject; Cubit as Relative.
  · "x" = Cubit as Subject; Stature as Relative
  · Line slope = 1 / 0.8, and line forced through 0,0.
  · One error corrected (see chapter text for explanation).

FIGURE 4.2
This figure is an *X, Y* plot of the transmuted Stature and Cubit values; the slope of this line, as seen from the vertical axis, is what Galton considered to be a measure of co-relation. (Adapted from the "figure" in Galton (1888c, p. 138).)

Regarding his figure, Galton said:

> *...we shall find the deviation of the* [transmuted Subject] *cubit as compared to the mean of the corresponding deviations of the* [transmuted Relative] *stature, to be as 1 to 0.8. Conversely, the deviation of the* [transmuted Subject] *stature as compared to the mean of the corresponding deviations of the* [transmuted Relative] *cubit will also be as 1 to 0.8.... The firm line in the figure is drawn to represent the general run of the small circles and crosses* ["x" *symbols*]. *It is here seen to be a straight line, and it was similarly found to be straight in every other figure drawn from the different pairs of co-related variables that I have as yet tried... the inclination is such that a deviation of 1 part of the subject* [Y-axis on this plot], *whether it be stature or cubit, is accompanied by a mean deviation of the part of the relative* [X-axis on this plot], *whether it be cubit or stature, of 0.8. This decimal fraction is consequently the measure of the closeness of the*

co-relation…… r [the measure of that co-relation] *is the same, whichever of the two variables is taken for the subject…* (pp. 136, 140, 145).

He next provided data on six other pairs of co-related variables, but he did not show plots for any of them. If the reader takes the time to plot them, Galton's assertions are discovered not to be true; that is, curiously, the small circles and crosses are *not* all seen to form a straight line, nor are *all* the converse *r* values close to being the same. A couple of example plots are shown here in Figure 4.3 (curved lines drawn with the help of Excel's "trend line" chart feature). Also provided here is Table 4.4, which shows the worst examples of *r* values not being the "same"; for example, the 0.72 value for Length of Left Middle Finger vs. Stature is more than 25% larger than the 0.57 value for Stature vs. Length of Left Middle Finger; Galton in his Table V claimed that both *r* values equaled 0.70 (rounded to the nearest 0.05).

Upon further analysis, I conclude that Galton's December 1888 paper's seven pairs of co-related variables *on average approximately* meet his claims. Therefore, it seems fair to conclude that he may not have been consciously trying to hide or alter results that did not conform to his already-formed conclusions (an accusation that many historians consider valid when leveled at Gregor Mendel (Columbia.edu-Mendel)). Galton's intuitive mind jumped to what turned out to be the correct conclusion, even though not all his data supported it. His claims in this regard should have been muted somewhat; he should have written something like this: "Plots of transmuted subject vs. transmuted relative (whichever of the two variables is taken as the subject) always appear approximately as straight lines and reciprocally always have

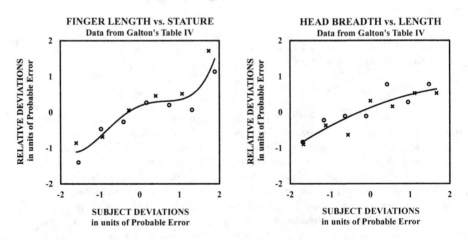

FIGURE 4.3

These charts (created using the exact same method that was used to create Figure 4.2) demonstrate that not all of Galton's transmuted data plotted as a straight line, contrary to what he claimed in his paper. (Derived from data in Galton (1888c).)

TABLE 4.4

Not All of Galton's Data Sets Produce Those Same "*r*" Values No Matter Which Variable is Plotted as the Subject and Which as the Relative; This Contradicts What He Claimed in His Paper

		Slope (= *r*) of Plots of Transmuted Deviations Based upon Data in Galton's Tables I, II, IV, and V	
Subject	Relative	Galton's Table V (He Rounded to Nearest 0.05)	Calculated by Excel's SLOPE Function
Stature	Cubit	0.80	0.71
Cubit	Stature	0.80	0.83
Stature	Middle finger	0.70	0.57
Middle finger	Stature	0.70	0.72
Stature	Knee height	0.90	0.81
Knee height	Stature	0.90	0.97
Cubit	Knee height	0.80	0.73
Knee height	Cubit	0.80	0.89

Source: Derived from data in Galton (1888c).
Note that the "Excel" values were calculated by putting the subject on the horizontal *X*-axis and the relative on the vertical *Y*-axis (Galton's values were calculated by his doing the reverse but then estimating the slope from the point of view of the vertical *Y*-axis).

approximately the same slope. When subject is plotted on the vertical-axis, the slope of the line as viewed from that axis is *r*, the measure of the co-relation."

One additional comment: Today's reader may wonder why Galton did not use the method of least squares to determine the best-fit line rather than to eye-ball draw it "to represent the general run of the small circles and crosses" (least squares had been known to mathematicians since the early 19th century). The likely reason is that (as previously discussed) he did not own or have access to a mechanical calculating machine – as explained by another author about a decade later, the huge task of calculating squares of hundreds if not thousands of data points is so laborious that for "most statisticians the method of least squares must always form an almost insuperable obstacle" (Crum, 1901, p. 85).

Galton's Concluding Paragraph

To conclude, the prominent characteristics of any two co-related variables, so far at least as I have as yet tested them, are four in number. It is supposed that their respective measures have been first transmuted into others of which the unit is in each case equal to the probable error of a single measure in its own series. Let y = the deviation of the subject, whichever of the two variables may be taken in that *capacity; and let x_1, x_2, x_3, &c., be the corresponding deviations of the relative, and let the mean of these be X. Then we find: (1) that $y = rX$ for all values of y; (2) that r is the same, whichever of the two variables is taken for the subject; (3) that r is always less than 1; (4) that r measures the closeness of co-relation* (p. 145).

Notice that his formula, $y = rX$, did not include an offset for what we today call the y-axis-intercept; the reason for that omission is that he was talking in terms of deviations from the mean or median, and therefore he manually forced his plot-line through the 0,0 point. Also note that this X was not the mean of all the x values, but rather was the mean of all the x (as Relative) values that corresponded to a given y (as Subject) value inside a single corresponding correlation matrix table cell.

That final-page formula, $y = rX$, is the most important one in his paper, and so it is astonishing that he wrote it blunderfully backward. As he had mentioned on previous pages, the "relative" is always less than the "subject" when both are measured in units of probable error (i.e., when both are transmuted). That regression-like relationship can be verified by simply looking at his figure (Figure 4.2, here): When the Subject value (on the vertical axis) is, for example, +1.0 or −1.0, the Relative values (on the horizontal axis) are +0.8 and −0.8, respectively. Thus, the correct formula, in words, is: *Relative = r(Subject)*. In his concluding paragraph, he clearly assigned y to be the transmuted Subject, x to be the transmuted Relative, and X to be the mean of the x values corresponding to a given y value. Therefore the correct formula, using his algebraic symbols, is $X = ry$, not $y = rX$. Unfortunately, Galton repeated that identical formulaic blunder in subsequent publications (Galton, 1889b, p. 238; 1889e, p. 408n; 1889f, p. 408n). I excitedly independently discovered all those blunders while writing this book, but I then learned that Karl Pearson had discovered the first of them a century earlier (Pearson, 1930a, p. 56n).

Analyses by Historians

Early in the 20th century, Karl Pearson undertook a mathematical analysis of Galton's December 1888 paper. After Galton's death in 1911, Pearson had inherited all of Galton's research-related records, notes, equipment, and personal books and papers. Pearson discovered that he could not check Galton's arithmetic because "Unfortunately I have not succeeded in discovering the original work and manuscript tables for [Galton's December 1888 paper]..." (Pearson, 1930a, p. 53). However, Pearson was able to find the original raw data (Pearson, 1930a, p. 53n). The output of Pearson's analysis covers almost eight large published pages (Pearson, 1930a, pp. 50–57). That analysis described Galton's correlation coefficient as the slope of the plot of two combined data sets. Pearson's description was:

> *He takes six or seven values of* [paired variates], *plots two sets of six or seven points and notes that the first and second series of points are nearly on one and the same straight line. He draws this straight line as closely as he can to the points and through the median, and reads off its slope. This slope is Galton's measure of co-relation.*

> *(Pearson, 1930a, p. 51; underlining added)*

Because Galton put Subject on the *Y*-axis and Relative on the *X*-axis, the slope of Galton's figure is not 0.8 but rather 1.2; the slope becomes 0.8 only if the figure is rotated by 90°. Pearson's described that plot as "the Graphical Process of finding...the Correlation Coefficient" (figure 11 caption). Pearson observed that...

> [Galton's] *results... depend essentially on assuming that all his data follow a normal (or 'curve of errors') distribution.... In the table* [Galton's Table II]... *it is very difficult to see any approximation to normality in the distribution of stature* (p. 51 and 51n).

However, using modern reliability-data-plotting techniques (Tobias and Trindade, 2012, chapter 6), the Box-Cox-transformed distribution ($\lambda = 3$) of Galton's stature data is found to be nearly perfectly normally distributed (i.e., nearly perfectly straight on the interval-version of a Normal Probability Plot). It is impressive that Galton's stature median of 67.2 and probable error of 1.75 compare well with the 67.105 and 1.746 that can be determined using a computer-based implementation of reliability data plotting (Zorich, 2021a).

After Pearson's analysis was published in 1930, it seems that no other historian performed an analysis of Galton's December 1888 paper until 1986, when two historians independently gave it another look. In that year, S. M. Stigler provided a mathematical analysis of Galton's Stature vs. Cubit data in his book *The History of Statistics* (Stigler, 1986, pp. 319f), in which he praised Galton's "intuitive and graphic approach" (p. 320). T. M Porter provided a similarly brief mathematical discussion; he praised Galton's "graphical virtuosity" for having solved the problem of how to measure the "degree of interdependence" of correlated organs in the same individual (Porter, 1986, p. 292). Porter further explained that...

> *Galton's invention of a method of correlation cannot be ascribed to mathematical acuity – though his intuitive grasp of simple mathematics was indispensable – but to his wide interests and his ability to use what he learned from one problem as an aid to the solution of others* (p. 294).

As previously mentioned, Pearson's view was this:

> *Like so much of Galton's work* [,] *the present paper* [i.e. from December 1888] *reaches results of singular importance by very simple methods.... It is the old*

experience that a rude [i.e. crude] instrument in the hand of a master craftsman will achieve more than the finest tool wielded by the uninspired journeyman.

(Pearson, 1930a, p. 50)

Chapter Summary

Galton's 10-page December 1888 paper introduced mathematics (as opposed to arithmetic) into statistical science, which until then had been simply a descriptive technique. It was a clumsy introduction with confusing sentences, misplaced sentences, incomplete explanations, some not-well-supported generalizations, false claims, an incorrectly plotted figure, an incorrectly stated formula, and many numerical errors. No explanation was provided for what his measure of correlation actually measured; for example, the reader had no idea how an r of 0.4 and an r of 0.8 should be interpreted and compared. His method of calculating r was purely graphical, even though (as discussed here in Chapter 1) he might have easily discovered a simple formula for it if he hadn't written and published so hastily.

Despite such criticism, I find it amazing that his brilliant insight on how to solve the "co-relation" organic-regression problem could be distilled into ten pages that were written in a time frame measured better in weeks than in months (lightspeed, in my view). I suspect that he didn't really know what he had until the standard deviation was invented and then used by Karl Pearson in 1896 to create the first exact r formula that required neither a graphical plot nor a correlation matrix table (Pearson's formula was discussed here in Chapter 1). Despite not quite knowing what he had, in 1888, he spent a significant amount of time the following year trying to convince his colleagues that they should be calculating r for their data – more about that, in this next chapter.

5

Francis Galton, after December 1888

Introduction

Karl Pearson's 1930 analysis of Galton's December 1888 paper was discussed in the previous chapter; that analysis ended by mentioning...

> [Galton made an] *attempt to introduce the conception of correlation to anthropologists in 1889. It was a hopeless task! Most physical anthropologists in this country lack a thorough academic training, and statistical methods will only penetrate here after they have been adopted in Germany and France as they are being adopted in Russia, Scandinavia and America. English intelligence is distributed according to a very skew curve, with an extremely low modal value; we have produced great men, who have propounded novel ideas, but our mediocrity fails to grasp them or is too inert to turn them to profit. Years later these ideas come back to England, burnished and luring* [i.e. alluring], *through foreign channels, and mediocrity knows nothing of their ancestry!*
>
> *(K. Pearson, 1930a, p. 57)*

That 1889 "attempt" to which Pearson referred included one abstract of his December paper, and two new papers, all three of which are discussed in this chapter. Unfortunately for the reader, it takes all four papers combined to convey Galton's two most important points, which I have stated here in my own words:

- The word "regression" applies to the linear slope of correlation matrix table data that have been plotted with Relative on the *Y*-axis and Subject on the *X*-axis. In such a table, there are two Subjects and two Relatives; their two respective regression values are not reciprocal (i.e., not equal). Those two regression values are measures of the two "relations" between the two variables involved.

- The word "correlation" applies to the linear slope of correlation matrix table *transmuted* data that have been plotted with Relative on the *Y*-axis and Subject on the *X*-axis. That transmutation involves dividing the original data by their respective probable errors. In such a table, there are two Subjects and two Relatives; their two respective correlation values are reciprocal (i.e. equal). In effect, there is only one correlation value, and it measures the one "co-relation" between the two variables involved.

DOI: 10.1201/9781003527893-5

The main subjects discussed in this chapter are:

- Abstract of the December 1888 paper; it was published on January 3, 1889.

- Paper by Galton that was given orally to the Royal Society on January 22, 1889; it was subsequently published in 1889 in four different versions in different journals at different times with different titles; in those versions, he tried to explain the difference in meaning that he gave to the words regression and correlation.

- Paper by Galton written in late 1889 but published in 1890; in this paper, he tried to explain the difference in meaning that he gave to the words relation and correlation.

- Galton's bewilderingly inconsistent use of the word correlation after 1889.

January 3, 1889

In a paper published in *Nature* on January 24, 1889, Galton mentioned that "a memoir [regarding correlation] read by me only a month ago before the Royal Society...will be published in due course in their Proceedings" (Galton, 1889d, p. 297). However, it would be several months before those Proceedings were published in a volume containing the papers read at Society meetings held November 1888 through April 1889. Because of similar routine delays in publication, "In the nineteenth century, contributors began using [the British journal] *Nature* and its weekly turn-around time... to give abstracts of longer forthcoming papers" (Baldwin, 2015, p. 14). It is therefore not surprising to find an abstract of Galton's late-December paper in the January 3 edition of *Nature*. Nor is it surprising that in his January 24 paper, after mentioning the not-yet-published December paper, he advised: "For abstract [of that December paper], see *Nature*, January 3. p. 238" (Galton, 1889d, p. 297n).

That half-page abstract was published in *Nature*'s "Societies and Academies. London" section. Surprisingly, it was not titled "Co-relations..." but rather "Correlations..." (Galton, 1889b, p. 238). Although his December paper had included more than a dozen instances of the hyphenated words co-relation or co-related vs. one instance each of correlation and correlated, this abstract contained the reverse, that is, only one instance of co-related vs. more than a dozen instances of correlation or correlated. That one instance was in the abstract's first words: "Two organs are said to be co-related...."

Did an editor at *Nature* change the spelling of co-relation to correlation after Galton submitted the abstract? It is unlikely that the spelling changes were editorial, because "...19th-century submissions to *Nature* were generally either rejected outright or rushed immediately into publication" (Baldwin, 2020, p. 8). The spelling changes may have been suggested by an editor and then pre-approved by Galton, but that cannot be proven because "Unfortunately...the *Nature* offices did not preserve much official correspondence before 1990" (Baldwin, 2015, p. 9).

Did Galton even write the January 3 abstract? Evidence that a *reporter* wrote the abstract is that it was written in the third person ("in his memoir"; Galton, 1889b, p. 238, underlining added) whereas the December 20 paper had been written in the first person ("in my paper"; Galton, 1888c, p. 143; underlining added). Evidence that a reporter did *not* write the abstract is that during 1888/1889...

- *Nature* was still a money-losing venture (Meadows, 2008, p. 215).
- Their staff was not large enough to send out people to report in-person on all the various societies (Zorich, 2019).

What evidence do we have that Galton himself wrote the abstract? The strongest evidence is in the abstract's last paragraph, all of which is given here:

> It should be noted that the use of the Q unit [i.e. probable error] enables the variations of the most diverse qualities to be compared with as much precision as those of the same quality. Thus, variations in lung-capacity which are measured in volume can be compared with those of strength measured by weight lifted, or of swiftness measured in time and distance. It places all variables on a common footing.
>
> (Galton, 1889b, p. 238)

That end-of-paper paragraph is an enlarged and modified version of this briefer text found on the third page of December's ten-page paper:

> It will be understood that the Q value is a universal unit applicable to the most varied measurements, such as breathing capacity, strength, memory, keenness of eyesight, and enables them to be compared together on equal terms notwithstanding their intrinsic diversity.
>
> (Galton, 1888c, p. 137)

If a reporter or editor were the abstract's author, I conjecture that is it highly unlikely that he would have understood the significance of that third-page sentence, i.e. understood it well enough to decide that it needed to be prominently highlighted as the final words in the abstract. I think it is also highly unlikely that in a paper that condensed ten pages to half a page (which is a 95% reduction) that a reporter or editor would have decided to *expand* December's single sentence of 41 words into January's three sentences totaling 68 words (which is a 66% *increase*). The author of that final paragraph in the abstract is much more likely to have been Francis Galton.

Other evidence of Galton's authorship of the abstract is found in a large footnote in a paper that Galton gave orally on January 22; there, he referred back to his December paper (the title of which he gave as "Correlations..." not "Co-relations...") (Galton, 1889e, p. 408n). The rest of the footnote comprised two summary paragraphs that the reader would assume were from the December paper; however, they are not—they are word-for-word copies of summary paragraphs from the January 3 abstract (minus one unimportant sentence). If Galton had not written the abstract, it seems reasonable to assume that in this January 22 paper of his, he would have quoted from his own December summary rather than from one written by someone else.

If the change from co-relation to correlation in the January 3 abstract had originated with a reporter (or with an editor or the type-setter/printer), and if Galton had considered those changes to be unauthorized and unwelcomed, he would have publicly said so. *Nature* published *errata* letters from authors, e.g. regarding a "typographical blunder...last week" (Galton, 1884b), and regarding a numerical "mistake" that had been published "last week" (Trouton, 1889); but no letters from Galton about spelling changes appeared in any 1889 issue of *Nature*.

The words co-relation, co-relate, or co-related are not found in any of the approximately 200 post-January-3-1889 books and papers authored by Galton that are currently available in Galton's "Complete Works" at www.galton.org (Zorich, 2020b). Those Complete Works include his near-end-of-life autobiography, which provided a list of his "Books and Memoirs," in which the title of his December 1888 paper was again given as "Correlations..." not "Co-relations..." (Galton, 1908b, p. 328).

Based upon all the above-mentioned evidence, it seems that...

1. Within about 2 weeks of December 20, 1888, Galton either authored, authorized, or reluctantly accepted the spelling change of co-relation to correlation.

2. He then used the word correlation rather than co-relation in virtually all his future publications.

3. During the rest of his life, in every reference he made to the title of his December 20 paper, he substituted Correlations for Co-relations.

It is interesting that Karl Pearson's 1920 paper titled *Notes on the History of Correlation* claimed that the word correlation had not yet defeated the word co-relation:

> ...the balance is still swinging between "co-relation" and "correlation" although it has ultimately fallen to the more weighty word.

(K. Pearson, 1920, p. 39)

On a related note...

In 19th century Britain, there was much debate regarding the most important criterion for establishing priority of discovery (Csiszar, 2018, p. 159f). Galton's first cousin Charles Darwin had successfully established priority over Alfred Wallace (for the theory of speciation by natural selection) on the basis of dated letters, dated draft manuscripts, and the testimony of distinguished colleagues who could date relevant conversations (Himmelfarb, 1968, p. 242f). However, according to Csiszar (p. 159f), the following three criteria were much more widely accepted than the ones used by Darwin, opinion varying as to which one of these was most important: (i) the date a paper detailing the discovery was delivered in hardcopy to a scientific body such as the Royal Society, (ii) the date that the paper was presented orally at a formal meeting of the members of such a body, or (iii) the date that the paper or its abstract first appeared in a reputable journal. It seems that Galton hastily addressed all three of those criteria, on December 5, December 20, and January 3, respectively.

January 22, 1889

On January 22, 1889, 1 month after his Co-relation presentation to the Royal Society of London, Galton gave his annual "President's Address" to the Anthropological Institute of Great Britain and Ireland. That speech was formally published later in the year; additionally, at least three other versions of it were published. The differences between those versions and how they are referenced are surprising:

- The original version given to the Institute (Galton, 1889e) was titled only "President's Address" and was (according to the copy published in the Institute's journal) given on January 22, 1889. Karl Pearson in his biography of Galton mistakenly dated this paper to January 2 (K. Pearson, 1930a, p. 57n). This paper is sometimes incorrectly referenced as having the title "Human Variety."
- What looks like an almost perfect photocopy of the original paper (including the same pagination and title) was published in 1889 as a pamphlet (Galton, 1889f).
- A slightly different version of the paper was published in *Nature* on January 24, 1889, under the title "Human Variety" (Galton, 1889d); in a footnote linked to the title, it stated that this paper was an "Address delivered at the anniversary meeting of the Anthropological Institute, on Tuesday, January 22"—however, the reader was not informed of the many slight wording changes, some rewritten sentences, the deletion of two very large footnotes, and the addition of one paragraph.

- Another version was included in a small book titled *Anthropometric Laboratory: Notes and Memoirs*, which Galton published in 1890 (Galton, 1890b, pp. 12–21). It too was titled "Human Variety"; a footnote linked to the title stated that it was a "Presidential Address delivered at the anniversary meeting of the Anthropological Institute, on Tuesday, Jan. 22, 1889." However, it actually was a copy of almost all of *Nature's* January 24 version (e.g. it included the additional paragraph that had appeared in *Nature's* version). There was a significant difference between this version and the one in *Nature*, the difference being the deletion of two paragraphs that had appeared on *Nature's* page 299 and in the original January 22 paper.

In the version published by the Institute, Galton introduced correlation as follows:

> *The various measurements made at the* [Galton] *laboratory have already afforded data for determining the general form of the relation that connects the measures of the different bodily parts of the same person. We know in a general way that a long arm or a long foot implies on the whole a tall stature—ex pede Herculem; and conversely that a tall stature implies a long foot. But the question is whether their <u>reciprocal relation, or correlation as it is commonly called</u>, admits of being precisely expressed. Correlation is a very wide subject indeed. It exists wherever the variations of two objects are in part due to common causes; but on this occasion I must only speak of those correlations that are of anthropological interest. The particular problem I first had in view was to ascertain the practical limitations of the ingenious method of anthropometric identification due to M. A. Bertillon, and now in habitual use in the criminal administration of France. As the lengths of the various limbs in the same person are to some degree <u>related</u> together, it was of interest to ascertain the extent to which they also admit of being treated as independent.*

> *(Galton, 1889e, p. 403; underlining added)*

That introduction is misleading in regard to what the "reciprocal relation" is "commonly called." Prior to Galton's December 1888 paper, there is no evidence that it had been commonly called anything other than a relation, even in Galton's own writings.

He continued:

> *Lastly, as regards the correlation of lengths of the different limbs.* [sic] *It has already been shown that the correlation connects the deviations, and has nothing to do with the mean or average values. Now, to express this relation truly, so that it shall be reciprocal, the scale of deviation of the correlated limbs, say, for example, of the cubit and of the stature of adult males, must be reduced to a common standard. We therefore reduce them severally to scales in each of which their own Q* [i.e. probable error] *is the unit. The Q of the cubit is 0.56 inch, therefore we divide each of its deviations by 0.56. The Q of the stature is 1.75 inch, so we divide each of its deviations by 1.75. When this is done the correlation*

is perfect. The value of regression is found to be 0.8, whether the cubit be taken as the "subject" and the mean of the corresponding statures as the "relative," or vice versâ (pp. 416–417).

The numerical values he mentioned here were identical to those presented in his December paper. Here he claims a "perfect" agreement between the values of "regression" of transmuted cubit and stature data, no matter whether he plots cubit as the Subject and stature as Relative, or vice versa. As previously discussed, that is not a true statement in every case presented by Galton in his December paper. For example, the values for Stature vs. Middle Finger are not both 0.70 (as he says in his Table V) but rather are 0.57 and 0.72; the values for Stature vs. Knee Height are not both 0.90 but rather are 0.81 and 0.97; and the values of Cubit vs. Knee Height are not both 0.80 but rather are 0.73 and 0.89.

In the next paragraph, he softened that "perfect" claim:

The value of the regression was ascertained for each of many pairs of the following elements, and a comparison made in each case between the correlated values as observed and those calculated from the ratio of regression. The coincidence was <u>close</u> throughout, quite as much so as the small number of cases under examination, 350 in all, could lead us to hope (p. 417; underlining added).

I conjecture that it may have been difficult for him to admit that the "coincidence was <u>close</u>," given his tendency to hyperbole (e.g. his claim in this paper that "the correlation is perfect" between cubit and stature, and his claim in December that his correlation plots were "seen to be a straight line…in every other figure drawn…."; that quote and the fact that those figures did *not* all produce straight lines was discussed here in the previous chapter). Despite the real-life imperfection of his data, his astounding mathematical intuition lead him to correctly claim that if he had a much larger number of cases, the reciprocal relationship of the regression coefficients of transmuted values would indeed be in perfect coincidence.

He continued on to a different topic:

The Q of the cubit is 0.56 inch, therefore we divide each of its deviations by 0.56. The Q of the stature is 1.75 inch, so we divide each of its deviations by 1.75. When this is done the correlation is perfect. The value of <u>regression is found to be 0.8</u>, whether the cubit be taken as the "subject" and the mean of the corresponding statures as the "relative," or vice versâ (p. 417; underlining added).

Compare that to what he said in his December paper:

In that of the cubit, the probable error is 0.56 inch = 1.42 cm.; in the stature it is 1.75 inch = 4.44 cm. Therefore the measured lengths of the cubit in inches will be transmuted into terms of a new scale in which each unit = 0.56 inch, and the measured lengths of the stature will be transmuted into terms of another new scale

in which each unit is 1.75 inch. After this has been done, we shall find the devia-
tion of the cubit as compared to the mean of the corresponding deviations of the
stature, to be as 1 to 0.8. Conversely, the deviation of the stature as compared to
the mean of the corresponding deviations of the cubit will also be as 1 to 0.8. Thus
the existence of the <u>co-relation</u> is established, and its measure <u>is found to be 0.8</u>.

(Galton, 1888c, p. 136; underlining added)

Thus in December he described his 0.8 as a co-relation; in this January paper, he described it as a regression. The point he is trying to make is that correlation is regression using transmuted values.

As previously mentioned, Galton added a paragraph to the January 24 version; that paragraph was:

In every pair of correlated variables the conditions that were shown to charac-
terize kinship will necessarily be present—namely, that variation in one of the
pair is on the average associated with a proportionate variation in the other, the
proportion being the same whatever may be the amount of the variation. Again,
when allowance is made for their respective scales of variability, the proportion
is strictly reciprocal, and it is always from 1 to something less than 1. In other
words, there is always regression.

(Galton, 1889d, p. 297)

For readers who were not familiar with Galton's work on kinship regression, that paragraph may have been confusing. For more than a decade, Galton had been researching his discovery that a correlation matrix table's data's linear plot of Subject (X-axis) vs. Relative (Y-axis) has a slope of less than one; e.g. height of Parents vs. that of Sons. In 1877, he gave that slope the name "reversion" (Galton, 1877a), which term he later changed to "regression" (e.g. Galton, 1885a). What he was trying to say in this added paragraph is that kinship regression and its corresponding correlation are similar in that both slopes never exceed a value of one.

Late 1889

Although his paper titled *Kinship and Correlation* was published in 1890, it was written in late 1889 (Stigler, 1989). On its third page was his first published use of the term "law of correlation":

I hope to be able to give in this brief notice a just idea of the <u>law of correlation</u>, but
it is quite out of the question to do more than explain its first and principal result.

(Galton, 1889a, p. 421; underlining added)

Near the end of the paper, he claimed that...

> *The purpose is now fulfilled that I had in view in writing this article, of giving a notion, that should be true as far as it went, of the chief <u>law of correlation</u>* (p. 430; underlining added).

In the nine pages between those two sentences, he did *not* explain what he meant by "law" of correlation. He did however give us the clearest explanation yet of the difference in his lexicon between the words relation and correlation.

> *I will take another of the same kind of examples in order to emphasize <u>the difference between relation and correlation</u>, of which <u>no explanation has thus far been attempted</u>.... There is not that <u>reciprocal</u> relation between them which is <u>conveyed by the word correlation</u>. So in respect to the lengths of two limbs or other bodily dimensions of the same person that vary on different scales. [sic] A long finger usually indicates a tall person, and a tall person has usually a long finger, but by no means to the same amount. <u>There is a relation between stature and length of finger, but no real correlation</u>. On the other hand, the scale of variation of symmetrical limbs, such as that of the right and left cubit, is so nearly the same that they can justly said to be correlated.... <u>Whenever the resulting variability of the two events is on a similar scale, the relation becomes correlation</u>. When it is not the same, and when the variations are...quasi-normal, a simple multiplication will be found to suffice...to <u>transform the relation into a correlation</u>. Thus we may speak of the length of the middle finger and that of stature being correlated together under a recognized understanding that the variations are quasi-normal, and that the multiplication in question shall be made. <u>Henceforth I will use the word correlation subject to these tacit understandings</u>* (pp. 425–426; underlining added).

That "henceforth" did not last forever, as evidenced by his 1901 and 1907 statements that correlation could be found in non-mathematical data (see quotes below, in the section titled "After 1889").

As discussed in a previous chapter, Galton had always described his new *r* as a measure of the closeness or nearness or degree of correlation...until now:

> *The unknown brother of the very tall man is probably only tall; the unknown thigh-bone of the very tall man is probably only long;...and so on. I have called this peculiarity by the name of regression. If there is no regression at all,—that is, if the regression is from 1 to 1,—then the correlation becomes identity. If the regression is complete,—that is, from 1 to 0,—there is no resemblance at all. In all intermediate degrees <u>the ratio of regression is an exact measure of the weakness of the correlation</u>.*
>
> (Galton, 1890a, p. 427; underlining added)

(and...)

The gain that has been now achieved is the discovery of the true and entirely unforeseen method of looking at correlation. The novelty of the idea is well exemplified by the question raised at the outset, of the thigh-bone and the probable stature of the man to whom it belonged. The old notion was that, the average length of the bone being so and so, and that of the stature of men of the same race being so and so, then if the bone were, say, a twentieth part longer than the average of such bones, the stature of the man to whom it belonged should be estimated at one-twentieth more than the average stature (subject to certain corrections). This we now perceive to be doubly erroneous in principle. We have nothing to do with twentieths or other fractional parts of the average length, and there exists no direct proportion between the total lengths of the bone and of the actually associated stature. The idea of regression being a factor in these relations has been hitherto quite unsuspected by anatomists. We now see that it necessarily plays an essential part in them, and that its value affords an admirable measure of the closeness, or weakness, of correlation between any two series that severally vary in a quasi-normal manner (p. 430; underlining added).

I have not encountered any other author who referred to the correlation coefficient as a measure of weakness, as Galton did in those quotations; the most common descriptor authors have used is "strength" (Zorich, 2018). Galton did not explain what he meant by weakness nor how r measures it. I speculate that what he meant, in Galton-speak, was this: If an r value is close to zero it is a measure of weakness, and if an r value is close to unity it is a measure of strength.

Galton ended this paper with a prediction:

There seems to be a wide field for the application of these methods to social problems. To take a possible example of such problems, I would mention the relation between pauperism and crime…. I can only say that there is a vast field of topics that fall under the laws of correlation, which lies quite open to the research of any competent person who cares to investigate it (p. 431).

Within the next decade, Galton's prediction became true, in the sense that the field of correlation topics became so vast that researchers were finding numerical correlation coefficients wherever they looked, including one researcher who found…

…an important correlation between mental and moral traits, [which was measured to be] *about.3* [0.3] *as worked out by Pearson's method. This is nearly the same as the correlation between strength of pull and weight.*

(Winslow, 1906, p. 116)

His *Kinship and Correlation* paper had great value for explaining correlation; however, it was published in America, not Britain. As previously mentioned, its existence and significance seemed to have escaped notice for almost 100 years.

After 1889

In the remaining 22 years of his life, Galton researched and published on topics other than correlation, such as anthropology, sociology, and criminal identification using finger prints. During these years, he wrote approximately 200 scientific works (books and papers), drafted a novel, and published an autobiography. Except for his autobiography, none of those 200 publications had any significant focus on correlation, although the word was used in them (Zorich, 2020b). Historian Stephen Stigler summed up the situation:

> *In the years immediately after 1889, Galton was preoccupied with other topics, notably the study of the classification of fingerprints, and he did not return to correlation, except in an advisory role to others.*
>
> *(Stigler, 1986, pp. 298–299)*

Regarding his preoccupation with the study of fingerprints, one of Charles Darwin's nieces wrote the following:

> *...Francis Galton was both pleasant and impressive, with his bushy, twitching eyebrows. We went to his house once to have our fingerprints taken for some experiment on the classification of fingerprints, on which he was working. He did not provide us with any means of washing off the printers' ink* [sic], *and we had to go about all day in London with sticky black hands.*
>
> *(Raverat, 1953, p. 274)*

Examples of the most relevant or interesting of Galton's post-1889 uses of the word correlation are given here next, in chronological order. As could be expected, such uses were mostly Co-Relational Correlations; but unexpectedly, some were Observational Correlations (all underlining added).

- *Owing to the large effect of <u>correlation</u>, an index* [of fingerprints] *based on all the ten digits is not much superior in efficiency to one that is based on six—namely, upon the first three fingers of both hands* (Galton, 1891, p. 141).
- *In the twelfth chapter* [of this book on Finger Prints] *we come to a branch of the subject of which I had great expectations, that have been falsified* [i.e., proven wrong], *namely, their use in indicating Race and Temperament. I thought that any hereditary peculiarities would almost of necessity vary in different races, and that so fundamental and enduring a feature as the finger markings must in some way be <u>correlated</u> with temperament* (Galton, 1892a, p. 17).

  That is another interesting use by Galton of Observational Correlation.
- *It is <u>hateful</u> to blunder in calculations of adverse chances, by overlooking <u>correlations</u> between variables, and to falsely assume them to be*

independent, with the result that inflated estimates are made which require to be proportionately reduced (Galton, 1892a, p. 109).

That may be the only time that the word "hateful" has ever appeared in a discussion of correlation.

- *Now the finger patterns have been shown to be so independent of other conditions that they cannot be notably, if at all, _correlated_ with the bodily measurements or with any other feature, not the slightest trace of any _relation_ between them having yet been found...* (Galton, 1892a, pp. 166–167).

  Galton here used both the root-word correlation and the word relation to mean exactly the same thing, namely an Observational Correlation.

- *...the ridges in the* [finger-print pattern of the] *middle finger should be counted as well. Their number is partly _correlated_ with those in the forefinger...* (Galton, 1895, p. 81).

  Here, Galton has extended mathematical correlation to counts, not just measurements.

- *The committee appointed by the Home Office to inquire into the two means of identification, that of measurements* [using Bertillon's system] *and that of finger-prints, and to report on their applicability to the detection of old offenders in England, strongly urged their use in combination, in which view I fully concurred. Severally, they are subject to so much _correlation_ that little is gained by using the measurements of many dimensions or the prints of many fingers, instead of a few of each* (Galton, 1900, p. 123).

- *The _correlation_ between youthful promise and performance in mature life has never been properly investigated. Its measurement presents no greater difficulty, so far as I can foresee, than in other problems which have been successfully attacked* (Galton, 1901a, p. 663).

  Here, Galton has applied mathematical correlation to subjective ordinal data, i.e. numerical scales of "youthful promise" vs. "performance in mature life."

- *The fifth and last lesson deals with the measurement of _Correlation_, that is, with the closeness of the relation between any two systems whose variations are due partly to causes common to both, and partly to causes special to each. It applies to nearly every social relation, and to environment and health, social position and fertility, the kinship of parent to child, of uncle to nephew, &c* (Galton, 1907, pp. 21–22).

  Here he redefined correlation and decided to capitalize it. Instead of it being a mathematical measure of the degree of co-relation between two variables that can be numerically measured (as he stressed it must be, in his 1888–1889 papers), it now applies to "any two systems...[and] nearly every social relation...."

- *One hears so much about the extraordinary sensitivity of the blind, that I was glad of an opportunity of testing a large number of children in an asylum.... I found afterwards a marked _correlation_ between at least this form of sensitiveness and general ability* (Galton 1908b, p. 249; this is another Observational Correlation).

- *It had appeared from observation, and it was fully confirmed by this theory, that such a thing existed as an "Index of Correlation"; that is to say, a fraction, now commonly written r, that connects with close approximation every value of deviation on the part of the subject, with the average of all the associated deviations of the Relative as already described* (Galton, 1908b, p. 303).

  This sentence appears in Galton's autobiography, which was published when he was 86 years old, a few years before his death. Apparently, an independent technical review was not conducted prior to publication. The problem is that neither this sentence nor any other in his autobiography mentions transmuting the deviations by dividing by the probable error. Therefore, this sentence does *not* describe his Index of Correlation but rather his Index of Regression. To be fair, in some cases (e.g. left arm vs. right arm), the probable errors would have been equal, in which case the Index of Regression and Index of Correlation would have equaled each other.

- *Mendel* [i.e. Gregor Mendel, of genetics fame] *clearly showed that there were such things as alternative atomic characters of equal potency in descent. How far characters generally may be due to simple, or to molecular characters more or less <u>correlated</u> together, has yet to be discovered* (Galton, 1908b, p. 308).

- *The harm due to continued interbreeding* [in humans] *has been considered, as I think, without sufficient warrant, to cause a presumed strong natural and instinctive repugnance to the marriage of near kin. The facts are that close and continued interbreeding invariably does harm <u>after a few generations</u>, but that a single cross with the near kinsfolk is practically innocuous. Of course, a sense of repugnance might become <u>correlated</u> with any harmful practice, but there is no evidence that it is repugnance with which interbreeding is <u>correlated</u>, but only indifference, which is equally effective in preventing it, but is quite another thing* (Galton, 1909, p. 54).

  Galton here described Observational Correlation. He may have been thinking of Charles Darwin, who married his own first cousin; they had many children, three of whom became noted scientists and fellows of the Royal Society (OSU. edu-Darwin).

The Word Relation

After 1888, Galton surprisingly sometimes used the word relation when the word correlation might have been better. For example (all underlining added):

- *The following observations are printed for the use of those who desire to investigate the <u>relations</u> of the various measurements of the same individuals...* (Galton, 1889c, p. 420).
- *The attempts of those who first experimentalised into Psycho-physics were mainly directed to ascertain the <u>relation</u> between the increase of stimulus and the corresponding increment of sensation* (Galton, 1893, p. 13).

Chapter Summary

In addition to whatever other projects Galton was pursuing during the 12 months following his December 1888 lecture on mathematical co-relation (soon renamed "correlation"), he also tried to promote the use of *r* among his colleagues. In that effort, he published six papers (one abstract, two new papers, and three other versions of one of those new ones). Unfortunately, neither the original paper in December nor any single one of those others clearly and fully explained what mathematical correlation was nor what value it had. In his last original paper in this series, he did at least clearly distinguished between what he meant by "relation" vs. "correlation." That distinction was worded with the precision of religious dogma, when he stated that *only* "Whenever the resulting variability of the two events is on a similar scale, the relation becomes correlation" (Galton, 1890a, p. 426). However, soon afterward, Galton became less dogmatic: He started using the word correlation for many types of data, even for Observational Correlations. Surprisingly, he sometimes reverted to using the word relation instead of correlation.

6

1889 to 1900

Introduction

In the wider scientific community, Francis Galton's correlation work may have at first been viewed simply as a slight extension of his well-known regression work, given the way that he described correlation in his December paper:

> ...the exactness of the <u>co-relation</u> becomes established when we have transmuted the inches or other measurement of the cubit and of the stature into units dependent on their respective scales of variability [i.e. Q].... When the deviations of the subject and those of the mean of the relatives are severally measured in units of their own Q, there is always a <u>regression</u> in the value of the latter. This is precisely analogous to what was observed in kinship, as I showed [two years ago] in my paper read before this Society on "Hereditary Stature".... The statures of kinsmen are co-related variables...<u>the index of co-relation, which is what I there called "regression,"</u>....

> (Galton, 1888c, pp. 136, 142–143; underlining added)

Among the late 19th-century authors discussed in this chapter are William Bateson (biologist), Charles Pidgin (statistician), Karl Pearson (mathematician), Walter Weldon (biologist), Francis Edgeworth (economist), Francis Warner (medical doctor), Franz Boas (anthropologist), Alphonse Bertillon (criminologist), C. B. Davenport (zoologist), John Hunt (reverend), George Yule (statistician), and Anna Thomas (educator). Additionally, the transcript of an interview that Galton gave in 1894 is discussed; it seriously misquoted him in regard to the size of correlation coefficients. During the 1890s, the idea that correlation could be viewed mathematically was adopted by relatively few scientists and mathematicians.

1889

On June 6, 1889, a paper by **William Bateson** (1861–1926) was read before the Royal Society; its title was "On some Variations of *Cardium edule*, apparently Correlated to the Conditions of Life." Bateson's paper reported on

DOI: 10.1201/9781003527893-6

"an investigation of the relation between the variations of animals and the conditions under which they live" (p. 297). He spoke of "the obvious correlation between the effects of the diminution in size of Shumish Kul [a dried up area that formerly had been connected to the Aral Sea] and the increase in the proportional length and thinness, &c., of the shells found there" (p. 309).

Unfortunately, he did not include X, Y graphs of any of his data nor any r values determined in the manner that Francis Galton had demonstrated to the Society just 6 months earlier. If Bateson had created graphs and calculated correlation coefficients per Galton's December 1888 paper, he might have been pleased to see, for example, a nearly perfect straight line and a correlation coefficient of 0.93 for his Table II's "First Terrace" data on length vs. breadth of shells (those results were obtained by me, using the software program MS Excel).

In December 1889, the American Statistical Society published a paper by **Charles Felton Pidgin** (1844–1923) that used the word correlation in reference to the process of cross-tabulating data. The paper was a review of a much larger paper on *The Mechanism of Statistics* that had been given earlier that year in Ireland by **Robert E. Matheson**; in Matheson's paper, in a section titled "Tabulation and Summarization," he described the "Process of Simple Extraction":

> In the simpler kinds of tabulation the forms may be ruled with columns for the several heads of information to be extracted, the width of the columns being proportioned to the probable number of marks to be made in each, a column for total being also provided.... In these cases strokes are the best for marks, each five strokes being kept together by making the fifth across the other four, to facilitate counting.
>
> (Matheson, 1889, p. 12)

Next, he described "Compound Extraction":

> In these cases there are headings not only at the top of the form but at the sides also. As an example, may be taken the tabulation of occupations at the Decennial Census of 1881 in combination with ages, sexes, religion, and education. The age-periods, names of religious bodies, and degrees of education were provided for by headings at the top of the form, while the occupations were ranged at the side, and the sexes distinguished by the use of different sheets (p. 13).

Pidgin explained in his review that...

> By "simple extraction" is meant the tabulation of one or more simple points, such as number of each sex. By "compound extraction" is meant what is here [in America] called "correlation," or "correlated tables," in which the tables show the statistical relations of several points of inquiry. For instance, in the Massachusetts Census of 1885, in one table, sex, native and foreign born, and age periods are correlated with color and race and conjugal conditions, so that

all possible relations of one of these points to all the other points can be seen on the same page.

(Pidgin, 1889, p. 478; underlining added)

The following year, Pidgin used the word correlation to refer to textual summaries of census data:

Each Special Agent of the Census deals with his own specialty, and each Bulletin and published Report will have the distinctive treatment of the department from which it comes. It will be the duty of the Special Agent for Abstracts and Items to <u>correlate this material, that is, to write abstracts and items containing results</u> drawn from various Bulletins or Reports, and brought together in such a way to show <u>correlated</u> results that the individual handling of Bulletins and Reports would not supply.

(Pidgin, 1890, p. 111; underlining added)

Karl Pearson (whose writings we've discussed previously many times) was so well known in the late 19th and early 20th centuries that he was off-handedly mentioned in a 1901 science fiction novel (*The First Men in the Moon*) as being "one of those great scientific people" such as "Lord Kelvin" (Wells, 1901, p. 21).

Pearson was *not* an early adopter of Galton's mathematical methods, despite having given a lecture on Galton's *Natural Inheritance* early in 1889; in that lecture, he actually warned the audience of the "considerable danger in applying the methods of exact science to problems in descriptive science, whether they be problems of heredity or of political economy" (quoted in Stigler, 1986, pp. 303–304). However, 35 years later, he recalled things differently: "It was Galton [in *Natural Inheritance*] who freed me from the prejudice that sound mathematics could only be applied to natural phenomena under the category of causation" (K. Pearson, 1934, pp. 298–299).

Pearson was similarly initially disinterested in Galton's concept of mathematical correlation as described by Galton in papers published in 1888–1890 (discussed here in the two previous chapters). Evidence of that disinterest was provided in Pearson's soon-to-be famous book *The Grammar of Science*. In the first edition of *Grammar* (1892), there was no mention of mathematical correlation on any of its nearly 500 pages; although the book's Index did not contain the word correlation, the word did appear many times in the Chapters, in reference to Observational Correlations—for example (all references to K. Pearson, 1892; all underlining added):

- *The scientific method is marked by the following features: (a) Careful and accurate classification of facts and observation of their <u>correlation</u> and sequence; (b) The discovery of scientific laws by aid of the creative imagination; (c) Self-criticism and the final touchstone of equal validity for all normally constituted minds (p. 45).*

- *The <u>correlation</u> of thought and consciousness seems to indicate that this complexity of the organism is to be sought in the inception and development of its capacity for storing sense-impressions* (p. 401).
- *In this Grammar...I have striven to indicate how natural law is a product of the* [sic] *human reason and how the <u>correlated</u> growth of the reasoning and perceptive faculties in man, assisted by the survival of the fittest, may possibly have left us with a normal type of man for whom only that is perception* [sic] *which can be reasoned about, and for whom the reason is keen enough to appreciate and analyze what is perceived* (p. 472).

At least one historian has spoken melodramatically of "Pearson's <u>conversion</u> to statistics in 1893" (Porter, 1986, p. 297; underlining added). Pearson soon became an admirer of Galton and a champion of mathematical correlation, to which a ten-page section was given in the 1900 second edition of *Grammar*; in the 1911 third edition, a 28-page chapter was added that was titled "Contingency and Correlation—The Insufficiency of Causation," which included one sub-chapter titled "The Measure of Correlation..." (pp. 152, 174). In that chapter, curiously, he discussed his own invention, the Correlation Ratio (η) but not Galton's correlation coefficient (r). Similarly, his end-of-chapter "Literature" list included only four publications, which he described as representing "the <u>original</u> memoirs on the subject" (p. 178; underlining added). That list included three of his own papers and a statistics book by W. P. Elderton, published respectively in 1903, 1904, 1905, and 1906—I wonder how many readers of that chapter concluded that Galton's December 1888 paper must *not* be one of the original memoirs on the subject of correlation?

Many of Karl Pearson's papers in the mid to late 1890's were about correlation or how to apply it (e.g. to the theory of evolution). He sometimes spoke about the "mathematical theory of correlation" (K. Pearson and Lee, 1897, p. 455), as if it were a separate branch of mathematics. Some of those papers dragged on for more than 80 pages of tiny font and hundreds of equations (e.g. K. Pearson & Filon, 1898). Given that extreme complexity of presentation, I wonder if Karl Pearson's papers on correlation actually delayed rather than promoted general adoption of correlation techniques.

In a published "Note" to the Royal Society in 1895, Pearson used the term "co-efficient of correlation" which he equated to "Galton's function" (K. Pearson, 1895, p. 241). About 6 months later, he expanded that Note into a lengthy paper (more than 60 pages), which interestingly did not use the term co-efficient of correlation but instead many times used "correlation coefficient"; that paper also twice used the hyphenated term "correlation-coefficient" and thrice used "Galton's function" (K. Pearson, 1896a). It is interesting to note that although some 19th-century authors (e.g. Galton, Pearson, and Edgeworth) sometimes spelled the word coefficient with a hyphen, they would typically not use a hyphen. I have been unable to find any dictionary that lists the hyphenated word "co-efficient." However, *NED* does give a definition for the hyphenated word "co-efficiency," which it defined as a "joint efficiency, cooperation; 'the state of acting together for some single end'"; *NED* indicated that use of this word was "rare" (*NED's* only historical-use quotation was from 1665).

As previously mentioned, Pearson's 1896 paper introduced a formula for what is now sometimes called the "Pearson Product-Moment Correlation Coefficient," which has since been used in many 20th-century textbooks to *define* the correlation coefficient. In Walker's *Studies in the History of Statistical Method* (1929), she stated that Pearson's 1896 paper "gave definitions of correlation and regression far more general than anything which had preceded" (pp. 110–111). Confusedly and amazingly, his definition of correlation carelessly neglected to mention dividing values by their respective probable error or standard deviation. Because of that oversight, the "Correlation...constant" (in his Correlation definition) is mathematically identical to the "coefficient of regression" (in his Regression definition), at least in the case of the "two organs" example that he mentioned in each definition. The complete texts of both definitions are given here next (all underlining added):

- *Correlation.—Two organs in the same individual, or in a connected pair of individuals, are said to be <u>correlated</u>, when a series of the first organ of a definite size being selected, the <u>mean</u> of the sizes of the corresponding second organs is found to be a function of the size of the selected first organ. If the <u>mean</u> is independent of this size, the organs are said to be non-correlated. <u>Correlation is defined mathematically</u> by any constant, or a series of constants, which determine the above function. The word "organ" in the above definitions... must be understood to cover any measurable characteristic of an organism, and the word "size," its quantitative value.* (Pearson, 1896a, pp. 256–257). (This definition is so poorly worded that the reader could easily mistakenly conclude that Pearson's emphasis on "mean" was in direct opposition to what Galton had described in his December 1888 paper where he said that co-relation "...*has <u>nothing whatever to do with the average</u> proportions between the various limbs....* (Galton, 1888c, p. 136; underlining added).)

- *Regression.—Regression is a term which has been hitherto used to mark the amount of abnormality which falls on the average to the lot of offspring of parents of a given degree of abnormality.... From this special definition of regression in relation to parents and offspring, we may pass to <u>a general conception of regression</u>. Let A and B be two <u>correlated</u> organs (variables or measurable characteristics) in the same or different individuals, and let the sub-group of organs B, corresponding to a sub-group of A with a definite value α, be extracted. Let the first of these sub-groups be termed an array, and the second a type. Then we define the <u>coefficient of regression</u> of the array on the type to be the ratio of the mean-deviation of the array from the mean B-organ to the deviation of the type α from the mean A-organ* (Pearson, 1896a, pp. 259–260).

1890

Walter Frank Raphael Weldon (1860–1906) was discussed in Chapter 1 in relation to the history of the nomenclature of the correlation coefficient (he proposed the name "Galton's function"). He was a seemingly tireless researcher, as evidenced by his famously generating random numbers by rolling 12 dice more than 26 thousand times (Wikipedia: Weldon). He was so well respected in the field of applied statistics that after his death, a Memorial Prize was established in his name, to be awarded to persons who have made important contributions to the development of statistical methods applied to biology; the prize has been given more than 50 times since 1911 (Wikipedia: WeldonPrize).

In a paper whose published version stated that it was "Received March 20, 1890" by the Royal Society, Weldon referenced Galton's December 1888 paper and its method; in it, Weldon spelled correlation without the hyphen:

> *I have attempted to apply to the organs measured the test of <u>correlation</u> given by Mr. Galton ('Roy. Soc. Proc.,' vol. 45, No. 274, pp. 135 et seq.); and the result seems to show that the degree of <u>correlation</u> between two organs is constant in all the races examined. Mr. Galton has, in a letter to myself, predicted this result.*
>
> (Weldon, 1890, p. 453; underlining added)

Two years later, he published "Certain Correlated Variations in *Crangon vulgaris*," in which he continued the long-standing custom of using the word "relation" in reference to the interaction of *X, Y* data sets. In that paper, he used relation in that sense many times, whereas he used the word correlation in that sense zero times; for example:

> *Having found a <u>relation</u> between the deviation of carapace lengths and that of post-spinous lengths…. the <u>relation</u> between the two organs, as measured by the value of r, was fairly constant…*
>
> (Weldon, 1892, p. 9; underlining added)

However, he did not clearly explain what difference in meaning he intended to convey (if any) when he used the word relation instead of the word correlation; his paper's concluding paragraph (given here next) is an example of such lack of clarity:

> *That is, the results recorded lead to the hope that, <u>by expressing the deviation of every organ from its average in Mr. Galton's system of units</u> [i.e. probable errors], a series of constants [i.e. r values] may be determined for any species of animal which will give a numerical measure of the average condition of any number of organs which is associated with a known condition of any one of them. A large series of such specific constants would give an altogether new kind of knowledge of the physiological connexion between the various organs of animals;*

> *while a study of those <u>relations</u> which remain constant through large groups of*
> *species would give an idea, attainable at present in no other way, of the <u>func-</u>*
> *<u>tional correlations</u> between various organs which have led to the establishment*
> *of the great sub-divisions of the animal kingdom.*
>
> *(Weldon, 1892, p. 11; underlining added)*

In that same paper, he exhibited what may have been the first published mathematical *formula* for calculating the correlation coefficient. After showing how the formula was derived, he said:

> *This constant* [r], *therefore, measures the "degree of correlation" between the two*
> *organs* (p. 3).

His formula was the equivalent of...

1. choosing a single plotted point on Galton's December 1888 "figure,"
2. determining the corresponding Y and X axis values of that point, and
3. dividing X by Y.

For example, one of Galton's plotted circles represented $Y=0.46$ and $X=0.36$. The ratio $0.36/0.46=0.78$, which compares very well to Galton's reported value of 0.8 (as previously discussed). Weldon's formula therefore at first glance seems to be satisfactory. However, the example just given was for a point that appears exactly on top of the straight line that Galton had drawn, the line whose slope he equated with r, i.e., the correlation coefficient. The very next point to the right in Galton's figure appears slightly off the line; its values are $Y=1.03$ and $X=0.53$, the ratio for which is $0.53/1.03=0.51$, which compares very poorly to Galton's value. Thus Weldon's formula produces at best a close approximation to the Galtonian value, and at worst an very inaccurate estimate.

Karl Pearson used the term "organic correlation" in reference to Weldon's 1892 paper (although that term did not appear in it). In Pearson's words:

> *In his* [Weldon's] *next paper, "On certain correlated Variations in Crangon*
> *vulgaris," Weldon calculated the first coefficients of organic correlation, i.e. the*
> *numerical measures of the degrees of interrelation between two organs or char-*
> *acters in the same individual.*
>
> (K. Pearson, 1906, pp. 16–17; note that the correct title of Weldon's paper
> is "Certain..." not "On certain...." It is interesting that in Walker's
> book on the history of statistics, she provided that same Pearson
> quotation (Walker, 1929, p. 109), but secretly substituted "that is"
> for the "i.e." in Pearson's original. Apparently she disliked "i.e.," as
> evidenced by her never using it in the body of the text of her entire
> book, and using it only once in a footnote, vs. many uses of "that is.")

Pearson's statement (*"Weldon calculated the first coefficients of organic correla-tion"*) is surprising, given that the first sentence in Weldon's own paper con-tradicted it:

> *The first successful attempt to find a constant relation between the variations in size exhibited by one organ of an animal body and those occurring in other organs was made some three years ago by Mr. Galton; and in a paper read before the Royal Society ("Roy. Soc. Proc.," vol. 45, p. 135) he determined this relation between several organs of the human body.*

(Weldon, 1892, p. 2)

Possibly Pearson meant that Weldon had "calculated" using a formula whereas Galton had only determined using a plot. Nevertheless, Pearson eventually fixed what seemed to be a mistake: After re-analyzing the data in Galton's December 1888 paper, he gave credit to Galton for "the first organic correlations ever published" (K. Pearson, 1930a, p. 53).

1892

In 1970, E. S. Pearson (son of Karl) and M. G. Kendall co-edited a book on the history of mathematical statistics, to which each of them contributed a chapter. It is humorous that Kendall's chapter claimed that **Francis Ysidro Edgeworth** (1845–1926) "was a great influence on his contemporaries and played a notable part in the development and acceptance of statistics as the subject was then understood," whereas Pearson's chapter claimed the oppo-site, namely that Edgeworth "seems to have had so little effect on the main line of development of mathematical statistics" (E. Pearson and Kendall, 1970, pp. 262 and 344, respectively).

Edgeworth obtained a college degree in classical literature, then studied law and became a barrister while also teaching himself higher mathemat-ics (Stigler, 1986, pp. 305–307). In 1881, he published a small book titled *Mathematical Psychics*—in a bookshop, such a title likely caught the eyes of engineering students (mistaking "Psychics" for "Physics"), until they read the subtitle: *An Essay on the Application of Mathematics to the Moral Sciences* (Edgeworth, 1881).

Edgeworth was a distant cousin of Galton's, and the fact that "their contacts continued over the years" may in part have led to his interest in statistical methods during the 1880s (Stigler, 1986, pp. 305–307). Edgeworth's pre-1888 statistical work has been given high praise: "In the 1880s Edgeworth was the only person in Britain, with the exception of Galton, doing anything approach-ing serious and sustained general work in statistical theory"; together, they produced "40 out of 78" of the papers on "statistical theory, error theory and actuarial theory" that were published during that time period (MacKenzie, 1981, pp. 98, 257n4). He was an early adopter of Galton's mathematical correlation concept, as evidenced by the syllabus for his "Newmarch Lectures" at University College London in 1892 (reproduced in Stigler, 1986, pp. 367–369). Those lectures were titled "On the Uses and Methods of Statistics"; one of the six segments of those lectures was titled "Types and Correlations," which included topics that he listed as:

- *Relation between the deviation…and the "correlated deviation"…as established by Mr. Galton*
- *Co-efficient which expresses the correlation*
- *Extension of the Galton-Dickson method to the correlation of three attributes*
- *Correlation between any number of attributes.*

In the 1890s, Edgeworth wrote several papers that included the word correlation in their titles. In those papers, he cited Galton's December 1888 paper as the source of the mathematical concept of correlation. He virtually always wrote about "correlation," except for an 1893 paper in which he mentioned "co-relations" and the "mathematical theory of co-relation" (Edgeworth, 1893, p. 674).

Edgeworth seems to have been the first in a long line of writers, including many 20th century textbook authors, who were more interested in discussing the *formula* for correlation than in explaining the meaning of the word. For example, the first sentence in his first paper on correlation began like this:

> *The "correlation" between…measurable attributes…may in general be expressed by the formula…*

> *(Edgeworth, 1892, p. 190)*

What followed was a formula that included almost 90 letters, numbers, commas, parentheses, and symbols. Instead of explaining what the word correlation meant, he footnoted that sentence's word "correlation"; the footnote advised…

> *See Galton…"Co-relations and their Measurement;"…[and] Weldon…"Certain Correlated Variations in Crangon vulgaris."*

However, as previously discussed, Galton's 1888 paper did not actually define mathematical correlation, other than to say it is the slope of a special line; likewise, Weldon's 1892 paper (previously discussed) did not define mathematical correlation, other than to say its coefficient could be calculated with a formula. Similarly, Edgeworth's 1892 paper defined correlation using formulas. An explicit example of his reluctance to define correlation in words is the very first sentence in his next-year's paper titled "Statistical Correlation between Social Phenomena" (Edgeworth, 1893): "An example may introduce my subject better than a definition." In the rest of that paper, he never defined mathematical correlation in words.

His explanations of correlation were so complicated that he, like Karl Pearson (previously discussed), likely delayed the wider acceptance of mathematical correlation rather than promoted it. Mercifully, in his 1892 paper, he did provide a simple formula for calculating *two*-variable correlation; that formula was a slightly enlarged version of the one published earlier that year by Weldon. As previously discussed, the data required by Weldon's formula was equivalent to what is found in a single column (or row) in a correlation matrix table. As illustrated by the example that Edgeworth provided in his 1892 paper, the data required by his formula was equivalent to what is found in *two* adjoining columns in a correlation matrix table (i.e., two neighboring "cross" marks on Galton's figure). The data he chose for his example were the last two cross marks on the right side of that figure, both of which lie almost on top of the straight line. That seems like a lucky choice because, as previously discussed in regard to Weldon's *single*-mark formula, the farther off the line a chosen mark is, the more inaccurate is the resulting correlation coefficient. However, in the case of Edgeworth's method, the inaccuracies were somewhat obscured by the fact that his chosen two marks were not on the same side of the line but rather on opposite sides of the line (thereby helping to cancel out their respective inaccuracies).

Stephen Stigler described Edgeworth's formula as being "computationally simple but unsatisfactory" (Stigler, 1986, p. 320). Additionally, Stigler discovered that Edgeworth's first example calculation (which used the cubit vs. stature data from Table II in Galton's December 1888 paper) was done blunder-fully incorrectly (a "comedy of errors" per Stigler)—and yet Edgeworth accidentally arrived at and reported 0.8 for the correlation coefficient (Edgeworth, 1892, p. 191), which was exactly the value that Galton had reported! As previously discussed, Galton's 0.8 was rounded up from 0.79; Edgeworth's 0.8 was rounded down from 0.86 (by my calculations). Stigler performed Edgeworth's formula calculations without blundering, yielding 0.68 for the correlation coefficient (Stigler, 1986, p. 321). In this case, the Edgeworth formula yielded a rough approximation to the Galtonian value; that approximation could have been much worse if he'd chosen a different pairs of marks on Galton's figure.

An interesting side-note is that when Edgeworth analyzed Galton's Table II data, he incorrectly referred to Galton's "Subject" as the "<u>independent</u> variable" (Edgeworth, 1892, p. 191; underlining added). Galton in 1888 would have disagreed with such a designation, since to him "co-relation must be the consequence of the variations of the two organs being partly due to <u>common</u> causes" (Galton, 1888c, p. 135; underlining added)—that is, two co-related variables are neither independent of nor dependent on each other but rather both partially dependent upon some other cause(s) that they share in common.

It is surprising (to a 21st century reader) that an 1892 book titled *The Dictionary of Statistics* contained no definitions of mathematical or statistical terms or methods, such as correlation; instead, it was a compilation of data summaries, e.g. on "Accidental deaths," "Asphalt paving," and "Balance of trade." The book was published in London; its author was **Michael G. Mulhall**.

1893

In 1893, **Dr. Francis Warner** published an analysis of data on the physical and mental condition of more than 50,000 children in more than 100 London-area schools. His paper focused on co-relations because "it is in the co-relation of conditions that we must seek for their relative importance" (p. 74).

Curiously, he also a few times used the unhyphenated word "correlation"; for example:

> *The correlation of individual defects and nerve-signs is given in Tables....* (top of p. 75)
>   *Despite that use of the un-hyphenated word, the titles of those tables each included the word co-relation not correlation* (pp. 86–88).

In contrast to those uses of the unhyphenated word correlation, every time he spoke about comparisons of defects, he described their inter-relationship using the hyphenated word co-relation. It seems that he reserved the unhyphenated word to reference a table itself, and reserved the hyphenated word for the relationships shown by such tables.

Warner's paper contained only Observational Correlations, e.g., counts of children who had various defects. The title of his Table III was "Showing Number of Children with each Defect in Development, also giving their <u>co-relations</u>" (underlining added). The body of that table provided only two types of numbers: Counts and percentages; the "Number of Children" must be those counts, and so the "co-relations" must be those percentages. Thus, although he did not state it explicitly in his paper, Warner had defined co-relation as a percentage; for example in Table III, a co-relation of 45.2% was given for "Children small for their age" who are also "With Low Nutrition." This is essentially the same numerical value for correlation that Bell had adopted in 1885 (previously discussed), where he too was hoping to determine the causes of health defects.

I am unsure of how to interpret the word correlate in the following text from a paper by **Edward Mussey Hartwell**; it was written at a time when he was "Director of Physical Training in Public Schools of Boston Massachusetts" (per the paper's title page):

> *The papers of Beyer, Enebuske, and Porter give evidence of a growing tendency to attempt to correlate the results of anthropometrical investigations, and the teachings of physiology as to the development of functional power* [i.e. physical strength in children].
>
> *(Hartwell, 1893, p. 555)*

The paper by **C. J. Enebuske** that Hartwell referenced summarized its own results in these words:

> *The* [mathematical analysis of] *anthropometrical data which we have presented above justify the opinion that the susceptibility of American women to gymnastic training is considerable.*
>
> *(Enebuske, 1893, p. 610)*

My best guess is that Hartwell used the word correlate to mean something like "examine...to determine if they could have had an influence on...." However, as I said, I am unsure.

Franz Boas (1858–1942) received a doctorate in physics and geography in Germany in 1881. Soon thereafter, he became interested in anthropology, which eventually led to his being hired for a series of anthropologically related jobs: In 1886 as an editor at *Science* magazine in New York, in 1889 as a professor of anthropology at Clark University in Massachusetts, in 1896 as curator of anthropology at New York City's Museum of Natural History, and finally in 1899 as professor of anthropology at New York City's Columbia University. His many scholarly, innovative contributions were mainly in the areas of linguistics, ethnology, and statistical anthropology. (Britannica. com-Boas)

His first paper on statistics explained his view of correlation:

> *I must add a few words regarding the subject of correlations. The admirable investigations of Mr. Alfonse Bertillon and those of Sören-Hansen, Bischoff, and others have proved that with increasing height all other measurements increased not proportionally, but at a slower rate. This law may be given a wider meaning by saying that whenever a group of people are arranged according to one measurement, with an increase of this measurement all others increase at a slower rate, the rate being the slower the slighter the correlation. This law leads us to establish the fact that we must consider each measurement as a function of a number of variable factors which represents the laws of heredity and environment. The correlation of two measurements will be close when they depend largely upon the same factor, slight when they depend largely upon distinct factors. This difference in the degree of correlation...is a well-established fact....*
>
> *(Boas 1893, p. 574; underlining added)*

Note that he made no mention of having transmuted the data into units of probable error. Curiously, he did not here mention Galton's work on correlation, even though elsewhere in the paper he mentioned Galton's contributions to other areas of statistical method.

In his next paper on statistics, he introduced a new method for estimating correlation and then applied it to three sets of measurements from several hundred human males. His data were *grouped* into *X, Y* correlation-matrix tables (like Table II in Galton's in December 1888), subsets of which (like Table III of Galton's) were provided in his paper; and therefore the word "grouped" was used in his explanation of correlation:

> *When any two biological measurements are considered as* <u>*correlated*</u> *and indi-*
> *viduals showing a certain value of the first measurement are* <u>*grouped*</u> *together;*
> [sic] *then the average of the values of the second measurement for this* <u>*group*</u> *of*
> *individuals will also be changed, but to a lesser* <u>*degree*</u> *than the first.*
>
> (*Boas, 1894, p. 320; underlining added;* again, notice that
> he makes no mention of transmuting the data.)

For each of three sets of paired measurements, he created two different curves that he showed on the same plot. For example, his Figure 1 included one curve based upon his having grouped the *breadth*-of-head data into evenly spaced intervals (Galton would have called these Subjects); on the *X*-axis, he plotted the averages of values found in each interval (rather than the mid-point of the interval as Galton had done). The *Y*-axis values were the averages of whatever *length*-of-head measurements corresponded to each grouped *X*-axis values (Galton would have called these Relatives). Similarly, the second curve in his Figure 1 was based upon his having grouped the *length*-of-head data into evenly spaced intervals (i.e. Subjects), the averages of each interval being plotted on the *Y*-axis; their *X*-axis values were the averages of whatever *breadth*-of-head measurements corresponded to each grouped *Y*-axis values (i.e. Relatives). Therefore, his Figure 1 was *not* equivalent to an untransmuted version of Galton's December 1888 "figure" (previously discussed). Galton put both Subjects onto one axis and put both Relatives onto the other axis (thus combining onto *each* axis the data from, e.g., head length and head breadth), whereas Boas combined Subject and Relative onto each axis, by putting all the head length values onto one axis and all the head breadth values onto the other. Boas then stated:

> *It is clear that if the breadth of head were a complete function of the length of head*
> *there could be only one curve expressing the interrelation between the two mea-*
> *surements* [i.e., the two curves would be superimposed onto each other,
> giving the appearance of only one curve]. *The fact that there are two curves*
> *shows clearly that the one measurement does not define completely the other, but*
> *that a number of factors influence each by itself.*
>
> (*Boas, 1894, p. 317*)

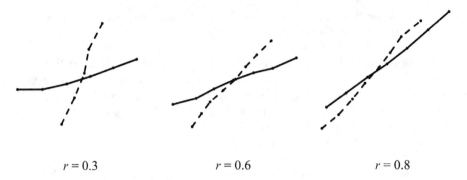

$$r = 0.3 \qquad\qquad r = 0.6 \qquad\qquad r = 0.8$$

FIGURE 6.1
These are simplified versions of the plots that Boas included in his paper; the correlation coefficient values shown here were not provided by Boas. In effect, solid lines in each chart are plots of Y on X, whereas the dashed lines are plots of the same data but X on Y. (Derived from data in Boas (1894, pp. 317–319).)

Using the same method with which he'd created his first plot, Boas created two other pairs of curves using the other two data-sets (he called those pairs Figure 2 and Figure 3). In regard to his three figures (reproduced here as Figure 6.1), he said:

> I have selected these <u>three pairs of measurements</u> in order to <u>illustrate the varying degrees of correlation</u>. It is clear that the correlation of the first pair [his Fig.1] is <u>very slight</u>, while that of the last pair [his Fig.3] is <u>very strong</u>—that is to say, the influence of z [i.e. shared causes] is very slight in the first pair and very strong in the last pair.
>
> (Boas, 1894, p. 320; underlining added)

Boas provided no guidance on how to mathematically interpret the differences between such plots. Should the reader measure and compare the angles formed by the two lines on each plot, correlation being some inverse function of angle magnitude? If the reader uses the provided data as input to Galton's graphic method, the correlation coefficients are found to be approximately 0.3, 0.6, and 0.8, for the data plotted in Boas's Figures 1, 2, and 3, respectively. (Note: I used standard deviations instead of probable errors when calculating those approximations; to obtain standard deviations, I used the "range" method explained in Wheeler and Chambers, 1992, p. 121.)

What is so very strange is that each of his three figures has major discrepancies between the plotted points and the table data; either his table data are wrong or his plots are wrong. For example, in his Figure 2, the "Breadth of face" table values of 138.9 and 160.4 are plotted at about 143 and 165, respectively; there are similarly serious mistakes in all three figures.

This method seems less than satisfactory as a measure of the degree of correlation; nevertheless, Boas's double-line method was included in some statistics textbooks for at least the next 95 years (e.g. Yule, 1911, p. 175; Davis & Nelson, 1937, p. 279; Snedecor & Cochran, 1989, pp. 177–179). That page in Yule's famous textbook recommended that "The student should study such [correlation-matrix] tables and ['two lines'] diagrams closely, and endeavor to accustom himself to estimating the value of *r* from the general appearance of the table" (Yule, 1911, p. 175).

Apparently Boas was not satisfied with that approach to correlation estimation, because within the pages of that same 1894 paper, he demonstrated a different measure of correlation, one that apparently he'd invented:

> *Whenever individuals showing a certain value of a measurement are grouped together, the variability of any second measurement of the group is smaller than the variability of the whole series. The smaller the influence of z [shared causes] as compared to that of x [causes not shared], the less the variability will be affected, and <u>we may consider the amount of the decrease in variability a supplementary measure</u> of the proportion between the influences of x and z or a measure <u>of the amount of correlation between the two measurements</u>.*

> (Boas, 1894, p. 320; *underlining added*; note that his "x" here has nothing to do with the X-axis)

Unfortunately, he did not demonstrate how to calculate such a measure, nor did he provide even one such value that he himself had calculated.

Almost 30 years later, he published a paper titled "The Coefficient of Correlation." Given that title and the new century, a reader might expect to find in it a positive review of the Galtonian/Pearsonian correlation coefficient; but instead the reader finds a warning in the paper's first sentence:

> *Students who are using the coefficient of correlation are aware that the value of the coefficient does not always express the biological, psychological, social, or economic relations which are the subjects of our studies; they realize that its value depends upon a good many extraneous conditions.*

> (Boas, 1921, p. 683)

In that paper, he defined correlation under the assumption that the data are arranged into the arrays of a typical correlation matrix table:

> *When two variables are interdependent so that a variant of one partly determines the correlated array of the other, we express their relationship by the coefficient of correlation* (p. 683).

Then he explained his main reason for discouraging the use of the correlation coefficient (note: Contrary to how he'd used the symbols *x* and *y* in previous text, he used them here to refer to the variables plotted on the horizontal and vertical axes, respectively):

If, for any reason, there is a <u>selection</u> of values of x, the coefficient of correlation will change its value, although the actual functional relation between x and y remains the same. In other words, in cases of this type, <u>the coefficient of correlation is an artificial value</u>, an algebraic convenience <u>which depends upon the composition of the series</u>…. In these cases, the essential problem will be not the determination of r but the determination of the two values of q [where q_1 is the regression coefficient for y on x and q_2 for x on y] (p. 685, underlining added).

As an example of "selection," Boas mentioned "military statistics, when individuals of underweight and of under-stature are eliminated…." (Boas, 1921, p. 685). Boas is correct that such censoring of data affects *r* values, but it is humorous to me that Boas ignored the fact that Galton's first-ever published correlation coefficient (which was between stature and cubit), was based on censored data. That is, Galton did not use stature Subject classes that were smaller than 64 inches or larger than 70 inches, nor did he use cubit Subject classes smaller than 16.75 inches or larger than 19.25 inches—as previously discussed, when plotting his "figure" and then using it to determine the first-ever published correlation coefficient, Galton did not use what he called the "flanking" data (Galton, 1888c).

1894

Another of the few times that the word co-relation was used post-1888 was in the title of a table in an 1894 report by a British "**Committee Appointed by the Secretary of State**" that was investigating the pros and cons of biometric measurements and/or fingerprints for identifying habitual criminals (Troup, Griffiths, and Macnaghten, 1894, p. 55). However, in the accompanying text, that table was described as showing a "correlation" (p. 50); that text was attributed to A. Griffiths, but the author of the table was not given. A footnote (p. 30) mentioned that "Dr. Garson" (i.e., **J. G. Garson**, who was *not* on the committee and not listed as one if its authors) had been consulted regarding measurements of the face. I speculate that Garson was the source of that use of the word co-relation, because (i) the table just mentioned was all about head and facial measurements and (ii) 6 years later, a follow-up report that *was* authored by Garson included another use of the word co-relation:

The absence of absolute similarity in morphological development, which occurs in all races of men, as in all animals, and is so important a factor in evolution, gives a certain range of variation in actual size to every part of the body, even in what are termed "pure races," that is to say, in communities which have for sufficiently long periods been isolated from their fellow-men to have acquired, in consequence, more or less similar morphological characteristics. The range of variation in such a community may be less marked than in people who have not

> *been so isolated. In these so-called pure races, also, there is a greater tendency for one part of the body to bear a more or less constant <u>relation</u> to another; thus we find the cephalic index, which expresses the percentage <u>relation</u> that the breadth of the head bears to the length, varies comparatively little in such races. In mixed communities, the range of variation of parts is considerable, and the <u>co-relation</u> of one part to another though increased is but slight.*

(Garson, 1900, p. 184; underlining added)

Note that Garson first talked of relations, and then of their co-relation, as did Galton in 1888 and 1889 (previously discussed).

That initial 1894 report included the text of an interview with Galton. He was asked to explain "the effect of correlation upon the successful classifica-tion [of finger prints]." Galton's *documented* response was this:

> *If correlation is close, the advantage of using the correlated elements becomes, [sic] considerably reduced. I have worked out the theory of correlation in a Memoir read before the Royal Society, in 1888, where I showed that there exists what may be called an index of correlation. This may be taken to range between 0, which signifies complete independence, and 10, which signifies the strictest intercon-nexion. Thus the index of correlation between head breadths and head lengths is as low as 4 or 5. Between the middle finger and the cubit, it is as high as 8 or 9.*

(Troup et al., 1894, p. 59)

Other than in a 1939 textbook that we'll discuss in Chapter 8, this is the only time that I've read of anyone describing the linear bi-variate correlation coef-ficient as having an absolute value larger than unity. I doubt that the text of the quote just given was written by Galton; what he and his interviewer said was likely recorded by a stenographer and then later transcribed. It seems beyond belief that Galton actually spoke of correlations equal to 4, 5, 8, 9, and 10. I speculate that the stenographer misplaced a decimal point or mis-heard what was said; that is, the values Galton actually spoke were most likely 0.4, 0.5, 0.8, 0.9, and 1.0. In support for that speculation, I point out that in Galton's December-1888-paper's Table V, Galton gave an r value of 0.45 for head length vs. breadth (vs. "4 or 5" in Troup's report) and an r value of 0.85 for middle finger vs. cubit (vs. "8 or 9" in the report); I speculate that Galton's words were "point four or five" and "point eight or nine" and that the stenographer did not understand or did not clearly record the "point."

1896

In the late 19th century, the world-wide standard for criminal identification was a body-measurements system that **Alphonse Bertillon** (1853–1914) had

invented in 1880 in France (Hacking, 1990, p. 182); his was the biometric system evaluated by the just-discussed 1894 "Committee."

In the 20th century, his system would be supplanted by the science of finger
prints (Fosdick, 1915), which was first systematized by Francis Galton in his
1892 book titled *Finger Prints*. Galton had previously been both an admirer
and critic of Bertillon's system:

> *Prisoners are now identified in France by the measures of their heads and limbs,*
> *the set of measures of each suspected person being compared with the sets that*
> *severally refer to each of many thousands of convicts. This idea, and the practical*
> *application of it, is due to M. Alphonse Bertillon.... The bodily measurements are*
> *so dependent on one another that we cannot afford to neglect small distinctions in*
> *an attempt to make an effective classification. Thus long feet and long middle-fin-*
> *gers usually go together. We therefore want to know whether the long feet in some*
> *particular person are accompanied by very long, or moderately long, or barely*
> *long fingers, though the fingers may in all three cases have been treated as long*
> *in M. Bertillon's system of classes, because they would be long as compared with*
> *those of the general population. Certainly his eighty-one combinations are far*
> *from being equally probable. The more numerous the measures the greater would*
> *be their interdependence, and the more unequal would be the distribution of cases*
> *among the various possible combinations of large, small, and medium values. No*
> *attempt has yet been made to estimate the degree of their interdependence.*

> > (Galton, 1888b, p. 348, 350; this paper was based upon a
> > lecture that Galton gave in May of 1888, several months
> > before his "Co-relations" lecture at the end of that year)

Galton's late 1888 discovery of mathematical correlation and the correlation
coefficient can in part be attributed to his earlier interest in Bertillon's system,
as explained by Pearson:

> *The idea Galton placed before himself was to represent by a single numerical*
> *quantity the degree of relationship, or of partial causality, between the different*
> *variables of our ever-changing universe.... I have said that Galton came to this*
> *fundamental conception from two aspects. The first problem was that of inheri-*
> *tance.... The second problem which impressed itself on Galton's mind was that*
> *of correlation in the narrow biological sense.... Galton's second idea of measur-*
> *ing the degree of relationship arose from the fact that he had recognised that two*
> *characters measured on a human being are not independent, they vary with each*
> *other.... Galton was driven to his second problem by Bertillon's system for the*
> *identification of criminals. Bertillon claimed, as I remember Dr [sic] Garson did*
> *at a much later date, that the measurements chosen were practically indepen-*
> *dent. Galton needed a criterion to show whether such measurements as head*
> *length, foot length, stature, etc. were or were not associated.*

> > (K. Pearson, 1930a, pp. 2, 3, 5)

Bertillon proposed to classify and identify criminals by measurements,
e.g. arm length and height. Galton seemed determined to demonstrate that

Bertillon's system was flawed; "Rivalry between Bertillon and Galton spurred the invention of the theory of correlation" (Hacking, 1990, p. 182). Although Galton's view in the 1880s was that (for example) arm length and height were *not* inherited independently, he had previously held the opposite view. For example, in *Hereditary Genius* he stated that although anatomical features and mental disposition are "in some degree correlated" (Galton, 1869, p. 348), human body parts themselves *are* inherited independently, not correlatively:

> *It is confidently asserted by all modern physiologists that the life of every plant and animal is built up of an enormous number of subordinate lives; that each organism consists of a multitude of elemental parts, which are to a great extent independent of each other; that each organ has its proper life, or autonomy, and can develop and reproduce itself independently of other tissues (see Darwin on "Domestication of Plants and Animals," ii. 368, 369). Thus the word "Man," when rightly understood, becomes a noun of multitude, because he is composed of millions, perhaps billions of cells, each of which possesses in some sort an independent life, and is parent of other cells.*

> *(Galton, 1869, p. 363; underlining added)*

That paragraph occurred at the start of a chapter titled "General Considerations." In the remainder of the chapter, he reinforced his claim about "all modern physiologists."

In the 1890s, after correspondence with and visits by Galton (Galton, 1908b, pp. 251–258), Bertillon began in a small way to include correlation into his thinking, as shown by the following example quotations taken from his identification-system's instruction book (the original French edition (Bertillon, 1893, pp. XLII, 19, 77, 140) did include the word *corrélation* or its related parts of speech, not just these English translations). All these quotations were taken from Bertillon (1896), and all underlining has been added:

> - *The drunkard often presents the maximum of sanguineous coloration; the mulatto presents an exaggeration of the pigmentary coloration. No correlation can be established between these two characters, each of which is the extreme of a special series requiring a special heading (p. 40).*
> - *There is a well known correlation between the reach and the height: the reach is ON AN AVERAGE about 4 centimetres greater than the total height. Thus these two indications check each other. Whenever the reach dictated is inferior by some centimetres to the height or exceeds it by more than ten centimetres, it is probable that a mistake has been made in one or the other of these observations and they should BOTH be verified (p. 104).*
> - *...a murderer recently arrested in Paris....is distinguished in an altogether exceptional manner, from an anthropometrical point of view, by the length of his feet, and correlatively, in a lesser degree, by the length of his fingers, and it is probably this last peculiarity, combined with relatively broad shoulders, to which the eccentricity by excess in his reach must be attributed (pp. 252–253).*

Such passages give no indication that Bertillon had found any value in calculating Galton's correlation coefficient.

Bertillon's system was still widely used in 1908, when Galton severely criticized it in his autobiography:

> *The subject was attracting much interest at the time* [1888], *and had received a great deal of off-hand newspaper praise. There was, however, a want of fulness* [sic] *in the published accounts of it, while the principle upon which extraordinarily large statistical claims to its quasi-certainty had been founded was manifestly incorrect, so further information was desirable. The incorrectness lay in treating the measures of different dimensions of the same person as if they were independent variables, which they are not.*

(Galton, 1908b, p. 251)

In 1896, zoologists **C. B. Davenport** and **C. Bullard** published a paper titled "A Contribution to the Quantitative Study of Correlated Variation and the Comparative Variability of the Sexes." They claimed that "To get quantitative [correlation] results" for their study of the number of Müllerian glands on the legs of swine, they would "employ a method devised by Galton" that was "explained in his [December 1888] paper" (Davenport and Bullard, 1896, p. 93 and p. 93n); however, that claim was far from true:

- As previously discussed, Galton's paper explained that his measurements needed to be "transmuted," i.e. divided by their respective measure of variability. Galton's measure of variability was the average of the first and third quartiles, which he referred to as the "probable error." Davenport and Bullard's measure of variability was the average absolute-value deviation from the arithmetic mean, which they referred to as the "Index of Variability" (p. 92).

- Galton's correlation coefficient equaled the slope of the best-fit straight-line that was eye-ball-estimated and hand-drawn through a plot of his transmuted X, Y paired deviations of X as the Subject and Y as the Relative and of Y as the Subject and X as the Relative (the meanings of the terms Subject and Relative were explained in Chapter 1 of this book). However, Davenport and Bullard's correlation coefficient was not derived graphically but rather formulaically, as explained here next:

  o Galton used the symbol r to represent the correlation coefficient. Davenport and Bullard used r to represent intermediate-calculation values leading to R, which was *their* symbol for the correlation coefficient (p. 94). In the remainder of this discussion, such intermediate-values will be represented by r with quotation marks around it, i.e. "r."

o One "r" value was calculated from the data in each of the 11 rows in their paper's correlation matrix table (p. 95). In effect, each "r" value was a row's average transmuted Relative deviation-from-mean divided by that row's transmuted Subject deviation-from-mean.

o Their R equaled the arithmetic average of those 11 "r" ratios. However, that average ratio was derived from only Subject=RightLeg vs. Relative=LeftLeg; that is, no "r" values were derived from the $n=11$ data of Subject=LeftLeg vs. Relative=RightLeg. The justification given for thereby doing only half as much calculation work as Galton's method required was that "Galton has shown that the same [correlation coefficient]... holds true when relative and subject are interchanged" (p. 96). As previously discussed, although Galton did make such a claim, some of his December 1888 data contradicted it (e.g. $r=0.57$ vs. 0.72, in the case of Stature vs. Middle Finger).

In effect, the Davenport-Bullard method for quantitating the correlation coefficient was a much-enlarged version of the formulas that had been published 4 years earlier by Weldon and Edgeworth (previously discussed). Weldon's and Edgeworth's formulas sometimes yielded r values that were wildly different from those derived using Galton's plot method. Might the much-enlarged Davenport-Bullard formula yield a much improved approximation, or would it too sometimes yield wild values? What follows here next is an investigation into that question:

Davenport-Bullard's formulaic method used on Galton's December 1888 paper's Table III data and its transmuted deviations (in that paper, Galton had reported $r=0.8$):

• Using the seven "r" values derived from Subject=Stature vs. Relative=Cubit, $R=0.327$, which is much lower than 0.8 because one of the "r" values showed negative correlation (i.e. $0.18/-0.11=-1.636$), whereas all the other "r" values were positive.

• Using the six "r" values derived from Subject=Cubit vs. Relative=Stature, $R=0.761$ (all "r" values were positive).

• Using all 13 of those "r" values, $R=0.527$.

That investigation demonstrates that the Davenport-Bullard formulaic method sometimes produces wildly inaccurate correlation coefficients.

It is interesting to see what Galton's method yields for the Davenport-Bullard data:

Galton's graphical method applied to Davenport-Bullard's 1896 paper's Table III data and their page 92 Indices of Variability (MS Excel's linear regression trendline feature was used to plot lines and calculate slopes):

- Using the 11 plotted points derived from Subject=RightLeg vs. Relative=LeftLeg, $r=0.782$ (Davenport and Bullard reported $R=0.772$).
- Using the 11 plotted points derived from Subject=LeftLeg vs. Relative=RightLeg, $r=0.705$ (Davenport and Bullard did not report an analysis of this data, but my calculation using their method outputed $R=0.756$).
- Using all 22 of those plotted points, $r=0.744$ (and $R=0.764$ by my calculation).

Davenport and Bullard's 1896 paper also provides an interesting example of a comparison of means that was performed years before the development of formal tests of statistical significance. The numbers of glands on the right and left legs were compared for 2000 males and 2000 females. The authors concluded (p. 91) that although the overall average number of glands in males vs. females was "tolerably close" at 3.544 and 3.511, respectively, the *ratio* of male to female (i.e. 1.009) was *large* enough to prove that "a real difference exists between the two [genders]." However, when comparing left and right legs, the authors concluded that because the overall average number of glands in left vs. right legs was found to be "so close" at 3.531 and 3.524, respectively, the ratio of left to right (i.e. 1.002) was *small* enough to prove that those averages "are about equal." In other words, a ratio of 1.009 proved significance, and a ratio of 1.002 proved equivalency; the authors did not reveal the criterion they used to reach those two conclusions.

In 1896, **Reverend John Hunt** wrote a book titled *Religious Thought in England in the Nineteenth Century*; in it he used an interesting mix of terms and concepts, including the word co-relation and the concept of conservation of forces—it's as if he'd read Grove's *The Correlation of Physical Forces* and took Grove's fifth-edition preface to heart (as discussed previously, Grove there had advised that "co-relation" was a more appropriate word than "correlation"):

> *God is everywhere present in nature. Everything in nature is His work. There are no second causes. God is immanent* [sic] *in the Universe, and whatever happens is by His immediate agency. An argument for immortality is drawn from the law of continuity, and the unity of the visible and invisible in the <u>co-relation of forces</u>. Miracles are not violations of <u>physical laws</u>. Continuity teaches that God's laws do not require revision, and that matter is not vile. <u>Co-relation</u> speaks of something behind and beyond matter and so an invisible order which will remain when the present system of things has passed away.*

> *(Hunt, 1896, p. 287; underlining added)*

1897

In a paper that was published in early 1897, **George Udny Yule** (1871–1951) explained the correlation coefficient, the formula for which he attributed to Bravais. The only thing he attributed to Galton was a graphical method of achieving essentially the same result. The following is interesting text from the paper's first two paragraphs:

> *The only <u>theory of correlation</u> at present available for practical use is based on the normal law of frequency, but, unfortunately, this law is not valid in a great many cases which are both common and important.... It seems worth while [sic] noting, under these circumstances, that in ordinary practice statisticians never concern themselves with the form of the correlation, normal or otherwise, but yet obtain results of interest—though always lacking in numerical exactness and frequently in certainty. Suppose the case to be one in which two variables are varying together in time, curves are drawn exhibiting the history of the two. If these two curves appear, generally speaking, to rise and fall together, the variables are held to be correlated. If on the other hand it is not a case of variation with time, the associated pairs may be tabulated in order according to the magnitude of one variable, and then it may be seen whether the entries of the other variable also occur in order. Both methods are of course very rough, and will only indicate very <u>close correlation</u>, but they contain, it seems to me, the point of prime importance at all events with regard to economic statistics. In all the classical examples of [economic] <u>statistical correlation</u> (e.g., marriage-rate and imports, corn prices and vagrancy, out-relief and wages) <u>we are only primarily concerned with the question</u> [of whether or not there] <u>is a large x usually associated with a large y (or small y)</u>; the further question as to the form of this association and the relative frequency of different pairs of the variables is, at any rate on a first investigation, of comparatively secondary importance.*

(Yule, 1897a, p. 477; underlining added)

It is interesting that after downplaying the correlation coefficient in that first-page text, Yule on subsequent pages provided a lengthy, detailed discussion of how to calculate correlation when there are three or four inter-dependent variables instead of just two.

He provided a lucid summary of the view that Galton labored to explain in his 1888–1890 papers:

> *...if we measure x and y each in terms of its own standard deviation, r becomes at once the regression of x on y, and the regression of y on x. The <u>regressions</u> being, in fact, the <u>fundamental physical quantities</u>, r is a coefficient of correlation because it is a coefficient of regression.*

(Yule, 1897a, p. 482; underlining added)

For more than a century, that claim of Yule's (i.e. that regressions are funda-mental and correlations derived from them) has in a sense been debated in textbooks (more about this, in upcoming chapters).

Later in 1897, Yule published a lengthy paper that investigated correla-tion theory. Although he claimed that "few of the results given [here]...are entirely new" (Yule, 1897b, p. 812), the historian S. M. Stigler concluded that "Yule's work on correlation reached full expression" in that paper because it "had a new and broader outlook" (Stigler, 1986, p. 348). In it, Yule provided definitions that would cause many a modern statistics instructor to cringe:

> *Instead of speaking of "causal relation," [and] "causally related quantities," we will use the terms "correlation," [and] "correlated quantities."*
>
> *(Yule, 1897b, p. 812; underlining added)*

I say "cringe" because it is difficult to rid students of the erroneous idea that correlation between variables implies causation by one of them over the other, and yet here Yule promotes that notion. That is, his definition promotes the view that "correlation" is partial causation (X sometimes causes Y, or is only partly responsible for causing Y), whereas a "causal relation" is full causation (X always causes Y, or is completely responsible for causing Y).

His data-plot also is reason to cringe. Of all the data that he could have possibly chosen to demonstrate regression and correlation, this (shown here in Figure 6.2) may have been the worst.

In that figure, he used a straight line (R–R) to model points that were obvi-ously not straight; and then he claimed that...

> *The equation to the line RR consequently gives a concise and definite answer to two most important statistical questions: Can we say that large values of x are on the whole associated with either large values of y or small values of y? And, What [sic] is the average shift of the mean of an x-array corresponding to a shift of unity in its type?*
>
> *(Yule, 1897b, p. 814; underlining added)*

It is no wonder that most of the scientific community did not immediately take correlation methods seriously.

1898

In 1898, **Anna B. Thomas** published a book that provided guidance on creat-ing lesson plans for teaching basic reading, writing, and arithmetic to the youngest school children. Its title was: *The First School Year: A Course of Study*

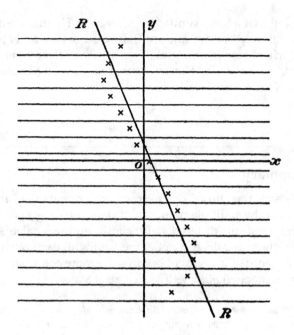

FIGURE 6.2
This chart shows a linear regression line through an obviously non-linear X, Y plot. (From Yule (1897b, p. 813).)

with Selection of Lesson Material, Arranged by Months, and Correlated for Use in the First School Year. Its author's preface stated:

> *Because of the child's physical surroundings and his love for living things, Nature Study has here been largely used as the basis of the course and other subjects have been correlated with it. The <u>principle of correlation</u> has not been forced. The child's surroundings in nature form an important and interesting part of his daily life, and hence should often have a controlling influence in the choice of lesson materials.*
>
> (*Thomas, 1898, p. 2, underlining added*)

You may ask "What could the term 'the principle of correlation' possibly mean in relation to the education of 5-year-olds?" It meant that the authors created a matrix table whose row headers were months of the year and whose column headers were major subject categories (e.g. Nature Study, Literature, Number [arithmetic], and Arts). In each cell in a row were detailed topics that were related to each other. For example: In the September row was listed a study of the golden-rod flower (in the Nature column), a poem titled Lady Golden Rod (Literature column), practice counting the parts of a flower (Number column), and examining paintings of flowers and singing songs about flowers (Arts column).

It is interesting that a book similar in purpose to Thomas's used the word "correlated" in its title but did not use that word or any of its related parts of speech anywhere else in the book. The title was: *Suggestions for Primary and Intermediate Lessons on the Human Body: A Study of its Structure and Needs Correlated with Nature Study* (Hallock, 1898).

Chapter Summary

I view the 1890s as the chronological frontier of correlation history, a time when r was called by multiple names, was determined graphically in multiple ways whose outputs differed significantly, and was calculated by multiple formulas whose results did not agree well with each other. In the first part of the 20th century, agreement began to be more common.

The view from Stephen Stigler's *The History of Statistics* is:

> *Before 1900 we see many scientists of different fields developing and using techniques we now recognize as belonging to modern statistics. After 1900 we begin to see identifiable statisticians developing such techniques into a unified logic of empirical science that goes far beyond its component parts.*

> *(Stigler, 1986, p. 361)*

The view from Theodore Porter's *The Rise of Statistical Thinking* is:

> *The intellectual character of statistics, however, had been thoroughly transformed by 1900. The period when statistical thinking was allied only to the simplest mathematics gave way to a period of statistical mathematics.... In the twentieth century, statistics has at last assumed at least the appearance of conformity to that hierarchical structure of knowledge beloved by philosophers and sociologists in which theory governs practice and in which the "advanced" field of mathematics provides a solid foundation for the "less mature" biological and social sciences.*

> *(Porter, 1986, p. 315)*

7

1900 to 1930

Introduction

Galton was well-known to the 20th-century scientific community. For example, in a book on evolution that was written in 1907 by two professors at Stanford University, in a chapter on "Heredity", he is mentioned only as "Mr. Galton", without even a first name or a literature reference (Jordan and Kellogg, 1907, p. 165). It was as if those authors assumed that anyone who'd be reading their book would have also read Galton's papers on kinship or his book *Natural Inheritance*. Similarly, a 1930 book on evolution referenced him first only as "Galton" without even a "Mr." in front of it; the book eventually (67 pages later!) identified him as "Sir Francis Galton" (Fasten, 1930, pp. 232 and 299). In a more recent publication, a book on the history of probability spoke of a "post-Galtonian statistical inference" without explaining what that meant or even who Galton was (Daston, 1988, p. 294).

Galton's December 1888 Co-relation paper may have represented the first injection of mathematics into what had previously been the science of arithmetic-based descriptive statistics (i.e., sums, totals, averages, charts, and plots). Nevertheless, the importance of that achievement was not obvious to all early-20th-century historians; for example:

- In H. S. Williams' 1909 book titled *The Story of Nineteenth-Century Science*, neither Galton nor mathematical correlation is mentioned, although Cuvier's "doctrine of correlation of parts" is.
- In D. C. Somervell's 1929 book titled *English Thought in the Nineteenth Century*, neither Galton nor mathematical correlation is mentioned, although Galton's "science of Eugenics" is.

The correlation coefficient was not commonly seen on pages of scientific journals during the 1890s. That situation changed starting in 1900, when Galton, Pearson, and Weldon founded a journal titled *Biometrika*, which had as its subtitle *A Journal for the Statistical Study of Biological Problems*. Its first issue was published in 1901; that issue contained a couple of anonymous editorials and an editorial-like paper by Galton. The first sentence in the first anonymous editorial was:

DOI: 10.1201/9781003527893-7

> *It is intended that Biometrika shall serve as a means not only of collecting under*
> *one title biological data of a kind not systematically collected or published in any*
> *other periodical, but also of spreading a knowledge of such statistical theory as*
> *may be requisite for their scientific treatment.*
>
> *(Anonymous, 1901, p. 1)*

The first sentence in Galton's paper was:

> *This Journal is especially intended for those who are interested in the application*
> *to biology of the modern methods of statistics.*
>
> *(Galton, 1901b, p. 7)*

That first issue of *Biometrika* included multiple papers that used the correlation coefficient, including one paper by Weldon that included the word correlation in its title: "Change in Organic Correlation of *Ficaria ranunculoides* during the Flowering Season".

In sharp contrast to the situation in *Biometrika*, there were the *Publications of the American Statistical Association*, in which W. M. Persons's (1910) was the first research paper to use a correlation coefficient to make its point, and C. E. Gehlke's (1917) was the second (Zorich, 2021b). Such a lack of focus on mathematical statistics is not surprising given that as late as 1926, the "Constitution of the American Statistical Association" stated that "The objects of the Association shall be to collect, preserve, and diffuse statistical information in the different departments of human knowledge" (ASA, 1926, p. 3). As previously mentioned, census and survey data were the 19th-century meaning of the word statistics.

Although it took a while longer for the American Statistical Association to embrace mathematical correlation, they quickly caught up to their British colleagues. Evidence of that is found in the *Index to the Journal of the American Statistical Association...1888–1939* (ASA, 1941); in it are references to dozens of articles on correlation applied to many areas of study, including agronomy, economics, scholastics, and anthropometrics. It listed a 1910 article by H'Doubler as being the first one related to the "correlation coefficient"; unfortunately, that paper does not contain any mention of nor formula for anything like a correlation coefficient—possibly the person who prepared the *Index* read only the paper's title, which was "A formula for drawing two correlated curves so as to make the resemblance as close as possible" (H'Doubler, 1910).

In 20th-century journals and textbooks, there was an almost century-long trend for correlation to be clearly explained with formulas but confusedly and sometimes even mistakenly explained with charts and text. It seems that statisticians were in agreement on how to calculate correlation but in disagreement on what it meant.

The topics covered in this chapter include:

- The affect that controlled vs. uncontrolled studies have on interpretation of correlational data analysis
- Acrimonious battles in journals about the meaning and value of correlation
- The writings of authors who worked primarily in just five fields of study:
  - *biology*: C. B. Davenport, E. Davenport, Henri Bergson, Sewall Wright, and Carl Jones
  - *economics*: Willford King, J. D. Magee, Horace Secrist, Edmund Day, Carl West, Harry Jerome, Frederick Mills
  - *mathematics*: Arthur Bowley, Henry Rietz, Edward Huntington, Karl Pearson, G. I. Gavett, Ronald Fisher
  - *psychology*: Charles Spearman, Percival Symonds, William Brown, Louis Thurstone, Clark Hull, Stuart Dodd
  - *statistics*: William Elderton, Reginald Hooker, George Yule, Harald Westergaard, John Koren, John Cummings, William Crum, Alson Patton, Robert Burgess, Helen Walker.

Controlled vs. Uncontrolled Study

Much of 20th-century disagreement regarding the meaning and value of a correlation coefficient can be traced to whether authors were talking about controlled or uncontrolled studies. In a simple *controlled* study involving only two variables, the focus is on the effect that a change in the magnitude of one variable has on the magnitude of the other. By definition, one of the variables is "controlled" or "fixed" in the sense that its values are chosen by the experimenter rather than allowed to vary randomly. For example, kilograms of fertilizer added per hectare vs. amount of vegetables produced per hectare; in that example, fertilizer inputs are controlled, and vegetable output is the random variable.

When analyzed by linear regression, the data resulting from a controlled study is (in modern times) typically arranged with the fixed variable (e.g., fertilizer) on the horizontal axis (also called the X-axis) and the random variable (e.g., vegetables) on the vertical axis (also called the Y-axis). Then the following are calculated:

- Standard deviation of the Y values ($= Sy$)
- Equation of the least-squares line through the plotted X, Y data (i.e., $Ye = a + bX$)
- Slope of that line ($= b$ in that least-squares line equation)
- Standard deviation of the difference between the Y values and their corresponding equation-derived Ye values; this is called the Standard Error of Estimate ($= See$)
- Correlation coefficient (r); or its square, the coefficient of determination (R^2).

In a controlled study, there is usually no purpose in calculating the standard deviation of the X values ($= Sx$), since they were chosen and as such are not a "result" of the study.

The values b, See, and r provide information about the size and/or consistency of the effect that X has upon Y (e.g., the effect that fertilizer has upon crop yield). The *size of the effect* is given by b, the slope of the line. That slope is known by various names, including *regression coefficient* and *effect coefficient*. Because slopes are in units of the original data, it may be difficult to compare results from different studies. In order to simplify such comparisons, the original data can be standardized before plotting, by dividing each X value by Sx, and dividing each Y-value by Sy; the slope of the resulting best-fit line is known by various names, including the *standardized slope, standardized regression coefficient, beta coefficient*, and (redundantly) the *standardized beta* (Miles and Shevlin, 2001, p. 19). Such a standardized value is what Galton described in 1888 (previously discussed); that is, it equals r, the *correlation coefficient*.

Unfortunately, some textbooks include sentences that do not clearly distinguish between the regression and correlation coefficients, in regard to which one of them indicates the *size* of the effect; for example:

> *If we measure the entire population from which we have randomly selected our sample, the size of the correlation <u>or</u> regression coefficient we would find is known as the effect size.*
>
> (Miles and Shevlin, 2001, p. 120; underlining added)

As measures of effect size and consistency, b and See, respectively, are unsurpassed in controlled studies. However, both of them are influenced by how the bivariate data is plotted; that is, a plot of Y on X produces different b and See values than does a plot of X on Y (i.e., where X values are plotted on the vertical axis, and Y values on the horizontal axis). Let's call these values b_1 and See_1, and b_2 and See_2, respectively. The correlation coefficient provides information about the geometric average value of b_1 and b_2, and about a standardize version of See_1 and See_2, as can be seen in the following formulas (in which all terms were defined previously):

$$|r| = \sqrt{(b_1 b_2)} = \sqrt{\left(1 - (See_1)^2 / (Sy)^2\right)} = \sqrt{\left(1 - (See_2)^2 / (Sx)^2\right)}$$

In a controlled experiment, the values of X are fixed, and so there is no purpose in determining b_2, which equals how much X varies for a given variation of Y. Similarly, there is no purpose in determining See_2, which equals the standard deviation of the difference between the X values and their corresponding equation-derived Xe values derived from $Xe = a_2 + b_2 Y$. In other words, when the X values are fixed, all useful information is represented by b_1 and See_1; and therefore there is no need to also calculate a correlation coefficient.

However, if the experiment or study is not controlled, then r has great utility for comparing the results from different studies. For example, when comparing human arm and leg length, which length should be plotted as X and which as Y? There is certainly a relationship between them, but how should it be measured? A report that includes b_1, b_2, See_1, and See_2 is unnecessarily complicated. A simpler approach is to instead report r, the correlation coefficient, because it averages, standardizes, and summarizes all four of those statistics for the purpose of comparison—or at least that is how many researchers have viewed the situation.

1899

In 1899, a book was published titled *Statistical Methods with Special Reference to Biological Variation*. Its author was an instructor in Zoology at Harvard University by the name of **C. B. Davenport** (previously mentioned for having published a paper with C. Bullard in 1896). As far as I have been able to discover, this was the world's first English-language book that focused completely on mathematical statistics. Because of that honor and because it was published in *late* 1899, it is included in this chapter rather than in the previous one.

Davenport's chapter on correlation was eight pages long, which represented 16% of the book's 51 pages of text (which was then followed by 97 pages of formulas and tables with titles such as "Table of ordinates of normal curve", "Squares, cubes, square roots, cube roots, and reciprocals", and "Logarithmic sines, cosines, tangents, and cotangents"). His explanations of correlation included the word "abmodality"; he also many times used the term "Index of Abmodality", which he defined with a formula rather than with words:

$$\text{Index of abmodality} = \frac{(X - M)}{S}$$

where *X* is a sample measurement, *M* is the mean of all sample measurements, and *S* is the standard deviation of all sample measurements (Davenport, 1899, p. 19). In effect, that is what today is called a Z-score. The only way his explanations make sense is if the reader equates the *word* abmodality with the *term* Index of Abmodality, but he does not say that anywhere in the book.

He spoke not of correlation but of "correlated variation":

> *Correlated variation is such a relation between the magnitudes of two or more characters that any abmodality of the one is accompanied by a corresponding abmodality of the others. The methods of measuring correlation depend upon the assumption that the variates of the characters compared are distributed normally about the mode. The method is approximately applicable to cases where the distribution of variates is slightly skew* (p. 30).

Like Galton, he used the terms Subject and Relative; unlike Galton, he defined them like this:

> *If we select individuals on the basis of one character (A, called the subject) we select also any closely correlated character (B, called the relative) (e.g. leglength [sic] and stature)* (p. 30).

That description leaves the reader clueless regarding the fact that the Subject values are mid-points of arbitrarily chosen intervals of "A", and that the Relative values are the averages of all "B" values that fall into a given "A" interval. Thankfully, the book provided a partial correlation matrix table in which for one Subject the corresponding Relative values were calculated (Davenport, 1899, p. 31); readers can then reverse-engineer the result, in order to resolve their confusion.

That matrix table was not as clear as it could have been in regard to what is Subject and what is Relative; Davenport too seems to have realized that problem, and curiously, his solution was to provide the clarification on the next page rather than on the same page as that table. There are a few numerical errors in that table; for example, he listed the standard deviations of Subject and Relative as both equaling 1.73, whereas elsewhere in his text (p. 34), he showed that they were slightly different from each other, namely 1.7195 and 1.7304.

In a small paragraph, he described how to determine the "correlation coefficient" using "Galton's graphic method":

> [section title:] *Methods of Determining Coefficient of Correlation*
>   [subsection title:] *Galton's Graphic Method*
>   *On co-ordinate paper draw perpendicular axes X and Y; locate a series of points from the pairs of indices of abmodality of the relative and subject corresponding to each subject class. The indices of the subjects are laid off as abscissae; the indices of the relatives as ordinates, regarding signs. Get another set of points by making a second correlation table, regarding character B as subject and*

character A as relative. Then draw a straight line through these points so as to divide the region occupied by them into halves. The tangent of the angle made by the last line with the horizontal axis XX (any distance yp, divided by xp) is the index of correlation (p. 32).

It is noteworthy that he used the term "Coefficient of Correlation" in that section title, but used the term "index of correlation" in the subsequent paragraph; and four pages later he thrice used the term "correlation coefficient" (p. 36). He continued to use different terms for the same correlation concept the next year when he published a paper titled "A History of the Development of the Quantitative Study of Variation", in which he used the terms "correlation-index", "index of correlation", and "coefficient correlations" (1900, pp. 866, 867, 868, respectively). Four years later, in the second edition of his book, he again used multiple terms: "coefficient of correlation" and "correlation coefficient" (1904, pp. 54, 55). His readers must have wondered… "Which term should I use?"

That "History" paper gave credit to Francis Galton for the initial discovery of "a method—somewhat rough, to be sure, because chiefly graphic—for measuring correlation…" and to Karl Pearson for placing it "on an analytical basis" (pp. 866, 867). His book, paper, and that acknowledgment might have been what helped to earn him an early editorship at Galton and Pearson's *Biometrika* (Cox, 2001, p. 3).

His book's description of Galton's graphic method was similar to Galton's December 1888 method, but there were a few important differences:

- Davenport put the Subjects on the X-axis and Relatives on the Y-axis, whereas Galton did the reverse. Davenport's method resulted in r equaling the slope as viewed from the X-axis, rather than from the Y-axis as Galton had done.

- Davenport *required* that the plot-line be drawn so that half of the plotted points be on one side and half on the other; Galton advised that the line be "drawn to represent the general run of the [points]" (Galton, 1888c, p. 140). It is humorous that the line drawn by Galton on the "figure" in his December 1888 paper does not meet Davenport's requirements because Galton's line resulted in seven points above the line, four below the line, and two points exactly on the line.

- Davenport measured the angle that the plotted line of transmuted values made with the X-axis, and then determined the trigonometric tangent of that angle (where r=tangent); Galton instead measured the X, Y slope directly. I wonder which method was more accurate, in the hands of the general public?

Some readers might not have a compass to measure that angle; in such cases, the user could use the large paper copy of a compass that Davenport included as the last page of each edition of his book—it may have been awkward to use, but it was better than nothing. It is interesting that the tables of tangents in the back of his book were given to six significant digits, but a user of the book's compass could effectively estimate angles to only two significant digits.

Immediately following his description of the Galton method, the book discussed a "more precise method...given by Pearson" (Davenport, 1899, pp. 32–35), namely Karl Pearson's product-moment correlation coefficient (previously discussed). Because the calculations for that coefficient are many and laborious, Davenport provided a short-cut method for use with correlation matrix table interval classes, a method that he copied from recent publications by a German scientist named **G. Duncker**. That short-cut method involved so many approximations that I wonder if its result might, on average, be *less* precise than Galton's method. The following are *some* of the steps in the Davenport/Duncker short-cut method:

1. "Separate the deviation from the mean of each class into its integral and fractional parts" (e.g., 3.45 is to be separated into 3 and 0.45).
2. "Draw rectangular co-ordinates" (i.e., divide the correlation table into four quadrants containing only the integral parts of the numbers).
3. Perform calculations on those integral numbers separately for each of the four quadrants.
4. Perform other calculations on the fractional numbers.
5. Subtract the fractional final result from the integral one.

Feedback from post-publication reviewers and readers regarding that short-cut method must not have been positive, given that Davenport eliminated it completely from his second edition and instead referenced other parts of his book where he'd given advice on performing calculations similar to those required by Pearson's formula. The second edition looks to have no other significant changes, at least in regard to correlation.

In 1914, he published a third edition. It included the first and second edition Prefaces but no new one; it was paginated virtually identically to the second edition, and the text of the sections on correlation looks to be identical to that in the second edition.

1901

Arthur Lyon Bowley (1869–1957) degreed in mathematics at Trinity College and by the 1890s began a decades-long career teaching statistics, most notably at the newly-formed London School of Economics (Encyclopediaofmath.

org-Bowley). He has been called F. Y. Edgeworth's "only statistical heir" (E. Pearson and Kendall, 1970, p. 262). Bowley's *Elements of Statistics* is considered by some historians to be the first statistics textbook in the English language (Wikipedia: Bowley); however, such a claim ignores the just-discussed 1899 book on *Statistical Methods* by C. B. Davenport.

Elements of Statistics went through six editions from 1901 to 1937, and the sixth edition went through at least three reprints, the last of which was in 1948 (as far as I have been able to determine). The preface to the first edition stated that...

> *There seems to be no text-book in English dealing directly and completely with the common methods of statistics. English writings on the various branches of the science are for the most part in the form of articles in the journals of learned societies.... The result is that there is no compact statement of principles acknowledged by statisticians, of the methods common to most branches of statistical work, of the artifices developed for handling and simplifying the raw material, and of the mathematical theorems by the use of which the results of investigations may be interpreted. This book forms an attempt to supply this want, so far as can be done without undue length.*
>
> (Bowley, 1901, p. v–vi)

A search of internet sites (e.g., Internet Archive) and of the bibliographies in books on the history of statistics reveals no earlier book-length English-language publication on mathematical statistics, other than Davenport's. Many other books were available with "Statistics" in their title, but they were collections and/or discussions of recent or historical *data* (i.e., "statistics"), such as census reports or economic surveys. In order to set itself apart from such books, Bowley first edition of *Elements* stated that...

> *No place has been given in* [this book] *to the history of statistics, and it does not contain any summary of the main groups of statistics extant; several tables, drawn from a wide range of subjects, are given, but only to illustrate particular methods, and their choice has been determined by their suitability for this purpose* (p. vi).

His first attempt at defining correlation did not indicate that it was a special concept but rather that it was a valuable way to arrange or interpret tabular data:

> *It may be supposed that the chief object of* [a study]... *was to find* [e.g.] *whether the labourers' families earned enough for their support, and what proportion was earned by the wives and children....The counties* [being studied] *might be taken* [i.e. arranged in a table] *in alphabetical order for convenience of reference, or in geographical order with subordinate averages for groups (e.g., Eastern : Norfolk, Suffolk, Essex).... For instance, if we wish to see the relation between total* [family] *earnings and the family's subsidiary contribution*

[i.e. the percentage contributed by the wife and children, exclusive of earnings by the husband]...*If we found signs of correspondence we should re-arrange the* [data in the table]...*in order of these subsidiary percentages, and see if they were approximately in order of total earnings also. This is an example of tabulation to show <u>correlation, the correspondence in the occurrence of two sets of phenomena</u>.*" (pp. 83–84; underlining added).

Later on in the book, he explained that not only tables but also diagrams can show correlation:

<u>*Diagrams may often be used to suggest correlation*</u> *between two series of figures, and this indeed is one of their chief merits, and they may be used to illustrate arguments on the subject, but at this point their utility ends, for* <u>*they cannot be made to prove much*</u>. <u>*Causal relations*</u> *are very difficult to establish, and the original figures must be critically consulted when theories are to be brought to the test* (p. 173; underlining added).

In a subsequent chapter titled "The Theory of Correlation", he defined correlation, first in words and then mathematically; and he applied it to virtually any paired data set:

When two quantities are so related that the fluctuations in one are in sympathy with the fluctuations of the other, so that an increase or decrease of one is found in connection with an increase or decrease (or inversely) of the other, and the greater the magnitude of the changes in the one, the greater the magnitude of the changes in the other, the quantities are said to be correlated. Correlation is a quantity which can be measured numerically.... Let individual values of X and Y...be grouped in pairs, as measurements of two quantities at the same date, or of two parts of the same organism, or <u>*in any other way*</u>.... *r is called the coefficient of correlation.... We see then that r measures the correspondence between deviations from their means of the two series of observations....* <u>*r serves as a measure*</u> *of any statement involving two qualifying adjectives, which can be measured numerically, such as "tall men have tall sons," "wet springs bring dry summers," "short hours go with high wages."* (pp. 316, 317, 320; underlining added).

At a time when mechanical calculators were rare and expensive, this subsequent statement by Bowley is surprising:

The <u>*calculation of r*</u> *is quite simple, and if we can assume normal dispersion, so that the probable error in a series is equal to.67* [i.e. 0.67] *of the error of mean square,* <u>*can be performed very rapidly*</u> (p. 320; underlining added).

Bowley then introduced what he called "The Galtonic Method": "An earlier method of estimating correlation, introduced by...Galton, is very useful for a rapid survey of two groups of figures" (p. 322). In a footnote to that sentence, Bowley told the reader to "*See Proceedings of the Royal Society,* 1886, vol. xi.,

Family Likeness in Stature". Unfortunately, the correct volume reference for that paper is *xl*, not *xi*; and neither Galton's concept of mathematical correlation nor even the word correlation can be found in it. That paper was published in early 1886, almost 3 years before Galton's December 1888 paper that introduced the mathematical concept of correlation. That 1886 paper (Galton 1886a) discussed the concept of regression not correlation. That mistake was not fixed until Bowley's fourth edition's "completely rewritten" section on correlation (Bowley, 1920, p. v); the fix was to remove the reference entirely.

Starting on page 322 of the 1901 first edition, Bowley claimed to be demonstrating how Galton graphically determined the correlation coefficient. For his demonstration, Bowley chose the minimum and maximum daily temperatures at some undisclosed location during 1898. Unfortunately, how he plotted the data was neither like Galton 1886 nor Galton 1888 but rather like a mixture of the two. Galton's 1886 paper's relevant diagram ("Figs. 5 and 6") had X and Y axes scaled in inches whereas Galton's 1888 paper's relevant diagram (on his page 138) was scaled in units of probable error. The axes of Bowley's relevant diagram (reproduced here as Figure 7.1) were indeed scaled in units of probable errors, but the axes were both labeled in degrees Fahrenheit; as a result, 10 degrees Fahrenheit on the X-axis confusedly had about the same length as 14 degrees Fahrenheit on the Y-axis. Unlike Galton, who had plotted Y=Subject vs. X=Relative, Bowley plotted Y=Maximum vs.

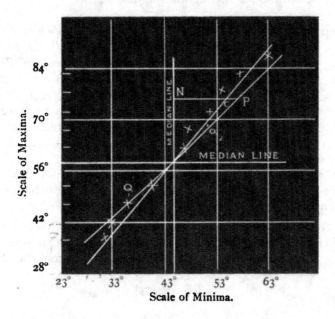

FIGURE 7.1

This is Bowley's plot (from his page 323) that he used to graphically determine that $r=4/5$, which agreed with the $r=0.8$ that he determined formulaically (on page 324). (From Bowley (1901).)

X=Minimum. To determine the correlation coefficient, Bowley first used the slope technique found in Galton 1888 and then used a version of Pearson's 1896 formula (previously discussed); his two results were "4/5" and "0.8", respectively. He concluded that "we obtain approximately the same value by either method" (p. 324). That is a true statement; but to anyone who has read Galton's 1888 paper, Bowley's claim to have used the graphical method "introduced by Galton" seems untrue.

Bowley subsequently proceeded to confuse his novice student reader by stating:

> ...*Galton applied this* [graphic] *method to the question of inheritance of stature. He found that the correlation between the statures of children and of their parents was 2/3. That is if a group of parents had an average stature x inches above (or below) the general average, the average for their sons was only 2/3x inches above (or below) the general average. This return towards the average is called in biological language "regression," and hence* the coefficient of correlation is often spoken of as the "coefficient of regression," *and such an equation as y = r(σ₂/σ₁)* x *is called the "equation of regression."* In words *this equation is: the ratio of the divergence of one quantity from its mean value to its standard deviation equals the ratio of the divergence of a correlated quantity to its standard deviation, multiplied by the* coefficient of regression (pp. 324–325; underlining added).

That paragraph contained two significant mistakes:

- When discussing correlation, r is *not* called the "coefficient of regression". He corrected this error in his second edition (1902), by identifying $r(\sigma_2/\sigma_1)$ and $r(\sigma_1/\sigma_2)$ as the "coefficients of regression" (Bowley, 1902, p. 325). However, while doing so, he introduced another error, when he said: "[Galton] found that the coefficient of correlation between the statures of children and of their parents was 2/3" (p. 324; underlining added). Those two words ("coefficient of") were not in the first edition; it is incorrect because the ratio of 2/3 for child vs. parental stature is what Galton had called the "ratio of regression", which today is called the regression coefficient (Galton, 1886a, p. 54). Bowley corrected this in his 1920 fourth edition, by removing the entire discussion about Galton and stature.
- The "in words" description of the formula did not match with the formula itself; what he described in words is this next formula (where x and y are paired deviations and b is the regression coefficient):

$$y/\sigma_2 = (x/\sigma_1)b.$$

Rearranging that formula gives the following incorrect one:

$$b = (y/\sigma_2)/(x/\sigma_1)$$

which is actually the formula for the correlation coefficient, as the reader could verify by rearranging the formula $(y=r(\sigma_2/\sigma_1)x)$ that Bowley had given in his text. The comparable and correct formula for the regression coefficient is:

$$b = y/x$$

That mistake was fixed in the second edition (the fix was to remove the entire "in words" sentence).

At the end of his first-edition's "Part II" (in which is found the chapter on "The Theory of Correlation"), he provided a 1-page, small-font list of "books and articles relating to the subject of Part II of this book which are most accessible and likely to be most useful to the English student" (Bowley, 1901, p. 327). That list contained three publications authored by Galton: *Inquiries into Human Faculty and its Development, Natural Inheritance,* and "*Family Likeness in Stature*. Proc. of Royal Soc., 1886, 1888". That list is interesting for two reasons: first, it does not include any of the papers that Galton used to introduce and explain correlation in 1888–1890; and second, none of Galton's 1888 publications was titled anything like "Family Likeness in Stature".

The fourth edition's preface stated that the book's introduction to correlation (in its Part II) had been revised:

> ...*Part II has been completely rewritten and considerably extended, both by the more detailed and extended treatment of theory and by the addition of a number of examples which illustrate the arithmetical use of the formulae and show the scope of the application of the theory....* [This revision was partly in response to] *the very loose reasoning often employed by writers who make too facile use of the standard deviation, of curves of frequency and especially of the <u>coefficient of correlation</u>. Very great care has been taken...to show as exactly as possible <u>the meaning of the measurement of correlation by this coefficient</u> and its implications, and very much more might have been said before the subject was too thoroughly explored. <u>No one should attempt to measure correlation till he has studied the theory closely and critically.</u>*
>
> (Bowley, 1920, pp. v, vi-vii; underlining added)

In all three previous editions, only a single chapter (each about a dozen pages long) focused on correlation, whereas in the fourth edition, four chapters totaling six dozen pages focused on correlation. His fourth-edition explanation of correlation began...

> *Suppose that we have pairs of observations* [X and Y].... *if there is anything common to X and Y in <u>causes</u> of their variations, the statement of the value of an X will presumably affect the probability of the deviations of the corresponding Y.... In the cases with which we have to deal, however, the connection is not one of direct relation; when X is given, Y is not determinate, but in a series of*

measurements (e.g. of height) we shall find for the same X varying values of Y. If
the average or shape of the frequency curve of the Y's associated with a given X
is not the same as that for all values of Y when the sorting by values of X is not
made, then there is something <u>*common*</u> *to the two quantities, and they are said to*
be correlated (p. 350–351; underlining added).

In a subsequent section titled *"Nature of r"* (p. 355), that explanation was
quantified with this formula:

$$r = p/(p+q)$$

where
  p=number of "causes" that X and Y have in "common"
  q=total number of "causes" affecting X and/or Y.
  In regard to that formula, he said:

This is the simplest conception of the numerical value of r; expressed in words
[,] it shows that the correlation coefficient <u>*tends*</u> *to be the ratio of the number of*
<u>causes</u> common in the genesis of two variables to the whole number of indepen-
dent causes on which each depends (p. 356; underlining added).

He said "tends" rather than "equals" because that formula is a theoretical
approximation without any practical value, given the fact that in most real-
life situations it is impossible to determine the numerical values of p or q; fur-
thermore, some causes may have a greater or lesser influence on the values
of X and Y, and yet this formula counts them all as equal. That formula leads
students to the conclusion that the correlation coefficient is a true proportion,
which (as we will see shortly) it is not (Spearman, 1904; Wright, 1921; Zorich,
2018). If students then take to heart Bowley's claim that "r is…a sensitive measurement of the amount of correlation" (p. 355), then they conclude that what
is being sensitively measured is the % of causes shared by X and Y—such
an erroneous conclusion seems inevitable, given that Bowley provided no
further textual explanation for the meaning of correlation and its coefficient.

His fifth and sixth editions (1926 and 1937, respectively) were not much
changed from his fourth, except for error corrections (according to their prefaces); the discussions and claims about correlation remained virtually the
same as in the fourth edition.

1904

In 1904, **Charles Edward Spearman** (1863–1945) was a British ex-military
officer who was two years away from obtaining his degree in experimental
psychology; he subsequently pioneered the use of mathematical statistics in

that field, and he invented the Rank Correlation Coefficient (Walker, 1929, p. 127). As far as I have been able to determine, what is not mentioned in any book or paper on the history of statistics is that he seems to have been the first person to *publish* the fact that r, the correlation coefficient, doesn't really measure anything precisely and that the true *measure* of correlation is r^2, the square of the correlation coefficient.

In April 1904, he published a 91-page paper titled "'General Intelligence,' Objectively Determined and Measured"; he stated its purpose:

> The present article, therefore, advocates a "Correlational Psychology," for the purpose of positively determining all psychical tendencies....
>
> (Spearman, 1904b, p. 205)

He included a brief discussion of a psychology-related book by Francis Galton that Spearman titled incorrectly (on page 206) as *Inquiries into the Human Faculty* (the correct title is *Inquiries into Human Faculty and its Development*). It is interesting to note that Galton's *Inquiries* continues to be difficult to reference correctly, as evidenced in 1996 by its being referred to as *Enquiries into the Human Faculty and its Development* in the editorial introduction to a 1996 reprint-edition of Galton's *Essays in Eugenics* (Galton, 1909; note the misspelled first word and the addition of the word "the").

Spearman's 1904 paper introduced a new type of correlation table that he introduced by saying…

> …we note that Discrimination has been tested in three senses and that Intelligence has been graded by three different persons; thus we have nine correlations… (p. 259).

He then described a 3×3 table composed only of those nine correlation coefficients. Years later, the invention of that table was mistakenly attributed to J. Garnett (more about this, later).

Throughout that paper, he used his own version of a correlation coefficient, which he derived from Pearson's Product-Moment version. What seems surprising is that 3months prior, in January 1904, he had published a paper that denigrated the value of such coefficients; relevant text from that earlier paper is:

> In psychology, more perhaps than in any other science, it is hard to find absolutely inflexible coincidences; occasionally, indeed, there appear uniformities sufficiently regular to be practically treated as laws, but infinitely the greater part of the observations hitherto recorded concern only <u>more or less pronounced tendencies of one event or attribute to accompany another</u>. Under these circumstances, one might well have expected that the evidential evaluation and <u>precise mensuration of tendencies</u> had long been the subject of exhaustive investigation and now formed one of the earliest sections in a beginner's psychological course. Instead, we find only a general naive ignorance that there is anything about it requiring to be learnt. One after another, laborious series of experiments are executed

and published with the purpose of demonstrating some connection between two events, wherein the otherwise learned psychologist reveals that his art of proving and measuring correspondence has not advanced beyond that of lay persons. The consequence has been that the significance of the experiments is not at all rightly understood, nor have any definite facts been elicited that may be either confirmed or refuted. The present article is a commencement at attempting to remedy this deficiency of scientific correlation.

(Spearman, 1904a, pp. 72–73; underlining added)

This following text is how that January paper introduced the correlation coefficient:

The most fundamental requisite is to be able to measure our observed correspondence by a plain numerical symbol…. The first person to see the possibility of this immense advance seems to have been Galton, who, in 1886, writes: "the length of the arm is said to be correlated with that of the leg, because a person with a long arm has usually a long leg and conversely." He then proceeds to devise the required symbol in such a way that it conveniently ranges from 1, for perfect correspondence, to 0 for entire independence, and on again to −1 for perfect correspondence inversely. By this means, correlations became comparable with other ones found either in different objects or by different observers….

(Spearman, 1904a, pp. 73–74; underlining added; it is noteworthy that
(1) in 1886, Galton did not use the word correlated in any publication
(as previously discussed); (2) the quote just given by Spearman was taken
from Galton's December 1888 paper, not an 1886 one; and (3) Galton did
not publish anything about inverse correlation in either of those years)

He continued in that January paper to discuss "The significance of the quantity" (i.e., of Galton's *r* coefficient):

…r is the measure of the correlation. But another—theoretically far more valuable—property may conceivably attach to one among the possible systems of values expressing the correlation; this is, that a measure might be afforded of the hidden underlying cause of the variations…. Hence A, in order to be correlated with B by $1/x$, must be considered to have only devoted $1/x^2$ (instead of $1/x$) of his arrangement to this purpose, and therefore to still have for further arrangements $1 − 1/x^2$, which will enable an independent correlation to arise of $\sqrt{(1 − 1/x^2)}$. In short, not Galton's measure of correlation, but the square thereof, indicates the relative influence of the factors in A tending towards any observed correspondence as compared with the remaining components of A tending in other directions.

(Spearman, 1904a, pp. 74–75; underlining added; the mathematical
symbols shown here have a slightly different appearance in
the original text; I made that change for the sake of clarity)

Spearman's January 1904 paper seems to have been the first publication (by anyone) to state that, in regard to correlation, *r* is *not* a measure in the sense that a meter-stick measures distance or a weight balance measures mass. His r^2, the square of Galton's measure, would in 1921 be named the Coefficient of Determination (more about that later).

1906

William Palin Elderton (1877–1962) was an actuary who became a colleague of Karl Pearson; his sister, **Ethel Elderton**, was a statistician who also worked closely with Pearson (Tappenden, 1962). The Elderton siblings collaborated to author an elementary statistics book called *Primer of Statistics*, of which Galton read enough to conclude (in the book's Preface) that it was designed for "familiarizing educated persons with the most recent developments of the new school of statistics" (Elderton and Elderton, 1910, p. vi).

Separately, William had previously written a more technical statistics book, which he titled *Frequency Curves and Correlation*. In its first edition, he said that his motivation for writing the book was "to bring before Actuaries the more practical methods of modern statistical work" (Elderton, 1906, p. vii). He included a 19-page chapter titled "Correlation", in which he defined correlation and explained the need for it:

> *Two measurable characteristics, A and B, are said to be correlated, when, with different values, x of A, we do not find the same value, y of B, equally likely to be associated. In other words, certain values of B are relatively more likely to occur with the value x than others.... it is required to find a method of measuring the amount of correlation statistically....* [What is needed is a] *measure of correlation suitable for comparison with the experiences of other* [life insurance] *offices or with the same office at a later date...* (pp. 106–107).

After presenting relevant mathematical formulas, he summarized by stating that "r...is a measure of the correlation. r is called *the correlation coefficient*" (p. 112; italics in the original).

That definition of correlation had nothing to do with cause and effect but rather only with occurrence, and therefore his subsequent definition of spurious correlation (given here next) seems incomplete:

> *...it is possible to obtain a significant value for a coefficient of correlation when in reality the two functions are absolutely uncorrelated. Such a result is called "spurious correlation"...* (p. 122).

As far as I can determine, the term "spurious correlation" was coined in 1897 by Karl Pearson, who applied it then to the realm of biology ("I term this a spurious <u>organic</u> correlation, or simply a spurious correlation"—K. Pearson, 1897, p. 490; underlining added). Although it can be inferred that he meant to apply it beyond the realm of biology, it was Francis Galton who shortly thereafter explicitly applied the term to "variables" in general (Galton, 1897, p. 499). Both Pearson and Galton defined spurious correlation in terms of lack of common causes, a meaning with which Elderton's definition was consistent. That is, what Elderton meant by "absolutely uncorrelated" was (in my words) "do not share any common causes of variability"; that interpretation is based upon his definition of correlation, which had included a probability statement, namely that "certain values of B are relatively more likely to occur with the value *x* than others" (underlining added). His definition of spurious correlation therefore incrementally improved upon those given by Pearson and Galton, in that spurious correlations cannot be said to have "likely" occurred (e.g., number of honey-producing bee colonies in the United States vs. number of nuclear weapons in Russia's stockpile, in the years 1990–2002: $r=0.979$ (Vigen, 2015, p. 45)).

His second and third editions (1927 and 1938, respectively) contained no important changes in regard to his presentation of correlation. However, there is one interesting sentence in the second edition in regard to how "statistical work" was starting to be viewed:

> *The object, in statistical work, is to find a measure; we have a scale for measuring probability and similarly we want a scale for measuring correlation.*
>
> (Elderton, 1927, p. 136; underlining added)

In contrast, as previously discussed, the "object" of statistical work *in the 19th century* had been the assembling, summarizing, and presenting of census and survey data.

I don't have access to the 1909 first edition of *Primer of Statistics*, but the 1910 second edition was published only 6 months after the first (based on information in the second edition). In the second edition, the two Elderton authors do not provide an explicit definition of correlation. The following is the closest that they come to it:

> *Is there any relation between the length and breadth of nuts or between the length and breadth of shells?.... Suppose we took a nut out of a collection of nuts, and found it was a long one: [sic] ought we, from this information only, to conclude it was a broad nut? If so, we are assuming that length and breadth are related, owing to some cause or other. If, on the other hand, we said it was impossible to estimate in any way the breadth of a nut from knowing its length, we should be assuming that length and breadth in nuts were not related. The same question can be asked about shells; and then, if we found that in both cases there was a relationship (correlation), we might ask whether there was a closer relationship between the lengths and breadths of nuts than between the lengths and breadths of shells, or vice versa. This suggests that what we must find is some way of comparing our correlations, so that, at the end of our work, we can arrange 'relationships' or 'correlations' in order [of magnitude], just as we have, up to the present time, been arranging lengths and breadths:[sic]*

> as we fix a scale of inches (say) for measuring heights, so <u>we must fix a scale of</u>
> <u>'somethings' for measuring relationships</u>.... It has been shown that, in order
> to make proper allowance, <u>we must take the standard deviation as the unit of</u>
> <u>measurement</u>; that is to say, we must measure each thing in the terms of its own
> standard deviation....

> (Elderton and Elderton, 1910, pp. 55–56, 63; underlining added)

In that explanation of correlation, the words relationship and correlation-ship are used interchangeably, whereas Galton's meanings for the base-roots of those words were very different from each other in his 1890 paper titled "Kinship and Correlation" (previously discussed).

The Eldertons described the correlation coefficient as giving "precision" to "rough ideas" about correlated variables (e.g., about stature and foot length in humans). The sibling brother, in his own book (previously discussed) that was published three years prior to this one did not use the word precision to describe the correlation coefficient; therefore, that word may have been contributed by his sister.

The third edition (1912) of *Primer* contained no correlation-related changes that I could find. Other editions were published, but I have not had access to any of them.

1907

In 1907, **E. Davenport** and **Henry Lewis Rietz** were both professors at the University of Illinois (USA); Rietz taught mathematics whereas Davenport taught plant and animal breeding and was director of an agricultural experiment station (Davenport and Rietz, 1907, title page). Their book titled *Principles of Breeding: A Treatise on Thremmatology* was published that year, with a subtitle of *The Principles and Practices Involved in the Economic Improvement of Domesticated Animals and Plants*. With such a title, it is surprising that any author who focused on general statistical methods would know about it; and yet it was mentioned in a footnote on page 414 of Secrist's 1917 book titled *An Introduction to Statistical Methods* (although Secrist mistakenly put a "The" in front of their title).

Their book was 713-pages long; Davenport devoted 20-pages to a discussion of the "Meaning" and "Coefficients" of correlation (in the main body of the book), and Rietz devoted 5-pages to "Correlation Theory" (in the book's appendix). Davenport explained that the "degree" of correlation can be expressed as a "ratio" between +1 and –1. Confusedly, inappropriately, and humorously, his "examples of perfect correlation" (shown below) are all non-mathematical Observational Correlations rather than Relational or Co-Relational ones:

The whole subject of correlation refers to that <u>interrelation between separate</u> <u>characters by which they tend, in some degree at least, to move together</u>. This relation is expressed in the form of a ratio. Thus, <u>if an increase of one character</u> <u>is always followed by a corresponding and proportional increase in a related</u> <u>character, the correlation is said to be perfect</u> and the ratio is 1. On the other hand, if an increase in one character is followed by a corresponding and proportional decrease in a related character, the correlation is said to be negative and the ratio is −1, or perfect negative correlation. Still again, if the characters in question are absolutely indifferent the one to the other, the correlation is said to be zero, indicating mere association under the law of independent probability, without causative relation of any kind. <u>Examples of perfect correlation</u> are furnished by such obvious relations as those between the <u>power of sight and the</u> <u>presence of eyes</u>; the giving of milk and the presence of an udder; the presence of sunlight and the fixing of carbon; and by such other relations as are involved in direct causation.

(Davenport and Rietz, 1907, pp. 453–454; underlining added)

He did provide one appropriate example of mathematical correlation:

In general, however, correlation falls somewhere between −1 and unity, and on one side or the other of the zero point; that is, a degree of relationship exists which is neither absolute, denoting direct causation, nor negative, signifying mutual exclusion. For example, a high degree of correlation exists between length of cob and weight of ear. It does not amount to unity, however, for the circumference also contributes to weight (p. 454).

Then he again invalidly applied mathematical correlation to Observational Correlation:

The student must distinguish clearly between correlation and mere association. For example, we might ask the question whether black pigs are more subject to cholera than are pigs of other colors. The first step would be to establish a ratio between the number of diseased pigs and pigs in general. This ratio would now express the chances that a particular pig, irrespective of color, will be afflicted with this disease,—that is, by that operation of independent probability which we call chance. <u>If now we find upon inquiry that under the same conditions the</u> <u>ratio of cholera subjects to black pigs is higher than the ratio to pigs in general,</u> <u>then we should conclude that an actual positive correlation exists between the</u> <u>black color and this particular disease.</u> On the other hand, if this ratio should be below the ratio of pigs in general, then we should conclude that black pigs are less susceptible to this disease than are pigs of other colors, and that a negative correlation exists, assuming always equal opportunities for infection (pp. 454–455; underlining added).

Note that he was *not* saying that there was a correlation between the likelihood of disease and the *intensity* of blackness, in which case this would have been a valid example of mathematical correlation. He eventually clarified his view by saying…

> *When the presence or absence of the characters in question is absolute, as red or black hair, presence or absence of horns, then the correlation is expressed by a single ratio* [i.e. +1 or −1] (p. 455).

Additionally, at the end of the chapter, he explained (more implicitly than explicitly) that for such attribute data, what is calculated is not a measure of the correlation but rather a "measure of the association" (p. 471), which is what others would call the Coefficient of Association (e.g., Yule, 1911, p. 38).

It is interesting that his explanation of the *regression* coefficient barely covered a single page. The formula he provided for that coefficient required the user to have *already* calculated the correlation coefficient (p. 466). He neither stated nor showed that the regression coefficient is the slope of Y vs. X in units of the raw data; instead, he described it only as being "useful for prediction" (p. 466).

In the book's Appendix, Rietz's definition of correlation conflicted with what his co-author Davenport had written on prior pages, in the sense that Rietz did not use the word correlation for *both* variable and attribute data but rather for *only* variable data:

> *Definition. Two <u>measurable characters</u> of an individual, or of related individuals, <u>are said to be correlated</u> if* [,] *to a selected series of sizes of the one* [,] *there correspond sizes of the other whose mean values are functions of the selected values. The word "sizes," here used, should be taken to mean "numerical measure."* (p. 703; underlining added)

Rietz then provided Pearson's 1896 formula for r, and stated that…

> *If $r = +1$, all the individual points of the population will lie on the line of regression, and we can therefore, when one character is given, tell exactly what the associated character is in magnitude. In this case the correlation is said to be perfect positive correlation. Similarly, if $r = -1$, the correlation would be perfect negative correlation* (p. 706).

Unfortunately for the reader, he provided no explanation of how to interpret intermediate values of r. The closest he came to such an explanation was to state that…

> *If correlation exists, these* [x vs. y values] *do not lie at random over the field* [plot], *but arrange themselves more or less in the form of a smooth curve called the "curve of regression." This curve is a crude picture of the function which defines the correlation of the y-character relative to the x-character* (p. 704).

He provided no guidance on how to interpret that "crude picture" in relation to intermediate values of r.

1908

Reginald Hawthorn Hooker (1867–1944) was praised for "developing methods for use in particular types of problems or illustrating its use by practical
examples" (Yule, 1944, p. 74). In 1908, when he was a Fellow of England's Royal
Meteorological Society, he wrote a paper titled "An Elementary Explanation
of Correlation: Illustrated by Rainfall and Depth of Water in a Well". The
paper's first paragraph explained:

> *I am desirous in the present paper of calling attention to recent methods of deal-
> ing with statistics, and more particularly with the process of correlation, and to
> illustrate, by working out a special example, what results can be obtained. This
> paper is therefore to be regarded rather as an explanation of the method than as
> a rigorous procedure to be followed in every case.... I want to enable all who
> read these remarks to appreciate the value of the method, so that they may give
> credence to facts ascertained by experts, even although [sic] they may not them-
> selves follow the whole of their work.... I have felt emboldened to attempt to make
> others understand what is the good of correlation.*

> *(Hooker, 1908, p. 277)*

His definition of the correlation coefficient was:

> *...I propose to regard the correlation coefficient as a "measure of the resemblance
> of two sets of observations," and as a <u>corollary</u>, as a measure of the dependence of
> one phenomenon upon another* (p. 277; underlining added).

He used that word "resemblance" in reference to having plotted rainfall and
well-water-depth vs. time: "...there is considerable resemblance between the
two 'curves' thus drawn" (p. 277).
  It is interesting that...

- The example he chose was a Relational Correlation, not a Co-Relational
  one; that is, the depth of water in a well is dependent upon rainfall.
- He admitted that he was "...not enough of the mathematician to do
  more than work out a few problems" (p. 277).
- His correlation coefficient definition's "corollary" implied causation.

His focus on causation is understandable, given that his primary references
were Yule's "On the Theory of Correlation" and Bowley's *Elements of Statistics*
(both previously discussed); Yule's paper had explicitly tied correlation to
causation (Yule, 1897b, p. 812) and Bowley's book had done so implicitly
(Bowley, 1901, pp. 173, 355–356).

1909

At an international conference of statisticians in mid-1909, **G. U. Yule** (discussed previously for a paper he published in 1897) presented a not-yet-published paper of his that was titled "The Applications of the method of correlation to Social and Economic statistics". The official conference copy wasn't published until the following year, but a version of it was published ahead-of-time in the *Journal of the Royal Statistical Society*. The Society's version claimed (in a footnote) that it was "Slightly condensed from a paper read before the Twelfth Congress of the International Statistical Institute at Paris, July, 1909" (Yule, 1909, p. 721). However, the seven pages of text in the Society's version is missing the first three pages (and all eight of the formulas on those pages) of the 12 pages of text in the Institute's version (Yule, 1910); therefore, whether that footnote's claim is considered true or false depends upon the definition of "slightly". Quotes from both versions are given here next.

The first sentences in the conference version provide a history lesson:

> *During the past five and twenty years [i.e. since 1884] there has been gradually developed and improved a general method, of great power and adaptability, for investigating the nature of the relations subsisting between statistical variables—such relations as those that subsists between the weather and the crops, between the marriage-rate and trade, between the mortalities in different districts and the various conditions of housing, employment etc [sic] in such districts. The method is generally known now as the method of correlation.*

> (Yule, 1910, p. 537)

It is important to note that all of the examples just given by Yule represent Relational Correlations, not Co-Relational Correlations; that is, in each case, one of the variables can be considered the independent and the other the dependent (at least in the mind of an economist).

The very first sentence in the Society version described the current extent of the application of the method of correlation:

> *The method of correlation has found many applications of recent years to the study of biology, more probably than to any other branch of science in which statistical methods are of service, but the applications to the problems that specially interest the student of social and economic statistics have, it seems to me, been relatively scanty.*

> (Yule, 1909, p. 721)

In the conference version of his paper, he described the formula for linear regression as...

...an estimation equation, or, as it is rather unfortunately termed, a regression equation.

(*Yule, 1910, p. 538*)

Linear regression is based upon the method of least squares; in the Society version of his paper, he had this to say about how that method related to correlation:

The method of correlation is only an application to the purposes of statistical investigation of the well-known method of least squares. It is impossible, therefore, entirely to separate the special literature of the theory of correlation from that of the theory of error or of the method of least squares.

(*Yule, 1909, p. 722*)

Yule defined correlation by stating that...

...the value of r measures the closeness with which (1) or (3) holds good. It is therefore termed the coefficient of correlation. [his (1) referred to the "regression equation" of Y on X, and (3) referred to that of X on Y]

(*Yule, 1910, p. 539*)

He did not define what he meant by "closeness" or "holds good".

In 1911, Yule published the first edition of a textbook titled *An Introduction to the Theory of Statistics*; during the next six decades, it went through 13 more editions, with the final 14th edition going through at least five re-prints. Based upon my admittedly unscientific survey, it seems that most other introductory statistical textbooks published during that time either quoted or referenced his textbook in their chapters on correlation. Yule's book "was to make his name known and respected all over the scientific world" (Kendall, in Pearson and Kendall, 1970, p. 420). A review of that 1911 edition was published by the American Statistical Association:

In this volume [i.e. Yule's first edition] have been brought together the recent contributions to the theory of statistics by such writers as Elderton, Pearson, Hooker, Edgeworth.... This volume possibly contains <u>the best study upon correlation</u> of any which can be put in the hands of a student of exceptional mathematical ability.

(*Bailey, 1911, p. 765; underlining added*)

Yule's influence on textbooks written by others was so great that in 1997 an historian stated:

His work on correlation and regression is now so standard that only history buffs would consult the original sources....

(*Johnson and Kotz, 1997, p. 169*)

More than 25% of Yule's book dealt with correlation. Surprisingly, none of those pages contain an explicit definition of correlation. Instead the book implied a definition by means of the correlation coefficient; that is, correlation is a characteristic of a paired X, Y data set if its correlation coefficient has an absolute value greater than zero—"Two variables for which r is zero are...spoken of as <u>un</u>correlated" (Yule, 1911, p. 175; underlining added).

He cautioned the reader against putting too much faith in the correlation coefficient:

> *The student—especially the student of economic statistics, to whom this chapter is principally addressed—should be careful to note that the coefficient of correlation, like an average or a measure of dispersion, only exhibits in a summary and comprehensible form one particular aspect of the facts on which it is based, and the real difficulties arise in the interpretation of the coefficient when obtained. The value of the coefficient may be consistent with some given hypothesis, but it may be equally consistent with others; and not only are care and judgment essential for the discussion of such possible hypotheses, but also a thorough knowledge of the facts in all other possible aspects.*
>
> *(Yule, 1911, p. 191)*

In the first edition, when he introduced the correlation coefficient, he said that "r is termed the coefficient of correlation, and the [Pearson formula for it]...should be remembered" (Yule, 1911, p. 174). Whereas in the 11th edition, he felt the need to change it to... "The coefficient r defined in equation...is of very great importance. It is called the coefficient of correlation." (Yule and Kendall, 1937, p. 211).

Starting in 1937s 11th edition, Yule provided a warning (given here next) that he had not clearly stated in prior editions:

> *It is important to note that the regression equations do not tell us whether a variation in one variate is <u>caused</u> by a variation in the other; all we know is that the two vary together, and so far as the regression equations show, either the feeding-stuffs price may exert an influence on the oats price, or <u>vice versa</u>, or their common variation may be due to some other cause affecting both. This is only one instance of a difficulty which pervades the theory of correlation and regression, namely, that of <u>interpreting</u> results in terms of causal factors.*
>
> *(Yule and Kendall, 1937, pp. 217–218;* in the original, all of that text was shown in regular font, but the words that are here underlined appeared in italic font)

Yule stated that his 14th (final) edition in 1950 was "a substantial revision" (p. v). However, it still did not discuss the importance of r^2 (or R^2) for measuring *linear* correlation; that is surprising, given the fact that such importance had been clearly explained in a 1904 publication by Spearman (previously discussed), had been given a name in 1921 by Wright (to be discussed later in this chapter), and had been prominently highlighted in 1930 by Ezekiel in the first textbook ever published solely on the topic of correlation (to be discussed here in the next chapter). In Yule's 1950 edition, the importance of R^2 was discussed only in relation to goodness-of-fit tests for *curved* lines (p. 361).

1910

Based upon my examination of every paper in every issue of the American Statistical Association's *Journal* from 1888 to 1917 (Zorich, 2021b), and upon the authoritatively published *Index to the Journal of the American Statistical Association, Volumes 1–34, 1888–1939*, that journal's first non-review paper containing a correlation coefficient appeared in 1910, was 36-pages long, and was titled "The Correlation of Economic Statistics". The author of that paper was **Warren M. Persons** (1878–1937), an assistant professor of economics at Dartmouth College. In later years, he would join the faculty at Harvard University and become a founding editor of the journal *The Review of Economics and Statistics* (Hetwebsite.net-Persons).

In that paper, Persons took no personal credit for the correlation coefficient, but neither did he give credit to anyone else, not even to Karl Pearson, whose 1896 formula for *r* he used.

> However, it should be noted at this point that the coefficient of correlation is not empirical but was <u>derived by a priori reasoning</u>. It was found by assuming that a large number of independent causes operate upon each of the two series X and Y, producing normal distributions in both cases. Upon the assumption that the set of causes operating upon the series X is not independent of the set of causes operating upon the series Y the value r...is obtained.
>
> (Persons, 1910, p. 298; underlining added)

Later in the paper, when he explained that *r* equals the slope of the regression line when *X* and *Y* are expressed in units of standard deviations, neither Galton nor Galton's December 1888 paper was mentioned (p. 306).

On the first page of his paper, the opening paragraph contained what may have been a paradigm-shifting challenge to many of its readers:

> It is rarely, if ever, possible for the economist to state more than "such and such a cause <u>tends</u> to produce such and such an effect." Events can only be stated to be more or less probable. He is dealing mainly, therefore, with <u>correlation and not with simple causation</u> (p. 287; underlining added).

His paper did not deal with biology, but his view was that...

> *The problems of economics are similar to certain problems of biology, such as the effect of environment and heredity upon the individual* (p. 288).

He went on to summarized the then current general practice:

> *The commonly used method of measuring the amount of correlation between any two series of economic statistics is to represent the two series graphically upon the same sheet of cross-section paper and then* [to visually] *compare the fluctuations of one series with those of the other* (p. 289).

His paper provided examples of such curves and of the vague conclusions that typically accompanied them (e.g., "the evidence of Chart I strongly supports the contention that there exists a close relationship between..."; and "The general movement of the two curves taken as a whole is the same" (pp. 290, 294, respectively)). He questioned the usefulness of that method and of such conclusions:

> *The graphic method of comparing fluctuations is well enough as a preliminary, but does it enable anyone to tell anything of the <u>extent</u> of the correlation between the series of figures being considered?...<u>The charts do not answer the questions proposed</u>. The painstaking collection of statistics to test correlation is useless if there be no more reliable method to measure correlation. <u>A numerical measure of the correlation must be found</u> if we wish to determine the <u>extent</u> to which the fluctuations of one series synchronize with the fluctuations of another series.... <u>Such a measure has been widely used in biological statistics and used to a limited extent in economic statistics.... This measure, the coefficient of correlation</u>...* (pp. 294, 298; underlining added).

Persons at first defined correlation by quoting from Bowley's first edition (previously given), but then offered his own view:

> *The straight line best fitting the points* [when plotted versus each other on an X,Y diagram] *is called the line of regression.... The coefficient of correlation (r) is a measure of the closeness of the grouping of the points about this line of regression* (pp. 301, 303).

Unfortunately, although that last sentence is true for some data sets, it is false for others (see Figure 7.2 here and a more extensive discussion in Zorich, 2017). The final paragraph in his paper summarized his opinion and recommendation:

> *The various illustrations which have been cited* [in this paper] *show the importance of questions of correlation in economics. The ordinary graphic method of measuring correlation is inadequate. The coefficient of correlation is simple and yet is sensitive to small changes. It has been used in many fields of statistics by Galton, Pearson, Yule, Hooker, Elderton and others. The experience of these writers warrants the adoption of the coefficient of correlation by economists as one of their standard averages* (p. 322).

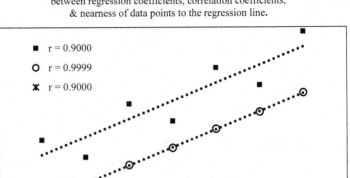

Three data sets that demonstrate the unpredictability of the relationships
between regression coefficients, correlation coefficients,
& nearness of data points to the regression line.

FIGURE 7.2
This chart shows that a large correlation coefficient (i.e., 0.9000) can be associated with either a
large or small regression coefficient (i.e., 10 or almost 0), and that a large regression coefficient
(i.e., 10) can be associated with either a large or very large correlation coefficient (i.e., 0.9000 or
0.9999). (Derived from Zorich (2017).)

1911

Early in 1911, Francis Galton died. Soon afterwards, the *Journal of the Royal
Statistical Society* published an obituary. The author of that obituary acknowledged that "By his invention of the method of correlation [,] Galton opened
an entirely fresh field of work…. [No one previously had thought] to employ
a single coefficient as a measure of the closeness of the relation between
two varying quantities." Curiously, that author did not mention a name for
that coefficient, other than to say that it was "termed 'Galton's function' by
Professor Weldon in his earlier papers" (Anonymous, 1911, pp. 316–317).

Another obituary of Galton was published that year, in the journal titled
Man, which was the recently renamed journal previously titled *The Journal of
the Anthropological Institute of Great Britain and Ireland*. This obituary claimed
that "In 1886 Galton made the great discovery of the Correlation table, and,
with the assistance of a mathematical friend, devised a method of calculating the coefficient of correlation which now plays so important a part in the
interpretation, not only of anthropometric, but of all kinds of statistics" (Gray,
1911). Unfortunately, that author was confusedly referring to Galton's discoveries regarding the "Law of Regression" (Galton, 1886b) and to the mathematical help from J. Dickson that facilitated those discoveries. As previously
discussed, Galton had discovered the regression (a.k.a. reversion) coefficient
in the 1870s. He would not discover the correlation coefficient until 1888;

however he never devised a method for "calculating" it, but rather only a method of graphically determining it.

In 1912, the Royal Society published a Galton obituary that unfortunately promoted confusion between regression and correlation by tersely connecting those two concepts in a single sentence that dealt with fractional biological inheritance. On the other hand, the obituary did clearly and correctly state that the correlation coefficient is an "average" of the relationship between paired sets of data (as previously discussed, the correlation coefficient equals the geometric average of the two relevant regression coefficients):

> [Because of the groundbreaking work by Galton and Pearson]...*it is now possible to assign a numerical value for the average degree of relationship or "correlation" between any pair of attributes in a large population.*
>
> *(G. H. D., 1912, p. xvi; underlining added)*

As the 20th century progressed, correlation in a mathematical sense (i.e., Relational or Co-Relational Correlation) became ubiquitous in scientific literature; however, the word was also still being used in a non-mathematical sense (i.e., for Observational Correlation). For example, the following is from a 1911 English translation of a widely-read 1907 French textbook on evolution that had gone through multiple editions (the instances of the words correlation and correlative given here also occurred in the 1910 French sixth edition). The author was **Henri Bergson** (1859–1941), a French philosopher who would be awarded 1927's Nobel Prize in Literature (Britannica.com-Bergson); his textbook on evolution is considered by some to be his "most important work" (A. Mitchell, "translator's note", in Bergson, 1911, p. v):

> *Let us assume, to begin with, the Darwinian theory of insensible variations, and suppose the occurrence of small differences due to chance, and continually accumulating. It must not be forgotten that all the parts of an organism are necessarily coordinated.... The law of correlation will be invoked, of course; Darwin himself appealed to it.... The examples cited by Darwin remain classic: white cats with blue eyes are generally deaf; hairless dogs have imperfect dentition, etc.... In these different examples the "correlative" changes are only solidary [solitary?] changes.... But when we speak of "correlative" changes occurring suddenly in different parts of the eye, we use the word in an entirely new sense: this time there is a whole set of changes not only simultaneous, not only bound together by community of origin, but so coordinated that the organ keeps on performing the same principle function, and even performs it better.... The two senses of the word "correlation" must be carefully distinguished; it would be a downright paralogism* to adopt one of them in the premisses [sic] of the reasoning, and the other in the conclusion. And this is just what is done when the principle of correlation is invoked in explanations of detail in order to account for complementary variations, and then correlation in general is spoken of as if it were any group of variations provoked by any variation of the germ. Thus the notion of correlation is first used in current science as it might be used by*

an advocate of finality; it is understood that this is only a convenient way of expressing oneself, that one will correct it and fall back on pure mechanism when explaining the nature of the principles and turning from science to philosophy. And one does then come back to pure mechanism, but only by giving a <u>new meaning to the word "correlation,"</u>—a meaning which would now make correlation inapplicable to the detail it is called upon to explain.

(Bergson, 1911, pp. 67–68, 70–72; underlining added; *paralogism:
A fallacious or illogical argument or conclusion)

Mathematician **H. L. Rietz** authored the "Correlation Theory" appendix to Davenport's 1907 book on plant and animal breeding (previously discussed). In 1911, Rietz published a paper whose title's first words were "On the Theory of Correlation". The paper's first sentence was a signpost on the road to correlation becoming an everyday word in all scientific fields:

The notion of correlation is of such importance in science that it seems it should become almost as familiar to the scientist as the notions of a mathematical function and of independence in the probability sense.

(Rietz, 1911, p. 187)

In 1911, **William Brown** published the first edition of his book titled *The Essentials of Mental Measurement*; at that time, he was a lecturer on psychology at University of London, King's College. The book's "Part I" was titled "Phsycho-Physics"; "Part II" was titled "Correlation". The first sentences in Part II were:

A somewhat detailed account of the mathematical theory of correlation and of the way in which it may be usefully applied to psychological measurements will be found in the later chapters of this Part. The object of the following introductory pages is to give the reader a general preliminary view of the method, free from mathematical complications, and to illustrate it by means of a simple example.

(Brown, 1911, p. 42)

The five chapter titles in Part II were:

- *Introduction* [to the mathematical theory of correlation].
- *The Mathematical Theory of Correlation.*
- *Historical* ["The history of the use of the theory of correlation in Psychology", as it said in the first sentence in that chapter].
- *Some Experimental Results* [on "what extent correlation exists between certain very simple mental abilities", as it said in the first sentence in that chapter].
- *The Significance of Correlation in Psychology.*

Two of the book's appendixes were titled:

- *Correlation Table Worked Out.*
- *Example of Multiple Correlation.*

Given that the book devoted more than 100 of its less than 150 pages to correlation, it seems that a more appropriate title for it would have been *Correlation in Psychology*.

Brown introduced correlation in Part I...

> ...*the interrelations of different mental abilities within any well-defined group of individuals situated within any definite environment may be determined by means of the <u>technical method of "correlation</u>." A correlation coefficient or other constant (e.g. correlation ratio) measures the <u>tendency</u> towards concomitant variation of two mental (or other) abilities within a group of individuals. The result may be transferred to any single individual within the group as <u>measuring the degree of probability</u> of connection or the closeness of connection of the two abilities in the particular case. <u>The correlation between the two abilities may be due to an actual direct relation of the abilities to one another</u>, or, indirectly, to the influence of a common external environment upon them both. <u>The first of these two cases is perhaps the more important</u>, but the possibility of the second should not be lost sight of, and it also has a special interest of its own* (p. 12; underlining added).

Such a view, that the correlation coefficient is a measure of probability, was elaborated by other authors decades later, as we shall see. He did not try again to define correlation until Part II, but curiously did so without mentioning his "probability" viewpoint:

> *Correlation may be briefly defined as "tendency towards concomitant variation," and a so-called correlation coefficient (or, again, correlation ratio) is simply a measure of such tendency, more or less adequate according to the circumstances of the case.... If the correspondence...*[between two sets of phenomenon measurements is such] *that the graphical representation of it (one phenomenon being measured along the axis of x, the other along the axis of y) is a <u>straight line</u>, the <u>correlation coefficient</u>, r, will be....* [If] *the graphical representation of it will be, <u>not a straight line, but a curve</u> of greater or less degree of complexity....* [it] *will be measured not by the correlation coefficient, r, but by the <u>correlation ratio</u>, η."* (pp. 42–43; underlining added).

In his chapter titled "The Mathematical Theory of Correlation", Brown provided only two scatter plots; he used one in his example for how to calculate a correlation coefficient, and the other in his example for how to calculate a correlation ratio. If the reader of Brown's textbook were to recall the advice given on his page 43 (namely, to use the correlation coefficient for data that plot as a straight line and to use the correlation ratio for data that plot as

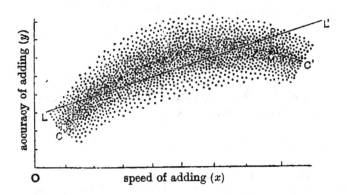

FIGURE 7.3
This chart shows a straight line fitted to an obviously curved plot. (From Brown and Thomson (1921, p. 107), which is identical to that in Brown (1911, p. 54).)

a curve), that reader would be very confused by his straight-line example, shown here as Figure 7.3.

He described that scatter plot as being a "broken curve" (p. 54) and then then said...

> Let us now find the "best fitting" straight line, LL', to this curve (p. 55).

He called the straight line LL' the "regression line" (p. 56), upon which basis he derived a formula for *r*. In other words, for the *linear* regression coefficient and for the *linear* correlation coefficient, he used a plot of data that is obviously *curved*. In virtually any other textbook from this era, this data set would have been used as an example of when *not* to use a linear correlation coefficient.

It is humorous to see the example curve that he provided (reproduced here as Figure 7.4) of when "It is clear that...the regression is not linear".

It is also humorous that what Galton considered to be true correlation in 1888 is what Brown called spurious correlation in 1911:

- *It is easy to see that co-relation must be the consequence of the variations of the two organs being partly due to common causes* (Galton, 1888c, p. 135).
- *Correlation is said to be "spurious" when it is due to extraneous conditions and does not arise <u>directly</u> out of the two functions under consideration* (Brown, 1911, p. 76; underlining added).

In other words, Brown viewed correlation as a measure of the *direct* connection between two variables: "*r* is a measure of the degree of dependence between *x* and *y*" (p. 56), but he viewed it as spurious if their apparent dependence is actually due to their mutual dependency on an "extraneous" third variable.

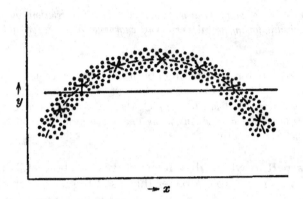

FIGURE 7.4
This chart shows a straight line fitted to a half-circle plot. (From Brown and Thomson (1921, p. 111), which is identical to that in Brown (1911, p. 57).)

Ten years later, Brown revised his book; he then had the assistance of a co-author, **G. H. Thomson**, who was a professor of education at Armstrong College. Some interesting quotes from the revised book's Preface are:

- *Owing to my own time during the War* [World War I] *being taken up entirely with army medical work, I have been unable to take any part in the further development of correlational psychology since 1914...* (Brown and Thomson, 1921, p. v).
- *My position is that hierarchical order is the natural order among correlation coefficients, that it only expresses the well-known fact that correlation coefficients are themselves correlated, and that the degree of perfection of hierarchical order found among psychological correlation coefficients is merely that which occurs by chance, and not, as Professor Spearman has been led to believe, extraordinarily high* (p. vi).

Compared to the first edition, some chapters in the second were "expanded and altered" and others were "entirely or almost entirely new"; however, the first-edition chapters titled "Introduction to Correlation" and "The Mathematical Theory of Correlation" were not so lucky, for they "remain with little alteration" (p. v). Nevertheless, this paragraph (a reflection of the just-discussed similar statement in the revised Preface) was added:

> *Correlation coefficients are themselves correlated, and n correlation coefficients form an n-fold or n-dimensional correlation-surface. The particular and convenient form of tabulation of correlation coefficients adopted by Professor Spearman and followed by most other psychological workers brings to light, in the form of "hierarchical order," one of the properties of this correlation-surface of the correlations* (p. 184).

Additionally, a strongly worded warning about sample size was added:

But it must be made quite clear that really to calculate correlations with only 10 cases is absurd, for the probable errors are enormous and moreover unknown... (p. 110).

The authors also include a warning about the term "correlation table":

It is very unfortunate that Mr [sic] J. C. Maxwell Garnett has recently [1919] used this term (already fixed in the meaning given in the text) for something quite different." (p. 121n)

The **Mr. Garnett** to whom they referred was a Fellow at Trinity College, Cambridge, when he wrote a paper that used the term "correlation table" in reference to a matrix of correlation coefficients, rather than in reference to a matrix of counts (Garnett, 1919, p. 100); as previously mentioned, such a table had been described in 1904 by Spearman. One such table of correlation coefficients mentioned in Garnett's paper was obtained from data on the performance of students who had taken "12 different...sensory, motor, sensori-motor, and association tests..." (Garnett, 1919, p. 91); he did not provide the table, but he described it as consisting of the values of r for each of the $(12 \times 12) - 12$ pairings. Despite criticism such as by Brown and Thomson, Garnett's "different" use of the term correlation table survived and can be found in 21st-century publications (e.g., Chiang, 2003, p. 338).

1912

Willford Isbell King (1880–1962) was a statistician with United States Public Health Service, a professor of political economics, and economist for the National Bureau of Economic Research (Uoregon.edu-King). In 1912, he published a textbook titled *The Elements of Statistical Method*. He published no new editions or revisions to that book (although there were at least eight additional printings of it during the subsequent four decades); thus, his was a textbook frozen in time during the golden age of discovery and invention in mathematical statistics.

His book's Preface explained and claimed that...

The purpose of this book is to furnish a simple text in statistical method for the benefit of those students, economists, administrative officials, writers, or other members of the educated public who desire a general knowledge of the more elementary processes involved in the scientific study, analysis, and use of large masses of numerical data. While it is intended primarily for the use of those interested in sociology, political economy, or administration, the general principles set forth are applicable likewise to every variety of statistical data.... So far as the author is aware, <u>there is no book published in America</u> which attempts

> to cover the field of statistical method in its present state of advancement.... it
> is believed that there is place for an elementary text of this nature. References
> are given to only a few of the principal works on the subject.... If more advanced
> study of any topic is desired, the student will find abundant references cited in
> those books (p. vii–viii; underlining added).

His statement that "there is no book published in America" is curious, given
that Davenport's 1899 *Statistical Methods* had been published in New York by
John Wiley & Sons, and that Bowley's 1902 and 1907 editions of *Elements of
Statistics* had been co-published in New York by Charles Scribner's Sons.

The previously mentioned 1912 Preface (date September 1911) was included
in all re-prints through at least 1947, without additional comment. That fact is
interesting because the book's claim to "to cover the field of statistical method
in its present state of advancement" rang less true as the decades passed:

- *E. S. Pearson considered the years from 1915 to 1930 to be the second
  of "two great formative periods in the history of mathematical statistics"* (E. Pearson and Kendall, 1970, p. 323).
- *Historians speak about the "Fisherian Revolution, 1912–1935"* (Hald,
  2007, Part V).
- *Davis and Nelson in their 1935 introductory statistics textbook,
  remarked that "Literature on the subject of probability and statistics
  has increased with bewildering rapidity in the last few years"* (Davis
  and Nelson, 1935, Preface, reprinted in Davis and Nelson, 1937,
  p. xi).

King's textbook included a 20-page chapter on Correlation, from which
comes the following:

- *Correlation means that between two series or groups of data there
  exists some causal connection.... it is seldom, especially in the field of
  social statistics, that any absolutely fixed mathematical relationship
  between two variables can be established. We very often must be satisfied if we learn that when one variable increases there is a certain
  tendency for the other to increase or vice versa. If it is proven true that
  in a large number of instances two variables tend always to fluctuate
  in the same or in opposite directions we consider that the fact is established that a relationship exists. This relationship is called correlation*
  (pp. 197–198; underlining added).
- *Correlation in two variables may be roughly illustrated in frequency
  graphs. Graphic methods, however, are all somewhat deficient since
  they cannot give a numerical measurement of the degree of correlation
  existing. For this purpose, we must compute a correlation coefficient,
  in other words a numerical measurement of the degree to which correlation exists between the subject and the relative as the two variables
  to be compared are called. By the term "subject" we mean the variable which is to be used as a standard or measure and, by the term
  "relative" we designate the variable which is to be compared with or
  measured in terms of the subject* (pp. 199–200; underlining added).

- *When two different characteristics of a given series of items...are to be compared, or when we wish to study the relationship between the long-time changes in two historical variables, the most satisfactory coefficient of correlation is that devised by the great biologist Karl Pearson* (p. 200).

The descriptions and definitions just given seem on first reading to be similar to those found in other publications. However, hidden within his definitions was a subtle difference compared to those of his contemporaries; I speak specifically of his view that a relationship is to be called correlation *only* if it is "proven true that in a large number of instances two variables tend always to fluctuate in the same or in opposite directions". In other words, if a researcher has relatively few data points, and no other proof of their co-fluctuation, then their relationship should not be called correlation. Such a viewpoint was off-target of the then current standard opinion, namely that correlation can be claimed based upon a single sample's high correlation coefficient.

Another viewpoint of his that differed from his contemporaries was his use of the terms "subject" and "relative" (shown in the just-given quote from his pages 199–200). Galton sometimes used those terms to literally mean a relative (e.g., a nephew) and a related subject (e.g., an uncle); however, as discussed here in Chapter 1, he and others mostly used those terms to summarize how data were arranged into a correlation matrix table: The "subject" values of *Variable A* were arranged into arbitrarily chosen intervals (e.g., 1–2, 3–4, and 5–6), whereas the "relative" values of *Variable B* were obtained from all its measurements that were found in a given interval of *Variable A*. Key to understanding how those terms *had* been used was the fact that for calculation of correlation, first Variable A would be considered the subject and *Variable B* the relative, and then B the subject and A the relative (e.g., the cubit vs. stature data that Galton analyzed in his December 1888 paper, previously discussed). Some authors performed only half the work that Galton did, i.e., by arbitrarily choosing one or the other variable as the subject (e.g., Davenport and Bullard, as previously discussed). In either case, they were dealing with Co-Relational Correlation; contrarily, King's definitions of those terms seem to relegate correlation only to instances of Relational Correlation; he would not be the last to do so.

As we shall soon see, a few years later King would publish derogatory statements about the value of correlation and its coefficient, which led to a war in print between him, Edmond Day, and Warren Persons.

King ended his chapter on correlation with a section titled "The Interpretation of the Coefficient of Correlation" (p. 215); that entire section is shown here next:

The following rules will assist in giving a general idea of the interpretation of r according to its relation to its probable error:

1. *If r is less than the probable error, there is no evidence whatever of correlation.*
2. *If r is more than six times the size of the probable error, the existence of correlation is a practical certainty.*

There might be added to the above the further statements that, in those cases in which the probable error is relatively small. [sic]

1. *If r is less than .30 the correlation cannot be considered at all marked.*
2. *If r is above .50 there is decided correlation.*

As we shall see, those rules were later recommended and quoted in full by other textbook authors (e.g., Jerome, 1924, p. 285).

As a relatively young man in 1912, **J. D. Magee** had a paper published by the American Statistical Association; the discussion of correlation in that paper would subsequently be incorporated into his doctoral dissertation in "political economy", which the University of Chicago would publish the following year. The titles of the paper and thesis were, respectively, "The Degree of Correspondence between Two Series of Index Numbers", and *Money and Prices: A Statistical Study of Price Movements*. In regard to correlation, both publications told the same story. I assume that the members of his doctoral thesis review committee were in agreement with his views on correlation, or they would not have approved it; that is important to note, given the negative critique of his dissertation that would be published by someone else in 1914 (more about this, soon).

In his paper, he wrote:

Our objections to the Pearsonian Correlation Coefficient as a means of testing the relationship between two series of index numbers are, to sum up: (1) it entirely disregards the element of time which in most problems in which index numbers are used, is of prime importance; and (2) the result obtained is not definite, if the relative changes are the same we get perfect correlation, but perfect correlation does not always mean that the relative changes are the same. It may be well to repeat that these objections apply only to the use of the correlation coefficient in connection with index numbers and not at all to the general use and especially not to its use in Biology.

(Magee, 1912, pp. 178–179)

In his dissertation, he phrased those objections more succinctly:

The objections to the use of the [Pearsonian] Coefficient of Correlation with index numbers are (1) that it entirely disregards the element of time, and (2) that it shows perfect correlation when the absolute changes are the same as well as when the relative changes are the same. When we are dealing with index numbers we are, of course, concerned with relative, not absolute, changes.

(Magee, 1913, p. 5)

In both of those publications, he proposed an alternative measure of index-number correlation that he called the Degree of Correspondence (1912, p. 179; 1913, p. 5). It is curious that neither of Magee's two publications mentioned previous papers by other authors on index-number correlation and rank correlation (discussion of such papers can be found in Walker, 1929, pp. 123f and 127f, respectively); it is curious because the calculation of Magee's Degree of Correspondence is similar to the calculation of rank correlation.

1916

Emanuel Alexsndrovich Goldenweiser (1883–1953) was born in Russia; as a young man, he emigrated to the United States, where he received a B.A. from Columbia University in 1903, and an M.A. and Ph.D. from Cornell in 1905 and 1907, respectively. His Ph.D. thesis subject was "Russian Immigration to the U.S." His first post-graduate work was for the US Immigration Commission, then for the US Census Bureau, and then (in 1914) for the office of farm management in the US Department of Agriculture. Starting in 1919, and for the rest of his career, he worked as an associate statistician for the US Federal Reserve Bank (Wikipedia: Goldenweiser).

In his opinion, even maps showed correlation:

> *The map, however, contains certain elements of correlation: the geographic location and concentration of the [agricultural] crop stands out; it appears whether the crop is in the South, the North, or the West; furthermore, a common knowledge of the location of the principal cities and rivers superimposes a mental set of correlations, even if the map is actually shown in bare outline.*
>
> (Goldenweiser, 1916, p. 206)

In that same paper, he expressed his views on the role that graphs played in correlation studies:

- *An <u>analytical graph</u> is one that is introduced into a discussion in order to show visually a relationship that the author wishes to emphasize. A graph consisting of two curves—one showing the value of the potato crop through a series of years and the other the production of the same crop—is analytical. The two curves are brought together in order to emphasize the circumstance that a large crop often results in so low a price that the total value of the large crop is smaller than that of a smaller crop harvested during another year. Such a graph is of distinct value because the <u>correlation</u> is made much plainer by the curves than it can be made by text and tables alone (p. 208; underlining added).*
- *A <u>research graph</u> is one that helps establish an unknown <u>correlation</u> as a result of plotting ascertainable data. This type of graph, used largely*

> *in the laboratory and in print, is more common in engineering and natural science than in statistics, but statisticians sometimes receive hints of possible relationships from experimental research graphs. These graphs are not, like the others, primarily a method of presenting, but of discovering a relationship* (pp. 208–209; underlining added).

In 1916, **Harald Ludvig Westergaard** (1853–1936) was a famous Danish statistician whose well-known textbook titled *Outline of the Theory of Statistics* was about to have its second edition published (Willcox, 1916, p. 227). Despite his fame then, his statistical credentials today are questioned by some historians: "Westergaard was in advance of his time in regard to the application of mathematics to economic theory, but somewhat behind it in appreciation of the tools of statistical inference—notably correlation analysis" (Johnson and Kotz, 1997, p. 319).

In 1916, the American Statistical Association published his 48-page paper titled *Scope and Method of Statistics*, which gave an unflattering view of mathematical correlation:

> *One of the methods of comparison which have of late become popular among statisticians with mathematical training is that of correlation based on Bravais's formula. A simple example is offered by the age distribution of brides and grooms.... r is a quantity defined by the equation* [for the least squares best fit straight line through the X,Y plot of bride age vs. groom age].... *If r = 1, all the points will lie on the straight line; the smaller r is, the greater will be the mean error; when there is no correlation, we shall have r = 0. This quantity r is called the coefficient of correlation.... The formula of correlation will prove useful in all cases where the points are grouped nearly around a straight line, as is the supposition, and this will very often hold good, as in the case supposed of the age distribution of brides and grooms. Still we must not forget that this formula removes us somewhat from the original data and it does not relieve us from the necessity of making a close investigation of these observations. On the whole, the formula of correlation does not introduce any new principle; by tabulating and grouping the observations we can easily establish as a rule the fact of correlation without the use of the formula. To take an example from Yule's Introduction to the Theory of Statistics (1911), the percentage of population in receipt of poor law relief in 38 English Poor Law Unions of an agricultural type is correlated with the average weekly earnings of agricultural laborers, and we find as the coefficient of correlation, r = −0.66. But it is unnecessary to make this calculation. Grouping the districts according to percentage of poor law relief, we find the following numbers which tell us, without any long computation, how relief and wages are related.... These numbers give us a perfectly clear idea of the connection between wages and poor law relief. The coefficient of correlation will tell us nothing which cannot be seen from an inspection of the original numbers. But in the illustration the amount of poor relief is influenced not solely by the average wages, but also by other influences, for in each group there are conspicuous deviations, and to explain them* [,] *other causes must be found. The coefficient of correlation teaches us this and no more.*

> (Westergaard 1916, pp. 267–268; underlining added)

In Westergaard's 1932 book titled *Contributions to the History of Statistics*, he mentioned Galton in regard to the initial work on correlation:

> *Thus, at the close of the nineteenth century an abundance of anthropometric observations was available. To this may be added observations with regard to heredity, for instance, Galton's investigations of the <u>correlation</u> of the stature of parents and their offspring.*
>
> *(Westergaard, 1932, p. 259; underlining added)*

That may seem like an endorsement of Galton's December 1888 *Co-relations* paper; however, Westergaard instead footnoted the end of that quote to "Fr. Galton, *Natural Inheritance*, London, 1889, pp. 87 sq.". Unfortunately, page 87 is partway through Galton's chapter titled "Discussion of the Data of Stature", which examined regression but *not* correlation. As previously discussed, the word correlation does not appear in *Natural Inheritance*; Galton's only use of what became the basis for his correlation method was in a chapter titled "The Artistic Faculty", in an analysis of the number of artistic children born to artistic parents (previously discussed). It is anyone's guess as to why Westergaard referenced the wrong publication; he was aware of Galton's *Co-relations* paper, as evidenced by his mentioning it in a footnote on his page 271. However, the rest of his history of correlation (see next quote) seems to show that he did not clearly understand the difference between regression and correlation as explained in papers by Galton and in textbooks by Bowley, Elderton, Yule, etc. (previously discussed); if that is a reasonable conclusion, then it is understandable that Westergaard would still consider the world-wide focus on correlation methods to be much ado about nothing. As previously mentioned, he'd declared in 1916 that "the formula of correlation does not introduce any new principle"; in his 1932 *History*, he expanded that declaration to cover not only the formula but also all of correlation theory:

> *There is another important subject which gave statisticians fresh impulses at the close of the past century, viz. the <u>theory of correlation</u> which since then has had a prominent part in the statistical literature, particularly in England and America, and most specially in biometry. The elements of the theory are contained in the… investigation from 1846 by the French astronomer <u>Bravais</u>…, but in its present form it is due to English statisticians who developed the theory without at first noticing Bravais' contribution. During his studies on heredity Fr. <u>Galton, 1877, measured the size of sweet-peas, and found a peculiar "regression" of daughter seed compared to mother seed towards the general mean.</u> Eight years later he returned to the problem. In connection with the health exhibition of 1884 he initiated an anthropometric laboratory where he collected observations on the inheritance of stature in man and in 1885–6 and later he published various articles on the results. Comparing, for instance, the stature of fathers and sons, he found <u>an evident relation, tall fathers having on an average tall sons</u>, etc., but there was in the latter group of observations <u>a step backwards, a "regression" to the mean</u>. These results could be represented by a straight line (a line of regression)*

showing how much the average stature of the adult offspring was reduced com-
pared to the parental stature. In dealing with this problem [,] J. D. Hamilton
Dickson reached some of Bravais' results. Also Edgeworth discussed the subject.
Later Pearson and Yule took up the question. Pearson gave the theory its present
form with its coefficient of correlation and its coefficients of regression, and Yule
treated the subject theoretically as well as practically, taking particularly the
pauperism in England as starting-point.

By these investigations the conception of the significance of the line of regres-
sion changed. In fact, the same correlation between two groups of observations
as with regard to heredity could be found in many other fields, as for instance
between old-age pauperism and proportion of out-relief, age at marriage of brides
and bridegrooms, stature and weight, etc. This again would influence the inter-
pretation of Galton's observations on regression from father to son, the ques-
tion being whether the regression partly, or wholly, was a consequence of the
unavoidable mixture of types, or whether a real stepping backwards had taken
place.

Strictly speaking, the theory of correlation did not introduce any new principle,
it being entirely based on well-known theorems of the Calculus of Probabilities.
But it gave statisticians easily-handled formulae, containing, so to speak, a com-
plete programme for their investigations, and the vast application of these for-
mulae to biometrical and other problems gave the statistical literature to a large
extent a new appearance. On the other hand, it cannot be denied that these for-
mulae contained a danger, tempting as they might to a too mechanical treatment
of the observations, and so to speak increasing the distance between the observer
and the facts, whereas the old-fashioned methods had the advantage of keeping
the observations themselves more in view, and therefore often made it easier to
draw safe conclusions from the material. It was left to the coming decades to find
the balance between old and new methods. It may be added that naturally the
theory of correlation with its numerous theoretical problems attracted mathema-
ticians, who were stimulated to deeper investigations, the latter again proving to
be a profit to statistics. Here again, at the close of the past century, we meet rich
possibilities of further evolution.

(Westergaard, 1932, pp. 270–272; underlining added)

1917

Horace Secrist (1881–1943) was a professor of economics and statistics at
Northwestern University (Illinois, USA) when his first book on statistics
was published, in 1917; it was titled *An Introduction to Statistical Methods: a
Textbook for College Students, a Manual for Statisticians and Business Executives.*
It included a chapter titled "Comparison—Correlation" that included a sub-
chapter titled "The Meaning of Comparison and What It Implies Statistically",
in which he *confidently* gave an introduction to correlation. Relevant quotes
from that introduction are:

- *Comparison is made between things possessing common qualities…. the purpose being merely to record a quantitative difference. But comparisons are rarely made for this alone. Generally, a more or less definite purpose of establishing causal connection lies in the background. A specific inquiry is to determine whether phenomena stand in the relation of cause and effect, or whether they are the result of a common cause* (Secrist, 1917, pp. 425–426).
- *The assignment of cause and effect must be in keeping with the fact that a single cause is rarely found, and if found cannot be said always to give rise to a single effect* (p. 426).
- *Comparison, therefore, involves the pairing of things or events which are not identical in all particulars as to time, place, and condition. Causation in fact becomes contingency or correlation. A study of cause and effect, whether of coincidence or sequence, becomes largely a study of association. The idea that a given effect is the result of a specific cause and that there can be no other, or that the result must in the nature of the case be uniform and absolute, does not apply to business and economic phenomena* (p. 428; underlining added).

The next subchapter was titled "The Meaning Of Correlation", in which he seems *unconfident*, as evidenced by his not using his own words to explain correlation but instead quoting multiple other authors such as K. Pearson, A. L. Bowley, E. Davenport, R. H. Hooker, and W. Brown. At one point, he did at last use a sentence without quotation marks, but it was footnoted:

> *But the presence of a high degree of correlation cannot logically be said to prove the relation between two phenomena* (p. 437; footnoted to a paper by Hooker titled "Correlation….").

In some cases, his habit of seeming to footnote everything is humorous. For example, he stated that "If $r=0$, no correlation exists, changes in the two phenomena being indifferent." (p. 453)—he footnoted that sentence to Yule's book titled *Introduction to the Theory of Statistics*. In no other textbook that I've encountered did an author need to reference a higher authority to justify saying that when r, the "measure" of correlation, equals zero, then no correlation exists.

In contrast to some of his contemporaries, his view of how to estimate correlation put high importance on the correlation coefficient and low importance on charts and graphs. His words were: "How nearly these phenomena are related is suggested but not measured by the graphic method. The most common *measure* is the Pearsonian coefficient…." (p. 452; italics in the original).

The second edition of his *Introduction* was published in1925; its title-page claimed that it was "Entirely Rewritten". It is interesting that the Preface to the new edition was much more philosophical that others of its era:

> *The book, it is hoped, is more than a "statistical arithmetic," or even a compendium of statistical practices. A conscious effort has been made to give it body and substance, and to state and illustrate the principles back of numerical calculation*

and manipulation. Mathematical formulae and descriptive methods of how to use statistics, while fully explained, are discussed in connection with the logical place which they hold in scientific thinking. Statistical analysis, requiring as it does observation of facts, their measurement, suitable analysis, and logical inference is treated broadly and fundamentally. <u>The book is concerned with the statistical ways in which each of the steps in constructive thinking should be carried out. It is intended to be an essay in applied logic.</u> While designed as an introduction to the subject, it is broad enough in scope, it is believed, to supply the basis for a thorough understanding of the elementary principles of statistics and statistical methods.

(Secrist, 1925, p. viii; underlining added)

The first edition's chapter on "Comparison—Correlation" became the second edition's chapter on "The Theory and Measurement of Correlation". In it, he continued to rely upon quotes from and references to other authors; that is, no sentences were "completely rewritten" so that they now explained correlation *in his own words*. Such deference was extended to the end-of-chapter "Conclusion", which in the first edition was all his own words, but which in the second edition became almost all a quote from A. L. Bowley; Secrist unabashedly stated that "Bowley's summary of his discussion of correlation may be used to close our own" (p. 434). The subsequent quote from Bowley consumed 2/3 of a page; the last sentence in it, and the last sentence in Secrist's chapter was:

In general, however, r [the correlation coefficient] may be said to <u>measure the amount</u> that is common in the systems of causation of x and y.

(Secrist, 1925, p. 436; underlining added)

It is important to note here that Bowley's text is from 1920, which is a year prior to the first publication of the Coefficient of Determination, which is the true "measure" of correlation (more about this, shortly, when S. Wright is discussed).

In 1920, Secrist published two other books on statistics that were much less technical than his *Introduction* book. One such new book was his *Readings and Problems in Statistical Methods*; its chapter on "Comparison—Correlation" included a subchapter on "The Coefficient of Correlation" that was "Adapted with permission" from W. G. Reed's 1917 paper by the same name (discussed here next); in many cases, entire sentences in Reed's paper were included word-for-word in Secrist's book, and both had sections titled "Limitations of the Coefficient of Correlation" (Reed, 1917, p. 677; Secrist, 1920a, p. 409).

Secrist's other new book, *Statistics in Business: Their Analysis, Charting and Use*, was "prepared primarily for business executives" (Secrist, 1920b, p. vii); surprisingly, it omitted any mention of correlation. His only allusion to correlation was in a single paragraph that was placed two pages before the end of the book:

> *Business phenomena are related; they do stand in the relationship of cause and effect to each other, but this is not so simple and so predictable as it is sometimes indicated to be. What the relationship is can rarely if ever be guessed; but it can be measured, and it is the function of scientific method to determine it.*

> *(Secrist, 1920b, p. 128)*

I cannot fathom why he there would avoid mentioning even the *word* correlation.

William Gardner Reed worked for the United States Weather Bureau in 1917 when he published a paper titled "The Coefficient of Correlation", in which he stated:

> *In many studies it is necessary or at least desirable to test the existence of concomitant variation between two series of variable qualities. A comparison of the plotted variables furnishes a rough, but for some purposes adequate, means of examining the relationship.... The usual tabular method is slightly more refined, but tables involve too many figures to give an adequate idea of the conditions and give no concise measure of the degree of relationship. The English biometricians have perfected a method of stating the degree of relationship.... The early statements of the use of the coefficient of correlation indicate clearly that the attempt to obtain such a coefficient from miscellaneous material is an abuse of this method of measuring relationship. The material in hand should be investigated carefully before any attempt is made to determine the relationship by the use of the coefficient of correlation.*

> *(Reed, 1917, pp. 670–671; underlining added)*

Reed's sentence that included the word "abuse" was followed by a footnote reference to pages 169 and 177 in Yule's 1912 second edition of *Introduction to the Theory of Statistics*. The casual reader would certainly conclude that Yule's textbook is the basis for Reed's statement; curiously, however, Reed's view is not supported by anything on Yule's pages 169, 177, or by any of the other 34 pages in Yule's chapter in which pages 169 and 177 are found.

Reed "investigated"(as he says in his just-given quote) by determining if an *X*, *Y* plot of the raw data looks to be best modeled by a straight line. He ended this discussion by saying...

> *The conclusion seems legitimate that the coefficient of correlation may be used strictly as a measure of relationship, when such relationship has been determined by other investigation to follow straight line relations. The use of the coefficient of correlation is to be recommended because it is independent of the personal equation of the investigator, and of the units employed.... If the relationship is not that of a straight line, it is obvious that...some other measure (e.g., the correlation ratio) must be used. Therefore, the coefficient of correlation should never be used to show relationship until after the phenomena have been investigated, at least far enough to show whether a straight line satisfies the relationship as well as any other curve (pp. 680–681).*

Battles in Print over the Meaning of Correlation

The third decade of the 20th century is commonly referred to as the Roaring Twenties (Britannica.com-Twenties). No common name has been assigned to the second decade, but we might refer to it as the Battling Teens because during most of those years the world was engaged in serious battles. Early in the Teens, the world was battling the disease of nationalism, which was raging most intently in Europe. In 1914, the Great War broke out between the two main European national alliances; by 1917, they'd fought to a stalemate, which was broken in 1918 when the USA entered the fray. As that war was ending, another was starting, against an enemy called the Spanish Flu, which killed tens of millions of people world-wide during the next 2 years.

During the Battling Teens, there seems to have been a dramatic increase in disagreements about the mathematical concept of correlation, some of which became discourteous, uncivil battles in print. One contemporary historian noted that "...there are at present not one, but several corps of statisticians, each trying earnestly to promote the science, but hardly able to coöperate for lack of mutual sympathy and sometimes acting in direct opposition to one another" (Westergaard, 1916, p. 229). Multiple examples of such battles are given here next. Some of these battles were as acrimonious as the ones that had occurred in medical journals in the mid-nineteenth century; those battles were between Joseph Lister and the prominent surgeons who rejected his germ theory of disease (Fitzharris, 2017, p. 173f). Lister eventually won his war, whereas there were no clear winners in these correlation wars.

In 1914, economics professor **Warren M. Persons** (previously mentioned) published a review of **J. D. Magee**'s 1912 doctoral dissertation that had in 1913 appeared in book form under the title *Money and Prices* (previously discussed).

Persons's paper's first sentence was:

> *Money and Prices is an <u>inconclusive study</u> based upon <u>unreliable statistical methods</u>.*
>
> *(Persons, 1914a, p. 79; underlining added)*

On subsequent pages, he said that Magee's second objection to the correlation coefficient...

> *...calls attention to <u>a virtue, not a defect,</u> of the coefficient.... <u>He does not appear to recognize</u>, however, that like growth elements in the two series make his [coefficient of] "Degree of Correspondence" <u>unreliable</u>.... Dr. Magee's entire study is based upon this <u>erratic</u> coefficient (pp. 80, 81; underlining added).*

Later that year, Magee replied in print:

In his review of my "Money and Prices" in the March number of the Quarterly Publications, Professor Warren M. Persons makes <u>several serious errors</u>. In the first place, <u>he confuses the arithmetic operations of addition and multiplication</u> in a most surprising way. I had criticized the Pearsonian Coefficient of Correlation as a means for testing the relationship between two series of index numbers, because the coefficient is not changed if a constant is added or subtracted from each term of one (or both) series. Professor Persons admits the fact, but claims that it is an advantage for the Pearsonian Coefficient. He says that it is analogous to moving a curve of index numbers up or down to facilitate comparison. <u>He seems to forget</u> that moving a curve up or down or adding or subtracting a constant from a series of index numbers changes the relationship expressed.

(Magee, 1914, p. 345; underlining added)

Persons's reply to Magee was published on the pages that immediately followed that reply by Magee:

The point at issue between Dr. Magee and myself concerns the reliability and accuracy of the second of the two methods which he uses to measure the correlation between two series of paired items.... I agree with his characterization of the first method but hold that the second method not only gives <u>erratic and unreliable results</u>, but that <u>it cannot be considered a measure of correlation</u> at all. <u>Our difference is not merely a difference of opinion</u> as to what constitutes perfect correlation.... It is clear, then, that <u>Dr. Magee uses the word "correlation" to indicate something quite different from the accepted connotation</u> of that term.

(Persons, 1914b, pp. 347, 348; underlining added)

That was the end of their battle, as far as I can determine.

Another battle started in 1916, when **Westergaard**'s 48-page paper (previously discussed) was followed by separate multi-page critiques of it by other statisticians, one of whom opposed Westergaard's view of correlation. That dissenter was the same just-mentioned Persons who had battled Magee:

With Westergaard's characterization of the Pearsonian coefficient of correlation I <u>emphatically disagree</u>....

(Persons, 1916, p. 277; underlining added)

Westergaard's paper had been published in 1916 in the *Quarterly Journal of the American Statistical Association*. The following year, the *Journal of the Royal Statistical Society* published a critical review of it written by Francis Ysidro Edgeworth (previously discussed):

...we [i.e. Edgeworth] attribute a more <u>serious error in defect</u> to our author's [i.e. Westergaard's] treatment of mathematical statistics. We quite agree with Professor Persons in his <u>emphatic disagreement</u> from Professor Westergaard's statement, "The coefficient of correlation will tell us nothing which cannot be seen from an inspection of the original numbers." <u>We protest</u> against the

*statement that on "the whole the formula of correlation does not introduce any
new principles". It is not too much to say with another of the commentators,
Professor West, that the author "is apparently not familiar with the properties of
the correlation ratio"…. his treatment of the statistics* [of contingency] *appears
to us <u>somewhat defective</u>. We do not, however, much complain of <u>his inability</u> to
trace correlation of the Galtonian kind between the degrees of temper… shown by
pairs of sisters. But <u>it strikes us with astonishment</u> that he shows no appreciation
of the correlation between physical characters….*

(Edgeworth, 1917, p. 408; underlining added)

That was the end of that battle, as far as I can determine.

In 1917, a few months after the publication of W. G. Reed's paper titled "The
Coefficient of Correlation" (previously discussed), **James Arthur Harris**
(1880–1930) published an unflattering critique of it. Harris was a botanist and
biometrician (Wikipedia: Harris) who, the previous year, had published a
paper on the history and current status of mathematical correlation, or as he
called it, "the theory of the measurement of inter-dependence" (Harris, 1916,
p. 53). Reed had a degree in climatology from Harvard and had then taught
meteorology and climatology at UC Berkeley, followed by employment at the
US Department of Agriculture (Brooks, 1932, p. 90). Reed in his paper had
quoted and referenced statistics textbooks by Davenport, Bowley, and Yule.
Despite such educational credentials, job experiences, and literary practices,
Harris considered Reed unqualified to be advising on the subject of correlation. Additionally, Harris seems to have considered himself superior to other
biologists, at least in regard to correlation:

*<u>Workers in the physical sciences realized long ago</u> that certain progress depended
upon the precision of their instruments of measurement and the adequacy of their
methods of mathematical description and analysis. <u>Biologists, here and there, are
beginning to see the importance of the analytical</u> as well as of the observational
tools. Among the analytical formulae none are of greater usefulness than those
for measuring interdependence….<u>biologists as a class still think</u> of correlation
as synonymous with the classical product-moment method. How <u>erroneous</u> this
impression is will appear in the following pages. The purpose of this review is
therefore to indicate in <u>non-mathematical terms easily comprehensible to biological readers</u> the lines of advances in the theory of the measurement of interdependence in order that they may the more easily select for dealing with their
actual data, formulae of <u>the existence of which they might otherwise be unaware</u>.*

(Harris, 1916, p. 53; underlining added)

In his 1917 critique of Reed's paper, Harris had this to say (all underlining
added):

- *Since the development of the modern higher statistical methods has
been largely in the service of biology, it is only natural that those
who adopt the biometrician's formulae <u>without a first hand</u>* [sic]

> *acquaintance* with the extensive biological literature should often labor
> under the disadvantage of *using methods of calculation which are not*
> *the best suited* to the practical needs of their work.... *When, however*
> *the beginner attempts*, as is surprisingly often the case, to elucidate the
> subject for others, real *harm may result*.... A case in point is afforded
> by a recent paper by Reed (p. 803).
>
> • *Reed lays entirely too great emphasis upon the necessity for a pre-*
>   *liminary testing of linearity of regression and his discussion of the*
>   *methods is very misleading. He says: "The correlations should never*
>   *be attempted without first investigating the relationship far enough to*
>   *see if it follows a straight line." Now while it is true that the correla-*
>   *tion coefficient is not strictly valid except in cases of linear regression*
>   *it always gives at least a minimum measure of the relationship between*
>   *two variables, and in the vast majority of cases this measure will be*
>   *the one finally adopted after linearity has been tested. Furthermore,*
>   *the determination of the coefficient of correlation is the first step in*
>   *the critical testing of linearity. It is quite unnecessary to plot the two*
>   *variables in units of their standard deviation or to use the method of*
>   *least-squares in adjusting the line as suggested by Reed* (p. 804).
>
> • *Thus the calculation of the correlation coefficient, which by a practical*
>   *method is a very easy task, should always be the first step in the analy-*
>   *sis of the data. The nature of the regression line can then be determine*
>   *in a convenient and really scientific manner.... This note is written*
>   *with no desire to criticize the work of an individual writer, but merely*
>   *in the hope of saving some beginner...[from] misconceptions* concern-
>   *ing the value or application of this most important statistical constant*
>   (p. 805).

Those words by Harris give the impression that he himself was very careful
to not make such "beginner" mistakes. Therefore it is interesting that he misquoted Reed: Harris's quote of Reed started with *"The correlations should..."*
but Reed's sentence actually started *"However, the coefficient should..."* (Reed,
1917, p. 677).

A couple years later, H. L. Rietz (previously mentioned) critiqued both of
the just-discussed papers by Reed and Harris. He started his critique with
uncomplimentary insinuations:

> *The recent papers of Reed and Harris in these Publications have brought to mind*
> *some simple cases in which a correlation coefficient is zero although the two*
> *variables are mathematical functions of each other represented by certain simple*
> *types of continuous curves. Those who have studied critically the theory of cor-*
> *relation are perhaps aware of the limitations of this valuable theory as well as*
> *they are aware of its useful applications. But those who are making applications*
> *without fundamental knowledge of the method are apt to overlook the limitations*
> *in applying a summary method of quantitative description such as is provided*
> *by the correlation coefficient.*
>
> *(Rietz, 1919, p. 472; underlining added)*

A couple pages later, he explained the basis of his negative view of those previous papers:

> *Harris holds that Reed put too much emphasis on the importance of testing the linearity of regression in the early stages of a correlation study. With this position taken by Harris, the writer [i.e. Rietz] is in agreement if the question is one merely of showing the existence of correlation rather than one of showing the degree of correlation; for, if regression is not linear, the value of r turns out to be smaller than the correlation ratio—a function appropriate to describe correlation without a limitation of linear regression. That is to say, the use of r does not lead us to infer a greater degree of correlation than exists, but in cases of non-linear regression it may lead us to infer a smaller degree of correlation than exists. The interpretation of the significance of the differences between two correlation coefficients cannot go far until careful inquiry is made into the form of the regression curve. The prediction of the mean value of y that corresponds to an assigned value of x is likely to be valuable only when the form of the regression curve is known* (p. 474; underlining added).

As far as I can determine, that was the end of that battle.

Possibly the most infamous correlation battle, at least in the USA, was precipitated by Willford I. King (previously discussed), whose paper titled "The Correlation of Historical Economic Variables and the Misuse of Coefficients in this Connection" was published in 1917; the next year, he would refer to it as "my article on correlation coefficients" (King, 1918, p. 171). He began his attack on correlation with this sentence:

> *He who reads statistical literature nowadays finds it literally teeming with studies in correlation.*
>
> (King, 1917, p. 847)

Karl Pearson would say something similar to that, in 1930:

> *Thousands of correlation coefficients are now calculated annually, the memoirs and text-books on psychology abound in them; they form, it may be in a generalised manner, the basis of investigations in medical statistics, in sociology and anthropology.*
>
> (K. Pearson, 1930a, p. 56)

That statement by Pearson was intended to be complimentary; King's was not, as demonstrated by the following (all from King, 1917; all underlining added):

> - *Apparently, the time has arrived when a statistician's ability is largely judged by his output of correlation coefficients. Unfortunately, however, many of the coefficients seem to be more valuable as mementoes of industry* [i.e., as proof of having completed many difficult and

lengthy analyses] *than for any new truth which they add to human knowledge. In fact, it may be said without exaggeration that only too often the statistician has been so entranced by the mathematical possibilities of the problem that he has lost sight of the real meaning of the operations involved and hence the conclusions presented have been utterly fallacious and entirely contrary to the facts. Is it not time to call a halt and ascertain just what ends may and may not be served by this newly popular mathematical aid?* (p. 847)

- *The first question is "What does the term correlation mean?" Evidently, it implies some relationship with things, and this particular relationship is always one of cause and effect* (p. 847).

- *One of the laws of physics is that a given force always produces identically the same effect whenever it is applied. Does this same rule hold good for economic or social phenomena? There is every reason to believe that it is exactly as valid here as in the other parts of the physical world. If this is not true, it is useless to hope for any material advancement of economic science. A law which only works part of the time is no law at all. Economics deals with physical beings and forces and it would be strange indeed if it should prove an exception to the general rules of the natural world* (p. 848).

- *...if economic and physical phenomena are governed by laws equally exact, there can be no such thing as imperfect or partial correlation. Every correlation to exist must be perfect. Every cause must produce its effect exactly and invariably. If the cause and its effect are once completely isolated, the coefficient must always be unity. In every case, there either is or is not correlation between variable A and variable B— there can be no middle ground. But it is common to obtain correlation coefficients of 0.3 or 0.6 or 0.7. What do these decimals mean? They merely indicate that the cause and its effects have not been completely isolated—that the effects of conflicting forces still enter in to mar the results. The low coefficients, then, in no sense indicate imperfections in the actual correlation, but they do show conclusively that the statistician has failed in his efforts to exclude some of the untoward forces which ought to have been eliminated. This failure may have resulted from a non-comprehension of the complexity of the forces involved; from a lack of sufficient information to render the elimination of the undesirable forces feasible; or from ignorance of the proper statistical methods which must necessarily be utilized in the process* (pp. 850–851).

- *Once the undesired forces have been completely eliminated, and the resulting figures have been plotted, two curves emerge* [e.g., for two variables such as rainfall and crop yield versus an index variable such as year]—*one representing the cause and the other the effect. If the elimination has been entirely successful, the fluctuations in the curves will correspond perfectly and the correlation will be evident at a glance. To compute a coefficient to prove whether correlation does or does not exist is, in this case, a manifest waste of energy. In attempting by graphic methods, therefore, the correlation of two supposedly related historical variables, two curves are derived which either do or do not resemble each other closely. If they are very similar, all the*

> *correlation coefficients in the world will add nothing to the knowledge*
> *gained simply by observing the curves. If they are dissimilar, it may be*
> *impossible to say whether there is or is not correlation. In this case, it*
> *is easy to tell in advance that a coefficient, if computed, will be low and*
> *will, therefore, prove nothing conclusively. In either instance, there-*
> *fore, the* [correlation] *coefficient has added nothing whatever to the*
> *knowledge obtainable from the graphic presentation. Too frequently,*
> *the coefficients presented are worse than useless for they lend the*
> *glamor of apparent erudition to a carelessly made study* (pp. 851–852).

- *Do coefficients, then, have any place in the correlation of historical*
  variables? So far as the present writer has been able to observe they
  have but one valid use. *The size of the* [correlation] *coefficient is*
  merely a means of determining the closeness of fit of two curves. If
  there is a lag, that is if there is a time interval between cause and effect,
  it is frequently difficult to determine by the eye the exact lag which has
  actually occurred. In this case, by trying coefficients with lags of dif-
  ferent lengths, it is possible to ascertain the approximate time interval
  which gives the best fit to the curves—in other words, the most com-
  mon lapse of time required for the cause to produce the effect. Aside
  from this purpose, the labor of computing coefficients for historical
  economic variables seems to be largely wasted. When coefficients are
  used to cover up careless work in failing to eliminate the causes not
  under consideration, their use becomes inexcusable (p. 852).

- *...no coefficient equals the graphic method for demonstrating whether*
  correlation does or does not exist (p. 853).

A formal rebuttal to King's 1917 paper was published in June 1918 by **Edmund E. Day**; the rebuttal was titled "A Note on King's Article..." (Day, 1918). Day was then a young Harvard-trained Ph.D. economist who would later become president of Cornell University (Cornell.edu-Day). A few months after Day's Note was published in the Association's official journal, King published a response in that same journal (King, 1918). A discussion of both papers is given here next.

Day started his paper with an aggressive sentence followed by a concilia-tory one:

> *Errors in the conception of correlation and the use of the correlation coefficient*
> *are so common a vice of pseudo-scientific research that any attempt to correct*
> *their influence would be a bootless* [i.e. ineffectual or useless] *task. Mistakes*
> *of this sort mark the beginnings of any newer branch of scientific knowledge and*
> *may be looked upon as natural growing pains.*
>
> *(Day, 1918, p. 115; underlining added)*

Subsequent sentences were again worded aggressively:

> *One can not* [sic] *but feel that much of Dr. King's bill of complaint is irrelevant*
> *to the work of statisticians of measurable professional standing.... But it is in*
> *his analysis of correlation and the correlation coefficient that Dr. King is most*

unfortunate. His discussion of the nature of correlation is so <u>extreme and inflex-</u>
<u>ible</u> as to be <u>wholly misleading</u>. To Dr. King, correlation invariably connotes
established causal relationship.... If correlation were being defined de novo, it is
possible that this view might prevail. But the term "correlation" has an estab-
lished meaning (p. 115).

Day then explained his own view of correlation, which he claimed was more
widely accepted than King's.

Correlation is not identical with causation, though closely connected with it.
Correlation connotes a tendency toward persistent association.... By established
usage and authority, correlation is nothing more than "one-to-one" correspon-
dence in paired items of selected variables (p. 116).

He described King's views as a series of unfortunate misconceptions:

If this is the nature of correlation, what is its relation to causation?...If correlation
were always perfect, the problem would be relatively simple. <u>Unfortunately for Dr.</u>
<u>King's exposition</u>, data drawn from experience—even artificially controlled expe-
rience—never exhibit perfect correlation. The essential task of correlation studies
lies in the interpretation of concrete evidence, not in the understanding of con-
ceptual limits.... It is clear enough that, when cause and effect are absolutely seg-
regated, correlation becomes complete and obvious. Such complete—or virtually
complete—segregation of forces is the ideal of causal analysis. Dr. King cites three
reasons for the failure of economic analysis to attain this ideal.... <u>Unfortunately he</u>
<u>neglects the reason which is most fundamental</u>: the practical impossibility of isolat-
ing economic forces completely. Dr. King, in suggesting the complete segregation
of forces in economic analysis, is not merely giving a counsel of perfection; <u>he is</u>
<u>proposing what is manifestly impossible</u> (pp. 116–117; underlining added).

Day also discussed the correlation coefficient and correlation curves:

Dr. King's argument leads him to discount the use of the correlation coefficient.
If all correlations were perfect, he could well afford to do so.... It is just because
correlation is never perfect that the coefficient is invaluable. As has been stated,
the significance of a given association of paired variables depends entirely upon
the extent to which the association exceeds that to be expected from chance. The
correlation coefficient enables an exact measurement of this all-important rela-
tionship. <u>For this purpose</u> [correlation] <u>curves are worthless</u>. Curves may yield
valuable suggestions for further study; they may prove effective graphic repre-
sentations of correlation. In general they are <u>a treacherous instrument</u> for prov-
ing correlation. For this purpose <u>nothing equals the correlation coefficient</u>....
In the analysis of time variables, <u>Dr. King would limit the use of the</u> [correla-
tion] <u>coefficient</u> to the measurement of lag. <u>Nothing could be more ill-advised</u>
(pp. 117–118; underlining added).

It is interesting that Day's view (just given) of the value of correlation curves
vs. correlation coefficients can be interpreted as being literally the opposite

of King's view (previously quoted) which was that *"no coefficient equals the graphic method for demonstrating whether correlation does or does not exist."*

Day ended his paper with more aggression coupled with a compliment:

> *It is difficult to say to what extent <u>the defects of Dr. King's article</u> arise from mere excess of statement in giving a warning, the need of which all might concede. Dr. King's contributions to the literature of statistics are a sufficient guarantee that we need fear no <u>abuse of statistical method</u> at his hands. Unfortunately <u>his December article</u> gives no equal assurance for others. Upon the contrary there is reason to believe that it <u>will add materially to that confusion regarding the nature of correlation</u> and the correlation coefficient which it has been Dr. King's intention to correct* (p. 118; underlining added).

King's two-page relatively mild response to Day's 3½ page harsh critique was published a few months later; the difference in style cannot be attributed to a difference in age, as both men were in their mid-30s. King began by acknowledging "Dr. Day's well stated criticisms", and then he explained:

> • *My <u>criticisms</u> were intended entirely for the benefit of those who did not use the [correlation] coefficient correctly.... The tendency to assume that the high coefficient shown by two variables with similar steady trends is a proof of correlation between the two is probably the most <u>flagrant misuse</u> of the coefficient. As a matter of fact, [Karl] <u>Pearson's coefficient is entirely without significance for comparing historical variables</u> unless [on a correlation curves chart] there are a considerable number of corresponding up-and-down swings in the variables. If such corresponding swings occur, the plotted curves usually reveal the fact quite clearly. My own rule is that when the correlation is not entirely evident from the curves, it is <u>unsafe</u> to assume that correlation exists. When an examination of the curves leaves one in doubt, the coefficient, <u>if computed correctly</u>, will normally be very low. If the curves seem distinctly to indicate correlation, it certainly does no harm to clinch the evidence by appealing to a high coefficient to prove how successful one has been in eliminating the extraneous factors* (King, 1918, pp. 171–172; underlining added).
> • *Dr. Day next takes exception to my definition of correlation. I still fail to see that I have in any way essentially changed the idea as it has been generally understood by statisticians. I have only sought to make definite and clear cut [sic] a concept which is frequently described in a <u>more or less nebulous and hazy manner</u>. And it still seems to me that "a tendency towards persistent association," as he puts it, between varying phenomena must be either chance or causal.... If a given cause always produces a definite effect [,] which proposition Doctor Day does not dispute, then correlation is always perfect no matter how hard it may be to isolate and detect it.... <u>I cannot agree</u> that the fact that correlation is always perfect, if existent, makes the problem of demonstrating the relationship between two variables necessarily a simple one. Extraneous factors sometimes are hard to eliminate, but only*

when these are disposed of successfully can the coefficient or any other method really prove correlation to exist. With other factors still present, <u>a high coefficient indicates merely spurious correlation</u>, due probably to the accident of similar trends. The use of coefficients, therefore, is <u>not in any sense a legitimate substitute</u> for the elimination of the effects of factors other than those which it is desired to compare. If the troublesome interfering causes cannot be scientifically disposed of, to the required degree, then <u>the use of coefficients gives no aid whatever</u>, and we are compelled to concede defeat and pass the problem on to our successors…. [F]ortunately my own experience with scores of typical economic problems demonstrates conclusively that it very often is perfectly feasible to eliminate all undesired influences to the extent necessary to show conclusively whether correlation does or does not exist (pp. 172–173; underlining added).

King's last paragraph summarized cordially:

In conclusion, I wish to thank Dr. Day for his courteous comments. I also wish to give assurance that while my <u>criticism</u> concerning the use of the coefficient was <u>not entirely confined to amateurs</u>, I have had no thought of <u>indicting</u> all statisticians indiscriminately. I also concede freely the fact that <u>the correlation coefficient, when intelligently used, is a perfectly satisfactory tool for measuring one's success in segregating a cause and its effect from a mass of confusing data</u> (p. 173; underlining added).

In partial support of his views, King's earlier (1917) paper had referenced recent papers by colleagues, I. Fisher and M. Persons. A relevant quote from I. Fisher is:

…close comparison [of correlation curves of two variables on the same chart vs. a third variable] *will usually give quickly, through the eye, a better practical picture, I think, of the degree of correlation and certainly of the location of the correlation, than can be obtained even by laborious calculations of coefficients of correlation.*

(I. Fisher, 1917, p. 592)

King had said that those papers by those authors "…discuss the merits of some of the various methods of eliminating the effects of the forces which must be excluded <u>before</u> correlating the particular cause and effect under consideration" (King, 1917, p. 851; underlining added). Being thus referenced by King apparently did not please Persons, as evidenced by his comments in support of Day's 1918 view, which Persons published in 1919:

<u>Some writers</u> on economic statistics have <u>confused</u> the notions of correlation (closeness with which the actual corresponding fluctuations obey any law which we may select) and of the law which we select for the test…. In an article on "The Correlation of Historical Economic Variables and the Misuse of Coefficients in

> *this Connection"....Dr. Willford I. King has concepts of economic law, causation,*
> *and correlation <u>quite different</u> from those defined by Karl Pearson and accepted*
> *by the present writer. Dr. King says, "A law which only works part of the time*
> *is no law at all.... There can be no such thing as imperfect or partial correlation.*
> *Every correlation to exist must be perfect. Every cause must produce its effect*
> *exactly and invariably".... Dr. E. E. Day has taken issue with Dr. King.... <u>The</u>*
> *<u>present writer agrees with Dr. Day that Dr. King's "discussion of the nature of</u>*
> *<u>correlation is so extreme and inflexible as to be wholly misleading" and that it</u>*
> *<u>"will add materially to the confusion</u> regarding the nature of correlation and the*
> *correlation coefficient."*
>
> (Persons, 1919, p. 134n7; underlining added)

King and Day may have agreed to disagree, since no further replies to each other appeared in ASA's official journal. As previously mentioned, King did not modify even one word of his textbook's treatment of correlation in any of the reprints he published during the next 29 years.

1918

The advice given by King in his December 1917 paper (previously discussed) may have been taken to heart by **Carl E. Jones** in a paper he published in that same journal, 12 months later. Jones's paper summarized the results of a study that had several aims, among which was "to obtain definite statistical data upon...the correlation between longevity and fertility...[and] between marriage age and fertility" (C. E. Jones, 1918, p. 203). Its 19 pages included 14 correlation tables; and he used the word correlation many times, including in a summary where he claimed that a "high correlation is shown between longevity or duration of life and fertility for both fathers and mothers...[and] between the marriage age and the fertility of both fathers and mothers" (p. 217).

In line with the spirit of King's paper, Jones in his paper did not provide even a single correlation coefficient to justify his claims of high correlation. However, neither did he provide any correlation *curves*, which deficiency goes *against* King's advice. Nevertheless, his Table III data can be plotted using MS Excel, whose third-order polynomial trendline feature produced the chart here in Figure 7.5. I am sure that many of his contemporary readers would have liked to have seen that plot and its R^2 value.

In March 1918, **Secrist** (previously discussed) published a 13-page paper on a topic unrelated to the ones featured in King's 1917 paper (previously discussed); although Secrist's paper did not explicitly mention King nor his paper, it included a paragraph which strongly downplayed King's claim that untoward and undesirable forces *can* and should be eliminated from an analysis in order to validate the value of a correlation coefficient:

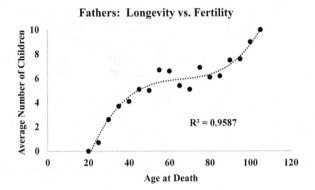

FIGURE 7.5
This chart shows that men who died at age 100 produced more children when they were young men than did men who died at age 80 when they were young men; a similar trend was seen in women. (Derived from C. E. Jones (1918, Table III).)

> *Since comparison involves the pairing of things or events which are not identical in all particulars, a study of cause and effect, whether of coincidence or sequence, becomes largely a study of Association. Causes never operate under exactly the same circumstances. Oneness of effect is only apparent, variation being evident the moment that the scale of measurement is reduced. Simply to assume the proviso "other things being equal" is not fully to atone for <u>the sins committed in statistical comparisons</u>. The "other things" are rarely if ever equal in actual life. Neither economic nor business phenomena go on indefinitely repeating themselves in one unending round of sameness. <u>To expect that an absolute cause will always result in an absolute effect or that the "other things" will automatically take care of themselves is futile.</u>*

> *(Secrist, 1918, p. 893; underlining added)*

In 1918, the American Statistical Association published a collection of papers by many authors; the collection was edited by **John Koren** and titled *The History of Statistics*. Despite its title, it was nothing like S. M. Stigler's 1986 book also titled *The History of Statistics*. Stigler's subtitle was "The Measurement of Uncertainty before 1900", whereas Koren's was "Their Development and Progress in Many Countries". Whereas Stigler's book discussed (among its many topics) Galton's correlation coefficient, Koren's book contained no discussion of mathematical correlation, and it mentioned Galton only in reference to having written a paper on "the graphic method in statistics" (Koren, 1918, p. 30n).

Koren's book is not surprising, in light of an earlier book by **Robert Giffen** that was titled *Statistics*. It had been written more than 20 years prior but was first published in 1913 (according to its title page). It contained nothing about correlation or Galton. Instead, its 477 pages contained advice on how to construct statistical tables. Mostly, it discussed data from "mass observations". Although Giffen admitted that "careful mathematical study" could have helped to explain statistical theory, he decided that regarding "the domain of mathematics...I propose rather to avoid it..." (Giffen, 1913, pp. 1–2).

Koren's book should have been titled *The History and Development of Official Statistics in Many Countries*, because like Giffen's book, it was about the analysis of government-collected data (e.g., from surveys and birth/death registries) and about the history of how such statistics had been collected, summarized, and presented. He hyperbolically claimed that "The science of statistics is the chief instrumentality through which the progress of civilization is now measured, and by which its development hereafter will be largely controlled" (Koren, 1918, p. 15).

His book contained a single instance of the word "co-relation" (in the middle of sentence, not as a hyphenated end-of-line instance):

> ...there still exist several fields in regard to which the difference between one country as compared with another are so great that some of the descriptions used frequently pertain to widely different things. It is precisely the <u>co-relation</u> of such heterogeneous numbers for which international statistics must have a care, and it is the more difficult to guard against them as only a thorough knowledge of the conditions in the individual countries make it possible to realize that there is danger.
>
> (Koren, 1918, p. 211; underlining added)

That instance is in a chapter that contained no instances of the word correlation, although it is found in other chapters.

On page 368, we find a mention of...

> ...life-tables in which mortality was correlated with occupation and age.

The correlation of mortality with occupation is an Observational Correlation, whereas the correlation with age is a Relational Correlation.

On page 447, the word correlation was used in reference to a classic correlation table:

> ...the annual statements in regard to marriages, living births and deaths, the data being transmitted by the pastors of the different parishes in the country to their respective bishops, by whom they were correlated in proper tables.

The word correlation was used there in place of the word "arranged" or "organized".

On page 595, a not-yet-common use of Relational Correlation in economics is found (underlining added):

> An obvious desideratum in the statistics of commerce and industry is a correlation of the statistics of foreign commerce with those of domestic production and consumption. Statistics measuring the volume of our foreign commerce acquire

significance largely in proportion as they can be so interpreted in terms of domestic manufactures, agriculture, and trade, as to measure with some degree of accuracy at least, in the case of each principal product, the relative amounts produced and consumed at home, the surplus exported, and the deficiency supplied from abroad. Specifically the question has been raised as to the possibility of correlating the statistics of foreign commerce with the census statistics of manufactures.

On the next page, that discussion was continued; and there can be found the term "close correlation", which in this case seems to have had nothing to do with mathematical correlation:

To the extent that the census scheme of classification is confined to a grouping of aggregate products of establishments by industries, no close correlation of the census data and of the foreign commerce data can be achieved except in those lines of industry in which the aggregate output of establishments is comparatively simple in character... (pp. 596–597; underlining added).

As we have seen here on previous pages, the value of correlation mathematics was slowly being realized by workers in fields other than biology. For example, in 1915, **Carl. J. West** at Ohio State University published a paper titled "The Value to Economics of Formal Statistical Methods", in which he claimed:

An important consideration from our point of view is that the introduction of somewhat complicated methods for determining correlation and variation has stimulated both the production of experimental data and the critical discussion of such data, which can but result in better and more accurate experimental work.

(West, 1915, p. 624; underlining added)

By 1918, West had become an assistant professor of mathematics; in that year, he published a textbook titled *Introduction to Mathematical Statistics*. For some unknown reason, it was published without adequate technical review; it contained many blunders (to use Galton's term) that appeared in a random pattern as if they were shot out of a blunderbuss (pun intended), and no errata page was provided. The text is generally well written, but some sentences and paragraphs are puzzling.

The following are several examples of obvious errors in the pages related to mathematical correlation; in those pages, there are a few dozen more errors, gross and small, which the reader can identify, as I have done, with an electronic spreadsheet and about 8 hours work:

- On page 67, a subset of a table containing counts of height is excerpted and discussed on that same page (the subset is for students whose weight was from 130 to 134 pounds); however, the count for height of

63 inches is given as 0 in the Table but 5 in the excerpt, and the count for height of 65 inches is given as 9 in the Table but 19 in the excerpt.

- On page 69, in a matrix table of temperature vs. precipitation, the axis for temperature is labeled precipitation, and vice versa.
- On page 77, we see that $7.9-4.6=3.1$, rather than 3.3.
- On page 84, a column of numbers is summed to 3018 instead of the correct total of 2813.

An interesting observation: On page 77, calculation of a correlation *ratio* was shown. On page 84, calculation of a correlation *coefficient* was shown. Those calculations involve a mean class deviation of weight and a mean class deviation of height; West used 7.9 for both of those values! The calculation of height deviation=7.9 is shown on page 33, but the weight deviation=7.9 calculation is nowhere to be found. Given all the other mistakes, and given how astonishing a coincidence it would be for those mean deviations to equal each other, the reader is left to wonder if weight deviation=7.9 is another blunder. In the copy of the book that I reviewed, some poor reader had (100 years ago?) put a very large question-mark in the margin of page 77. In fairness to Professor West, when an electronic spreadsheet is employed to perform the simple but extensive arithmetic used to calculate those two mean deviations, they do both round to 7.9, coincidentally.

Because it is the most confusing and error-filled exposition of correlation that I've ever encountered, West's book is now the first-ever recipient of my Sir Francis Galton Memorial Blunder Award.

West's view of the value of the correlation coefficient was shown by the closing remarks in his chapter on the correlation coefficient:

> *Statistical Properties of the Coefficient of Correlation: The coefficient r is, as the preceding discussions show, a conservative measure of correlation. In periodic data exhibiting a sinusoid form for the regression curve [,] the correlation may be high but because the departure of the regression from linearity is so wide [,] the value of r underestimates the correlation [,] and hence its applicability in such data is not of importance. The characteristic importance of the coefficient r is in defining the slope of the regression lines. It furnishes the most convenient method for defining the general tendencies in the data…. Therefore for the single purpose of measuring correlation [,] the coefficient of correlation is distinctly inferior to the correlation ratio both in convenience and reliability. It should never be used as a measure of correlation without first carefully testing the form of the regression. It does have however the highly useful property of giving the slopes of the [linear] regression lines.*

(West, 1918, pp. 85–86; underlining added)

What West meant by the underlined sentences is that one formula for a linear regression slope includes the correlation coefficient, namely: Slope$=r (Sy/Sx)$.

1919

One attempt to promote agreement in print regarding correlation (at least in the United States) was a paper published in a mathematics journal in 1919 by **Edward Vermilye Huntington** (1874–1952), who was a Harvard University professor in mathematics and statistics. At various times in his career, he was vice president of the Mathematical Association of America, vice president of the American Mathematical Society, and vice president of the American Association for the Advancement of Science (MAA.org-Huntington).

He considered his role in promoting agreement to be a duty:

> *The* [Mathematical] *Association* [of America] *should accept as perhaps its primary obligation the duty of interpreting the results of pure mathematics to the workers in the field of applied mathematics. This does not mean the "degradation of pure mathematics to utilitarian purposes." It means rather the search for identity of essential form among apparently diverse problems.... This search for identity of form among the diversities of practical problems is then the task of the interpreter.... I desire to bring to the attention of the Association the opportunities for such interpretive service presented in a comparatively new field of mathematics, namely, the field of mathematical statistics.*
>
>   *Mathematicians as such seem to me to have been slow to enter this field. Of the professional mathematicians in this country* [,] *only about a dozen have thought it worthwhile to join the American Statistical Association...[which now has] over 800 members.... Of the published papers read before the American Mathematical Society during the last five years, only three or four have had any relation to statistics. The very terminology of modern statistical method is unfamiliar to the great majority of professional mathematicians.... Most of the development of the science has been left to the economists, the actuaries, the biologists, the psychologists, and, more recently, the pedagogues. The result has been a wide scattering of the literature of statistical theory; many theoretical results have been first developed in articles having miscellaneous titles like "Family likeness in stature," "The trend of the stock market," or "The reliability of spelling scales"; any unification of effort was clearly lacking.*

(Huntington, 1919, pp. 421–422; underlining added)

Those quotes are from Huntington's paper titled "Mathematics and Statistics, with an Elementary Account of the Correlation Coefficient and the Correlation Ratio", which was an attempt at such "unification". Although it was extremely mathematical compared to most other papers on this subject, it was written simply enough that most statisticians would likely have understood it. His paper focused on correlation because:

> *In the field of statistical method and theory, the most characteristic single problem is the problem of correlation. The establishment of the existence or nonexistence of correlation between two things is the final goal of most statistical work* (p. 422).

In 1919, it is doubtful that the 800+ members of the American Statistical Association would have agreed that correlation was the goal of most statistical work, especially since (as previously discussed) the ASA Constitution as late as 1926 still stated: "The <u>objects</u> of the Association shall be to <u>collect, preserve, and diffuse statistical information</u> in the different departments of human knowledge" (ASA, 1926, p. 3; underlining added).

Huntington's paper focused on both the correlation coefficient and the correlation ratio because "Of the several mathematical measures of correlation that have been proposed...[those two] are perhaps the most fundamental" (pp. 422–423).

The third section of his paper was titled "The Central Problem of Statistics. Correlation between Two Functions, $x(i)$, $y(i)$." His explanation of that problem began with...

> *In the problem of correlation, what is sought is for some measure of agreement or disagreement between two series of paired quantities, x_1, x_2, x_3, ... x_n and, y_1, y_2, y_3, ... y_n (p. 423).*

He then provided a representative plot of 20 pairs of such quantities. The appearance of his plot (shown here as Figure 7.6) would have seemed familiar to his statistician readers but would have seemed strange to mathematicians. Instead of plotting y vs. x on the vertical and horizontal axes, respectively, which is a mathematician's "ordinary way" (p. 426), he...

- Sorted the paired quantities so that the x quantity was in numerical order.
- Assigned $i = 1$ to the smallest x value, $i = 2$ to the next smallest, and so on until $i = 20$.
- Plotted x (on the vertical axis) vs. i (on the horizontal axis).
- Then, on the same graph paper, plotted y (on the vertical axis) vs. the i of its paired x value.

His next statement duplicated what many prior statistical authors had written, but he wrote in a unique way—no author discussed here so far used such a simple, concise, clear, mathematical statement:

> *what we seek is then some measure of agreement or disagreement between the two curves or functions $x = f(i)$, $y = \omega(i)$ (p. 423).*

He then proposed something radical, at least to the mind of a statistician:

> *The case $r = +1$, called the case of perfect positive correlation.... The case $r = -1$, called perfect negative correlation.... <u>Either case will occur...when and only when the original y's and x's are connected by a linear equation</u>.... We see, therefore, that in the Pearsonian sense, <u>perfect correlation</u> (positive or negative)*

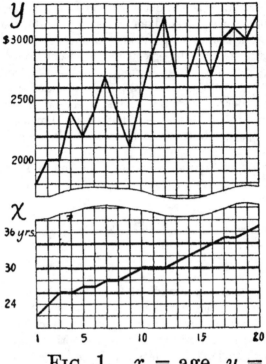

FIG. 1. x = age, y = income, of a certain group of men (n = 20). Means: $\bar{x}$ = 30 years, $\bar{y}$ = 2600 dollars. Standard deviations: σ_x = 4.025 years, σ_y = 417.1 dollars. Correlation coefficient: r = .855.

FIGURE 7.6

This chart is similar to a typical 19th-century correlation plot (i.e., two variables plotted vs. a third variable that the two variables shared in common); however, in this chart, the third variable is nothing more than counting numbers assigned to one variable that has been sorted in order of magnitude. (From Huntington (1919, p. 423).)

> between two sets of quantities x and y <u>means</u> <u>nothing more nor less than the</u> <u>existence of a linear algebraic equation connecting those quantities</u>. Indeed a bet-
> ter name for the coefficient of correlation might be the "coefficient of linear rela-
> tionship". <u>In general, the given sets of values will not be linearly related, and the</u>
> <u>value of r will be less than 1</u> (pp. 424–425; underlining added).

In other words, he proposed that unless an x, y plot is *perfectly* linear, the x, y relationship is non-linear, and that therefore the correlation ratio rather than the correlation coefficient should be used (although he does not state that clearly at this point in his paper). He understood that such an extreme view was impractical, and so he allowed that if the plot were *approximately* straight, then r was a good measure of correlation; in his words:

> ...*the value of r is a suitable criterion of the approach to the linear relationship of the variables x and y* (p. 427).
> ...*the correlation ratio,* η_{yx}, *will be equal to the correlation coefficient, r,* <u>*when and only when*</u> *the regression of y on x is linear...* (p. 432; underlining added).

That proposal obviates the need for arguments over whether or not a data plot is straight enough to justify calculation of a correlation coefficient; his view is that *theoretically* it never is straight enough. That is, if r is less than 1, then by definition the plot is non-linear; however, r is "suitable" if the plot is approximately linear. Based upon my review of textbooks published during the remainder of the 20th century, it seems that Huntington's view was not widely adopted. For example, neither he nor his paper were mentioned in the first-ever entire book on correlation (Ezekiel, 1930, to be discussed here later).

It is interesting that in his paper he made a major mistake, namely to say: "A typical graph of y as a function of x...is called a *correlation graph* (Galton, 1888)" (p. 426). Any casual reader would assume that the paper he referenced (i.e., Galton, 1888c) is the source of the term "correlation graph", or at least includes an example of a correlation graph; however, Galton's paper does *not* contain the term correlation graph, nor even the word graph, nor does it include a graph of (untransmuted) y as a function of (untransmuted) x.

In 1919, **Warren M. Persons** (previously mentioned) published a 95-page paper (book!?) titled "An Index of General Business Conditions"; it appeared in the journal *The Review of Economics and Statistics*. That paper was divided into four sections, one of which was titled "The Method Used"; in the 23 pages of that section, the word correlation was used about 190 times. He included a brief history of the biometric origins of mathematical correlation, and then he explained how the correlation coefficient could be applied to economics. He introduced the method of correlation by describing it as a problem to be solved:

> In the January [1919] number of [the journal titled] The Review of Economic Statistics [sic] fifteen monthly series of business statistics were presented and analyzed. Each item was conceived to be a compound or composite, that is, each actual figure was conceived to be <u>a magnitude determined by the concurrent action of distinct causes</u>. A method was developed of measuring and eliminating those constituents of the items of time series ascribable to secular trend and seasonal variation. The method developed was applied to the fifteen series of business statistics covering periods of from sixteen to twenty-six years....

<u>*The problem now before us is the measurement of correlation between the cycles*</u>
of the various series.

(Persons, 1919, pp. 117, 120; underlining added)

He assumed that the reader understood that the yearly values of series were to be plotted versus date, two-series-at-a-time, on the same chart (i.e., a "time series"); he provided such a chart for pig-iron production vs. interest rate on commercial paper. The two plotted lines seemed to go up and down together, with a lag time of several months. He asked: What is the best estimate of that lag time? He first answered using a "graphic method", by plotting each series on its own separate "translucent tracing cloth", placing the cloths on top of each other, shining a light through them from below, and having "three independent observers" each separately slide the cloths back and forth until "the observer [could] estimate the relative positions of the two graphs which would secure the best fit of one curve to the other" (p. 121). However...

Such conclusions need to be checked up by a more objective method of measuring correlation. There is need of a quantitative measure, of a coefficient of correlation.... The results obtained [using the graphic method] *give a valuable guide for further investigation; but a quantitative measure of correlation is needed to supplement and test the tentative conclusions which the graphic method has provided* (pp. 121–122).

His second answer to "What is the best estimate of that lag time?" was to calculate the correlation coefficient between the two series, at monthly lags of 0 through 12 months. He plotted the resulting r values versus lag time (i.e., as a correlogram, as discussed here in Chapter 1), and the resulting curve (shown here as Figure 7.7) had a peak correlation coefficient of 0.75 at about 5.5 months.

He then asked an important question:

The limiting values of the scale of correlation are then +1, 0, and −1. The first value, +1, indicates that the two series of cycles are identical; the second value, 0, indicates a complete lack of correspondence between the series, such as would occur from pairing items at random; the third value, −1, indicates that the items of the two series are numerically identical but opposite in sign. <u>What do intermediate values indicate? In other words, what is the meaning of such coefficients as</u> +<u>.04,</u> +<u>.37,</u> +<u>.48,</u> +<u>.80,</u> +<u>.96 or their negatives?</u> (p. 124; underlining added).

His answer was:

To answer our specific questions fully and rigorously would lead us into the intricacies of the mathematics of the theory of probability, an unnecessary digression. Some further consideration of the general question, however, is advisable.... Expressed otherwise, <u>the larger the coefficient of correlation the smaller is the probability that it has resulted accidentally</u>.... In judging the significance of

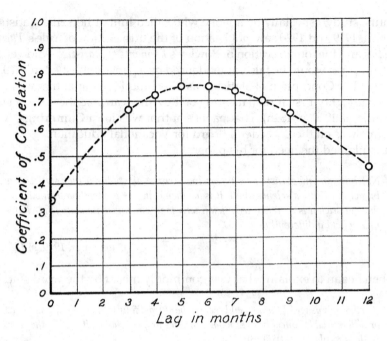

FIGURE 7.7
This correlogram shows that the strongest correlation between pig-iron production and the subsequent interest rate on commercial paper occurs 5–6 months post-production. (From Persons (1919, p. 124).)

> *coefficients, then, the fundamental notions are the size of the coefficient and the number of pairs of items utilized. Expressed in mathematical terms, the probability that a given coefficient has resulted from totally unconnected causes is a function of the coefficient and the number of pairs of items used in its computation. The form of this function has been derived. The value of the function which is most widely used is called the "probable error."* (pp. 124–125; underlining added).

Thus, he consider r to be a probability statement about whether or not there is correlation, whereas most other statisticians (and Galton) considered it to be a measure of the strength or degree of correlation itself. The two views can be bridged by a discussion of probable error, but that bridging was not made clear in Persons's paper.

1920

One might assume that by 1920, more than three decades after Galton's December 1888 paper, the new concept of *mathematical* statistics had advanced to the point where it was familiar to and practiced by workers in every field of scientific study, especially by anyone who called him or herself a statistician. After all, 1919 and 1920 saw publication of the fifth edition of Yule's *Theory of Statistics* and the fourth edition of Bowley's *Elements of Statistics*. And yet, such an assumption would be false, as evidenced by a paper published in 1920 with the title "The Conception of Statistics as a Technique", a paper that was "Read before the eighty-first annual meeting of the American Statistical Association" (Cummings, 1920, p. 164n). That paper's author was **John Cummings**, who at that time was on the US Federal Board for Vocational Education.

He explained the need for his paper:

> *Not infrequently statisticians themselves express the opinion that statistics is in fact not really a science at all. It is, as they conceive it, merely a method, and the statistician is merely a technician who may hire out to do statistical chores in any field indifferently.*
>
> (Cummings, 1920, p. 166)

He then began to explain what was commonly meant by the word statistics:

> *By common usage [,] there would seem to be at least four fairly well differentiated or differentiable concepts as to what statistics are, or what it, if we use the term in the singular, is…* (p. 166).

The first three of those concepts related to "compiled" or "compiling" of "numerical data"; and the fourth was…

> *Statistics is a social science, with a method or technique of its own, a defined field of operation, and a cumulative fund of systematized data or knowledge relating specifically to social conditions and phenomena…. [This] concept comprehends the data with the method or technique of compilation, but restricts the range of professionally statistical research to social conditions and phenomena, which is traditionally the special field or preserve of statistical inquiry…. statistics is the mother of the social sciences…* (pp. 166–167; underlining added).

He described the common view that a statistician is a "technician" or "skilled artisan", who is…

> *…expert in working out tabulation fields for punching cards, expert in manipulating various types of punching, sorting, tabulating and calculating machines, expert in reading his own tables into text, and in filling in blanks in standardized texts, preserved inviolate from decade to decade, or from year to year* (p. 168).

He then asked whether a statistician should also be viewed as a practitioner of…

> *…a separate and distinct science…the subject matter of that science [being] the method, technique, and mathematics of compilation, including of course, the mathematical determination of correlations…?* (p. 168; underlining added).

He admitted that most statisticians are unprepared for such a role:

> *It will be freely admitted by statisticians that occasionally the statistician him-*
> *self, for the precise interpretation of his data, must seek aid of the mathematician,*
> *who understands better than the statistician commonly does the harmonics and*
> *intricacies of mathematical calculations (p. 170).*

In his concluding paragraph, he gently prodded his audience into the world of mathematical statistics:

> *It is the statistician's function <u>to measure social tendencies precisely</u>.... to deter-*
> *mine precisely rates and frequencies and <u>correlations</u> and probabilities... (p. 176;*
> underlining added).

Also in 1920, **Karl Pearson** (discussed previously) wrote a 21-page paper titled "Notes on the History of Correlation". In it, he discussed a paper that Galton had written in 1885, a paper that contained a "Diagram" that Galton had created from what Pearson described as "the first correlation table...as we should now call it" (p. 36). That table and diagram are shown here as Figures 7.8 and 7.9.

After discussing the mathematical analysis of that Diagram's curves and lines (an analysis conducted in 1885 by Galton's mathematician friend

TABLE I.

NUMBER OF ADULT CHILDREN OF VARIOUS STATURES BORN OF 205 MID-PARENTS OF VARIOUS STATURES.
(All Female heights have been multiplied by 1·08).

Heights of the Mid-parents in inches.	Heights of the Adult Children.														Total Number of		Medians.
	Below	62·2	63·2	64·2	65·2	66·2	67·2	68·2	69·2	70·2	71·2	72·2	73·2	Above	Adult Children.	Mid-parents.	
Above ..	..	..	..	..	..	..	..	..	..	..	..	1	3	..	4	5	..
72·5	..	..	..	..	..	..	..	1	2	1	2	7	2	4	19	6	72·2
71·5	..	..	..	..	1	3	4	3	5	10	4	9	2	2	43	11	69·9
70·5	1	..	1	..	1	1	3	12	18	14	7	4	3	3	68	22	69·5
69·5	..	..	1	16	4	17	27	20	33	25	20	11	4	5	183	41	68·9
68·5	1	..	7	11	16	25	31	34	48	21	18	4	3	..	219	49	68·2
67·5	..	3	5	14	15	36	38	28	38	19	11	4	..	..	211	33	67·6
66·5	..	3	3	5	2	17	17	14	13	4	..	..	..	..	78	20	67·2
65·5	1	..	9	5	7	11	11	7	7	5	2	1	..	..	66	12	66·7
64·5	1	1	4	4	1	5	5	..	2	..	..	..	..	..	23	5	65·8
Below ..	1	..	2	4	1	2	2	1	1	..	..	..	..	..	14	1	..
Totals ..	5	7	32	59	48	117	138	120	167	99	64	41	17	14	928	205	..
Medians ..	..	..	66·3	67·8	67·9	67·7	67·9	68·3	68·5	69·0	69·0	70·0	..	..	..	..	..

NOTE.—In calculating the Medians, the entries have been taken as referring to the middle of the squares in which they stand. The reason why the headings run 62·2, 63·2, &c., instead of 62·5, 63·5, &c., is that the observations are unequally distributed between 62 and 63, 63 and 64, &c., there being a strong bias in favour of integral inches. After careful consideration, I concluded that the headings, as adopted, best satisfied the conditions. This inequality was not apparent in the case of the Mid-parents.

FIGURE 7.8
This table from an 1886 publication is what in later years would be called a correlation table.
(From Galton (1886c, p. 248).)

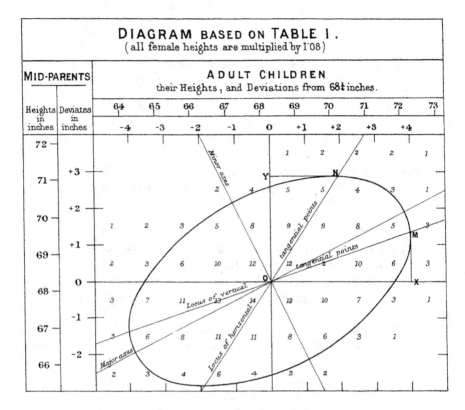

FIGURE 7.9
This diagram (based upon the data in Figure 7.8) shows the approximate distribution of heights of parents vs. children; Galton used this type of plot to demonstrate generational reversion (a.k.a. regression) to the mean, i.e., that children of tall parents tend to be shorter than their parents, and that children of short parents tend to be taller than their parents. (From Galton (1886c, Plate X).)

Dickson), Pearson then makes a valid claim but with an unfortunate choice of words:

> *Thus in 1885 Galton had <u>completed</u> the theory of bi-variate normal <u>correlation</u>.*
>
> (K. Pearson, 1920, p. 37; underlining added)

I say "unfortunate" because if he had substituted the word "distribution" for "correlation", it would have been an accurate statement; his sentence as written gives the reader the false impression that Galton's 1885 paper applied to the concept of "co-relation" that he'd discovered in late 1888. That 1885 diagram demonstrated regression, not correlation, at least not the correlation whose discovery Galton described in his papers written in 1888/1889. As previously discussed, Galton in those papers did not consider a table, diagram, or *X,Y* plot to exhibit correlation unless they had been created with data transmuted into units of *Q* (the respective probable error)—neither the Table nor Diagram just mentioned were created with such transmuted data.

Also unfortunate was Pearson's summary statement that...

> *In 1889 appeared Galton's book Natural Inheritance embodying most of the work we have discussed in the earlier memoirs of* [Galton's that he'd written from] *1877 to 1888. Beyond this Galton did not carry the subject of correlation.*

> *(K. Pearson, 1920, p. 40; underlining added)*

I say "unfortunate" because Pearson was apparently unaware of or had forgotten about the two lengthy papers (previously discussed) that Galton had written in 1889 (which date is "beyond" 1888), in which Galton had clarified his own thinking on the subject of correlation.

1921

In 1888, a problem was created that was partially solved in 1904 and fully solved in 1921. The problem was that Francis Galton had not provided a way to interpret intermediate values of his measure of co-relation; he had said only that "two variable organs are said to be co-related when the variation of the one is accompanied on the average by more or less variation of the other...*r* measures the closeness of co-relation" (Galton, 1888c, pp. 135, 145). He did not provide any more clarity in his follow-up papers published in 1889 and 1890 (previously discussed).

As previously discussed, an 1896 paper by Karl Pearson had introduced the product-moment formula for *r*. In that paper, he repeatedly compared *r* values from different data sets, thereby giving the reader the mistaken impression that the magnitudes of intermediate *r* values can be compared directly. The rest of his early statistical papers in which correlation was a major topic were similarly *r*-focused (see papers in E. Pearson, 1956).

In 1911, in Pearson's third edition of *The Grammar of Science*, he philosophized that mathematical correlation and mathematical functions are essentially the same concept. His view was that a mathematical function is a perfect correlation but that such perfection is absent from nature. A mathematical function predicts a precise *Y* value for each *X* value (e.g., $Y=a+bX$), whereas nature's *Y* values are merely correlated with *X* and so cannot be predicted precisely, i.e., $Y=a+bX\pm natural\ variation$ (if Pearson had been describing an engineering study instead of a biological one, he would have talked about $\pm process\ variation$ or $\pm measurement\ variation$ instead of $\pm natural\ variation$). The precision of the prediction is indicated by the magnitude of the correlation, which ranges from 0.00=no correlation and no precision, to 1.00=perfect correlation and perfect precision. Pearson's view has been interpreted to mean that...

> ...*the correlation coefficient allowed <u>precise</u> measurement of the <u>strength</u> of the [X,Y] association, between [r =] zero (independence) and [r =] one (strict dependence).*
>
> (Desrosières, 1998, p. 107; underlining added)

However, in all 27 pages that Pearson devoted in *Grammar* to this topic, he provided no mathematically rigorous definition for such strengths, except for the extremes of 0.00 and 1.00 (K. Pearson, 1911, pp. 152–178). Similarly, in 1919, the just-published fifth edition of Yule's *An Introduction to the Theory of Statistics* did not provide any help to a reader who wanted to rigorously interpret intermediate *r* values. The best Yule said was:

> ...*r is of very great importance.... If the two variables* [whose correlation is being calculated] *are independent, r is zero.... The numerical value* [of r] *cannot exceed ± 1.... r is termed the coefficient of correlation....*
>
> (Yule, 1919, pp. 173–174)

Were Pearson and Yule unaware of, or in disagreement with, what Spearman had written in 1904 (previously discussed), namely that "not Galton's measure of correlation, but the square thereof" was "theoretically far more valuable" as a "measure of the correlation"? Possibly part of the problem was that the square of *r* had no name; that part of the problem was solved by S. G. Wright in 1921, when he named it the "coefficient of determination" (to be discussed here in the next section).

NOTE: I have pieced together the following series of formulas from various sources, starting with Pearson's correlation function metaphor (just discussed); I've found this series to be useful for introducing correlation to novice students:

$Ye=a+bX=$values predicted by the function (i.e., the linear regression equation)

$Y=a+bX\pm natural$ *(or other) variation*=values observed in any experiment or study

$Sye^2=$Variance of the predicted Ye values

$Sy^2=$Variance of the observed Y values

$See^2=$Variance of the values obtained by $Y-Ye$

$Sy^2 = Sye^2 + See^2$

$r^2 = Sye^2 / Sy^2$

That last formula shows that the coefficient of determination (i.e., r^2) is the ratio of Sye^2 and Sy^2, where Sye^2 is the variation "caused" by X variation (i.e., Ye is an exactly predictable function of X, under the assumption of no "natural variation"), and where Sy^2 is the total variation of Y (which is caused by a combination of X variation and "natural variation"). Therefore, the coefficient of correlation is the fraction of total Y variation that is explainable by (or due to, or caused by) the linear relation between Y and X. If there were no "natural variation", then all the Y values would fall on the linear regression line, and then $Y=Ye$, and $See^2=0$, and $Sy^2=Sye^2$, and $r^2=Sye^2/Sy^2=1$ exactly. If there were no linear relationship between Y and X, then the "linear regression line" would be horizontal (i.e., $Ye=a$, and therefore $Y=a\pm natural\ variation$), and $Sye^2=0$, and $r^2=Sye^2/Sy^2=0$ exactly.

In 1921, **Sewall Green Wright** (1889–1988) was a senior animal husbandman in animal genetics at the United States Department of Agriculture (Wright, 1921, p. 557). He is remembered today as one of the founders of the science of population genetics (Britannica.com-Wright).

From the point of view of the history of the correlation coefficient, he is best known for describing r^2 as the "coefficient of determination" and for expanding on what Spearman had said about r^2 in 1904 (previously discussed). Wright accomplished all that in a 1921 paper titled "Correlation and Causation", in which he stated:

> Relations between variables which can be measured quantitatively are usually expressed in terms of Galton's coefficient of correlation...or of Pearson's correlation ratio.... Use of the coefficient of correlation (r) assumes that there is a linear relationship between the two variables.... For many purposes it is enough to look on it [r] as giving an arbitrary scale between +1 for perfect positive correlation, 0 for no correlation, and −1 for perfect negative correlation. The correlation ratio (η)...does not, however, depend on the assumption [of linearity]...and is always larger than r when relations are not exactly linear. It can only take values between 0 and +1, and it can be looked upon as giving an arbitrary scale between 0 for no correlation an 1 for perfect correlation.... The numerical values of both coefficients, however, have significance in another way. Their squares (η², or r² if regression is linear) measure the portion of the variability of one of the variables which is determined by the other....

> (Wright, 1921, pp. 557–558; underlining added)

Curiously, he did not there or elsewhere reference Spearman or his 1904 paper. After further discussing the correlation coefficient, he said:

> Another coefficient which it will be convenient to use, the coefficient of determination of X by A...measures the fraction of complete determination [of X] for which factor A is directly responsible...

> (p. 562; underlining added; in this example, X is the effect-value plotted on the vertical axis, and A is the cause-value plotted on the horizontal axis).

In his paper, Wright tended to speak generically about "degrees of determination" and numerically about "coefficients of determination", but he was not consistent about that distinction.

His goal was to quantitate the correlations in a series of causes that led to a final result; he called the sequence of causes a "path", and he used the term "path coefficient" for the correlation value applicable to a single step on a path (p. 563). Even after reading his paper twice, I do not understand why he calculated each path coefficient as a correlation coefficient rather than as a coefficient of determination:

> *A cause has a linear relation to the effect and is combined additively with the other factors if a given amount of change in it always determines the same change in the effect, regardless of its own absolute value or that of the other causes. The conclusion is that, under these conditions, the <u>path coefficient</u> equals the <u>coefficient of correlation</u> between cause and effect, and the <u>degree of determination</u> equals the square of either of the preceding coefficients* (p. 563; underlining added).

In his Fig. 1 (shown here as Figure 7.10) each arrow represents a step on the path to the "Weight [of guinea pigs] at 33 days"; his goal was to inscribe a path coefficient next to each arrow, as he did in his Fig. 8 (shown here as Figure 7.11).

The reader might assume that Wright considered correlation to prove causation, but that would be an incorrect inference. Wright used path coefficients not as a way to prove the presence of a cause-and-effect relationship, but rather to help quantitate a relationship that he already knew existed; for example, he considered it obvious that the length of the "gestation period" influenced "weight at birth". He sought "a method of <u>measuring</u> the direct influence along each separate path" (Wright, 1921, p. 557; underlining added).

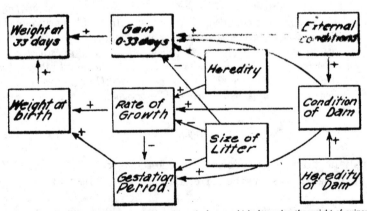

Fɪɢ. ɪ.—Diagram illustrating the interrelations among the factors which determine the weight of guinea pigs at birth and at weaning (33 days).

FIGURE 7.10
This flow diagram shows the factors that influence the weight of 33-day-old guinea pigs. (From Wright (1921, p. 560).)

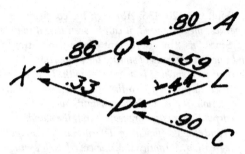

FIG. 8.—Path coefficients measuring the relations between birth rate (X), rate of growth (Q), gestation period (P), size of litter (L), and other causes (A, C).

FIGURE 7.11
This flow diagram shows path coefficients (i.e., correlation coefficients) derived from factors that influence birth rate. (From Wright (1921, p. 570).)

His search led him to discover that the correlation coefficient was an inaccurate measure of that influence but that the square of the correlation coefficient (i.e., the coefficient of determination) was an accurate measure.

At least one contemporary textbook author disagreed with Spearman and Wright about the *relative* importance of the coefficient of determination *vis à vis* the correlation coefficient, but he did so without providing a supporting argument: "The quantity $b_1 b_2$ [i.e., r^2] then makes a good measure of correlation in the case of linear regression.... [However] r becomes an excellent measure of correlation in the case of linear regression." (Gavett, 1925, p. 228; underlining added). Other than Gavett, it seems that all more recently published statistics textbooks that discussed r^2 described it as a *better* method than r for measuring correlation. The following is a chronologically arranged example-list of quotes from authors who have promoted the importance of r^2 and the coefficient of determination and/or who have demoted the importance of r and the correlation coefficient, starting with Spearman in 1904:

- 1904a, Spearman, p. 75: *In short, not Galton's measure of correlation, but the square thereof, indicates the relative influence of the factors in A tending towards any observed correspondence as compared with the remaining components of A tending in other directions.*
- 1921, Wright, p. 558: *For many purposes it is enough to look on it* [r] *as giving an arbitrary scale between +1 for perfect positive correlation, 0 for no correlation, and −1 for perfect negative correlation. The correlation ratio* [η]*...can be looked upon as given an arbitrary scale between 0 for no correlation and 1 for perfect correlation.... Their squares (η², or r² if regression is linear) measure the portion of the variability of one of the variables which is determined by the other....*
- 1927, Burgess, pp. 208–209: *Because of the vagueness in the interpretation of r..., r should not be relied on to give the complete and only analysis and summary of relationships.*

- 1930, Ezekiel, pp. 120, 136: *Although the coefficient of correlation was the earliest measure used, it can be seen that it may be misinterpreted.... Since this* [the coefficient of determination] *is the most direct and unequivocal way of stating the proportion of the variance in the dependent factor which is associated with the independent factor, it may be used in preference to the other methods.... Particularly in linear correlation, there are three constants which summarize nearly all that a correlation analysis reveals. First, the standard error of estimate.... Second, the coefficient of determination.... Finally, the coefficient of regression.... (notice that the correlation coefficient was not among those "three constants").*
- 1938, A. E. Waugh, p. 275: *In our example...r = −0.664. We might,* [sic] *then interpret it by computing the coefficient of determination....* [which] *tells us the percent of variation in the dependent variable that can be explained in terms of the independent variable.*
- 1939, Treloar, p. 127: *It is sometimes taken as a basis for condemnation of the correlation coefficient that a difference of 0.1 (or any fixed increment) between two values of r means an increasing difference in intensity of association, from the prediction point of view, as one passes from low to high values of r. This deficiency is regrettable, but no wholly satisfactory way of avoiding it by using a function of r, such as r^2* [i.e., the coefficient of determination]*...has as yet found general appeal.*
- 1944, Blair, pp. 264–265: *Correlation* [coefficient] *is the square root of a percentage, and is, therefore, quite misleading to the beginner in statistics. A more easily understood, and in many respects a better measure, is the coefficient of determination, which is a true percentage of the portion of one variable that is associated with another. Determination is the square of correlation, and is coming into general use as the more accurate and easily understood measure.... For instance, one might think that r = 0.3 is one-half of r = 0.6, but it is in fact but one-fourth. One might think that r = 0.2 is one-fourth of r = 0.8, but actually it is only one-sixteenth of 0.8. (Note: Leading zeros were not present in the source article; they have been added here to promote legibility.)*
- 1960, Freund, p. 333: *Values of r falling between 0 and +1 or between 0 and −1 are more difficult to explain; a person who has no knowledge of statistics might easily be led to the erroneous idea that a correlation of r = 0.80 is "twice is good" as a correlation of r = 0.40, or that a correlation of r = 0.75 is "three times as good" or "three times as strong" as a correlation of r = 0.25.*
- 1961, Kozelka, p. 132: *Is r_{xy}* [the correlation coefficient] *merely a notational convenience, or does it really tell us something?.... It is customary to use r_{xy} as a measure of the strength of the linear tendency...* [but the] *safest way to interpret r_{xy} is in terms of r^2_{xy}....*
- 1979, Schmidt, p. 157: *Because correlation coefficients look like proportions, many people respond to them as if they were proportions, thinking that a correlation of 0.70 is twice as high as a correlation of 0.35. This is unfortunate, because r itself is not a proportion.... r^2 is a*

> *proportion, specifically the proportion of variation in the scatter plot accounted for by a linear equation.*
>
> - 1981, Selkirk, p. 17: *The correlation coefficient r is not a proportion and one cannot talk about one correlation coefficient being twice another, nor about one correlation coefficient being 0.2 more than another; its scale must be regarded as ordinal.*
> - 1984, Healey, p. 267: *While a* [correlation coefficient] *value of 0.00 indicates no linear relationship and a value of ±1.00 indicates a perfect linear relationship, values between these extremes have no direct interpretation.*
> - 1988, Bohrnstedt and Knoke, pp. 269, 271:*...the coefficient of determination...is the proportion of total variation in Y "determined" by X.... The usefulness of the correlation coefficient lies in its communication of directionality as well as magnitude of association, unlike* [the coefficient of determination], *which conceals whether the variables are directly or inversely related.*
> - 1998, Holland, p. 47: *The square of the correlation coefficient, or r^2, can be shown mathematically to represent the proportion of variability "explained" by the correlation. That is, r^2 answers the question, "What proportion of the variability in x is explained by its association with y?"*
> - 2000, Gravetter and Wallnau, pp. 536, 565: *When judging "how good" a relationship is, it is tempting to focus on the numerical value of the correlation* [coefficient]. *For example, a correlation of +0.5 is halfway between 0 and 1.00 and, therefore, appears to represent a moderate degree of relationship. However, a correlation should not be interpreted as a proportion.... [However] squaring the correlation* [coefficient] *provides a measure of the accuracy of prediction: r^2 is called the coefficient of determination because it determines what proportion of the variability in Y is predicted* [determined] *by the relationship with X.*
> - 2017, Triola, (Kindle section 10–3): *The value of r^2* [the coefficient of determination] *is the proportion of the variation in y that is explained by the linear relationship between x and y.... [to] compute r^2...we can simply square the linear correlation coefficient r.*

I have not been able to find any published succinct explanation of the meaning of the Coefficient of Determination as it applies explicitly to Co-Relational Correlation; Wright's paper does give such an explanation but uses many paragraphs and many pages to do so. As seen in the just-given list of quotes, the meaning has been explained succinctly only for Relational Correlation; for example, *"the coefficient of determination...tells us the percent of variation in the <u>dependent</u> variable that can be explained in terms of the <u>independent</u> variable"* (Waugh, 1938, underlining added). Such an explanation fails to help the reader understand what meaning the Coefficient of Determination has in regard to X, Y data sets for which the concept of dependent and independent variables does not apply, i.e., in regard to Co-Relational Correlation. Let me now attempt a more helpful explanation:

As previously discussed, Galton in December of 1888 stated that *"It is easy to see that co-relation must be the consequence of the variations of the two organs being partly due to common causes"* (Galton, 1888c, p. 135). Let's use the symbol Z to represent those common causes. Galton incorrectly claimed that *r* quantified the part that Z played in determining the values of X and of Y, whereas Spearman in 1904 and Wright in 1921 discovered that the square of *r* (i.e. *r²*, the Coefficient of Determination) is the correct quantity. Galton in 1888 had determined *r*=0.8 for the Co-Relational Correlation between cubit and stature; the corresponding Coefficient of Determination is therefore 0.8×0.8=0.64, which means that Z caused 64% of the variation in cubit length as well as 64% of the variation in stature. Generally speaking, the Coefficient of Determination for Co-Relational Correlation is the proportion of variation in both X and Y that can be explained in terms of one or more causal variables that X and Y have in common but whose identity might not be known.

Wright may have given *r²* a name, but T. L. Kelley (1923; p. 155) assigned it a Greek letter ("$\lambda = r^2$"). By 1939, Wright's name for *r²* was included in dictionaries (e.g., Kurtz and Edgerton, 1939, p. 28), but Kelley's Greek letter was not.

Eight decades later, a formula for the "Signed Coefficient of Determination" was developed (Zorich, 2018); the output of that formula indicates whether or not the original *r* value is negative or positive:

$$\text{Signed coefficient of determination} = {_s}R^2 = r^3 \big/ |r|$$

It is interesting that in some introductory statistics textbooks, the correlation coefficient *r* is introduced and/or *defined* in terms of *r²*, rather than the other way around (e.g., Ezekiel and Fox, 1959, p. 127; Crow, 1960, p. 157; Spiegel, 1961, p. 243; Bohrnstedt and Knoke, 1988, p. 271). That is…

Correlation coefficient = Square root of coefficient of determination

rather than…

Coefficient of determination = Square of correlation coefficient.

1923

In 1923, **Truman Lee Kelley** was a professor of education at Stanford University when he published a textbook titled *Statistical Method*. It is interesting that the book devoted more than 150 pages to correlation and measurement of relationship, but yet spoke disparagingly about the topic:

*The derivation of the correlation coefficient shows it to be the regression coef-
ficient in the case of standard measures. The regression coefficient is statistically
the more fundamental [;] and in all actual problems involving the estimate of
one variable knowing a second, the regression coefficient and not the correlation
coefficient is the essential measure. A wider use of regression coefficients in place
of correlation coefficients would lead to a more accurate and detailed understanding of the situations portrayed.*

(Kelley, 1923, pp. 181–182)

One concrete example of his view was:

*That there is a correlation between the votes of men and women is of quite sec-
ondary interest to the fact that there is a wide difference in the regressions of
the two sexes. The interpretation of the correlation table given hinges upon the
slopes of regression lines in a much more fundamental sense than upon the value
of the correlation.*

(p. 185; he was referring to a correlation matrix table of percent votes by men
vs. percent votes by women for a particular candidate, precinct by precinct,
in the municipal election of April 6, 1915, in the city of Chicago, USA)

1924

In 1924, **Harry Jerome** was an assistant professor of economics at the
University of Wisconsin when he wrote an introductory textbook on statistics titled identically to the just-mentioned book by Kelley, namely *Statistical
Method*. In Jerome's mind, the value of the correlation coefficient lay in its
ability to predict the future, or so he said in his "Correlation" chapter's opening paragraph:

*Just as the study of history is valuable largely as a guide to the future, so like-
wise, to a considerable extent, may statistics be looked upon as an historical
method of study, by which, out of past experience, we formulate statements
of the most probable future occurrences. By statistical analysis the economist hopes to obtain forecasting formulas which will afford practical aid in
anticipating coming changes in economic conditions. He wants to be able to
anticipate the most probable change in the price of corn with a given change
in the quantity produced, or the most probable change in interest rates with a
given change in bank reserves. Measures or <u>coefficients of correlation are bases
for such forecasting formulas</u>, in that <u>they express the relations which have
existed in the past</u>, in concise quantitative form convenient <u>for use in estimating future probabilities</u>.*

(Jerome, 1924, p. 263; underlining added)

He defined correlation on his next page:

> We may tentatively defined correlation as the typical amount of similarity, in direction and degree, of variations in corresponding pairs in two series of variables (p. 264).

He also provided a definition of the Persons's formulaic correlation coefficient but surprisingly does so in terms of Galton's graphic method:

> ...the Pearsonian coefficient of correlation is the slope of the straight line which best represents the plotted points portraying the associated deviations of two series when these deviations are expressed in multiples of their standard deviations (p. 280).

An extrapolation of his opening paragraph's view was given in the opening sentence in the section titled "The standard error of the regression equation". That sentence was:

> We have seen that the coefficient of correlation enables us, through the regression equation, to state the most probable change in one variable, given a certain change in another (p. 281).

The last section in his Correlation chapter was titled "The Use and Interpretation of the Coefficient as High or Low". There, he stated that...

> In many cases...the statistician wishes to know whether he should consider a [correlation] coefficient of .25, or .40, or .50, or .60, etc., as low, moderate, or high. It is somewhat hazardous to venture the formulation of a guiding principle for the interpretation of the coefficient, for its significance will vary somewhat with the type of data involved. The meaning of the term is more definite when the regression equation is given, but it is more customary to use the coefficient as the expressed measure of correlation. The following rules have been suggested for its interpretation:" (p. 285)

On that same page, he then quoted rules from King's *Elements of Statistical Method* (previously discussed). Those rules stated that any correlation coefficient greater than 0.50 is "decided correlation" whereas if it is less than 0.30 it "cannot be considered at all marked". Neither Jerome nor King explained what "decided" or "marked" means, and so the reader is left to wonder. Whatever the case, Jerome did not include a discussion of Wright's Coefficient of Determination (previously discussed). Jerome's book therefore is one of the many post-Wright examples where the author misleads the reader into thinking that the correlation coefficient is a rigorous *measure* of co-variation, rather than it being (as shown by Wright) an approximate, ordinal-scale indicator.

In 1924, **Frederick Cecil Mills** (1892–1964) was a professor of economics at Columbia University when he published a "new" coefficient of curvi-linear correlation (Mills, 1924); 3 months later, **Mordecai Ezekiel** (a statistician at the U.S. Department of Agriculture) published the same coefficient, which he had developed independently (Ezekiel, 1924). They both had noted that the classic correlation ratio (Karl Pearson's coefficient of non-linear correlation) involves calculation of y-data deviations by taking the difference between a given y-value and the mean of the correlation-matrix-table cell in which it was found. Because a line drawn from mean to mean to mean in such a table typically produces a zig-zag pattern, Mills and Ezekiel recommended instead plotting the best fit *smooth* curve through the means and then calculating the y-deviations as the vertical difference between a given y-value and its corresponding point on that curve; Ezekiel considered the final resulting value to be a "corrected correlation-ratio" (Ezekiel, 1924, p. 434).

When a correlation ratio is calculated in that way, Mills suggested that it be called the *index of correlation* and that it be symbolized by ρ, which is the Greek letter *rho* (Mills, 1924, p. 273). Bowing to the fact that Mills had published first, Ezekiel agreed to use the same name and symbol as had Mills, although Ezekiel also called it the *correlation-index* (Ezekiel, 1924, pp. 434, 434n2, 435).

Mills used ρ (without any subscripts) in the "general formula for the index of correlation" when y is dependent on x, and also for when x is dependent on y (Mills, 1924, pp. 274, 274n); but two pages later he used ρ_{yx} for y dependent on x—the reader is left to infer that ρ_{xy} must be for x dependent on y (that inference is validated by Mills' subsequently published textbook on *Statistical Methods Applied to Economics and Business*, in which he did use ρ_{xy} for x dependent on y (Mills, 1938, p. 408n)).

From among all the letters in the Greek alphabet, and from among all the nouns in the English language, it is surprising that Mills chose a symbol (i.e., ρ) and name (i.e., index of correlation) that had previously been assigned other meanings in the field of correlation, as Mills acknowledged in a tiny-font footnote in his paper:

> This symbol [i.e., ρ] has been employed by Spearman [in 1904] to represent a coefficient of correlation based upon the squares of differences in rank. It has not been so widely used in this sense, however, that confusion should arise from its adoption for the general purpose here suggested. It is of interest to note that Galton [in December, 1888] used the term "Index of Co-relation" for what is now called the coefficient of correlation.
>
> (Mills, 1924, p. 273n2)

However, Mills was wrong: The symbol ρ was *still* being widely used for rank correlation, as evidenced by statistical textbooks that had been published very recently (e.g., Brown and Thomson, 1921, p. 130; Kelley, 1923, p. 192); and it would continue to be used as such for decades—for example, in a mid-20th-century reference book whose sole topic was rank correlation (Kendall, 1962, p. 8). Although the term "index of correlation" had only infrequently been used to refer to the Galton/Pearson linear correlation coefficient (as discussed here in Chapter 1), it had been used relatively recently (Whipple, 1907, p. 323). Also confusing to the reader is the fact that the symbol ρ had been used by Edgeworth and others since 1892 for the population parameter *linear* correlation coefficient (as mentioned here in Chapter 1).

Nevertheless, in later years, Mills continued to use that term and symbol for his new correlation coefficient, e.g., in his *Statistical Methods*, but there at least he assigned ρ_r (instead of just ρ) to Spearman's coefficient of rank correlation (Mills, 1938, pp. 377, 408ff). Ezekiel initially followed Mills's lead but eventually began using the symbol i_{yx} instead of ρ_{yx} for their index of correlation (Ezekiel, 1930, p. 119; Ezekiel and Fox, 1959, p. 128).

A brief discussion of the Mills/Ezekiel index can be found in a 1925 statistics textbook by **William Leonard Crum** and **Alson Currie Patton** (pp. 249–250). Apparently, not enough time was taken to examine the final printer's proof of that book, as evidenced by the fact that although in *one* instance it was stated and shown correctly that the symbol for that index is ρ, in four other instances it was shown upside down (it looks as if the printer mistakenly used an italics version of the lower case letter b from the Russian alphabet).

A half-page discussion of that term was also included in a 1927 statistics textbook by **Robert Wilbur Burgess**, who at that time was a "Senior Statistician, Western Electric Company"; it is interesting that Burgess reverted to using the symbol R for it (Burgess, 1927, p. 228).

Given the fact that Mills' 1924 paper hijacked what he considered to be Galton's term for linear correlation and hijacked Spearman's symbol for rank correlation (and Edgeworth's symbol for linear correlation) and then applied them both to a new type of correlation ratio, it is humorous that a 2014 paper hijacked a version of their term and used it to introduce a new type of *rank* correlation coefficient, which they called the "weighted correlation index" (Vigna, 2014).

1925

William Leonard Crum and **Alson Currie Patton** (previously mentioned in the discussion of Frederick Cecil Mills) taught statistics at Stanford University and Yale University, respectively, in the 1920s. Students who were planning to take a course in introductory statistics from either of them may have worried about how difficult it would be to understand the lectures. Such worry might have been based upon a reading of the seemingly incomprehensible first sentence in the Preface of the assigned textbook (*An Introduction to the Methods of Economic Statistics*), which those two professors had recently published. That first sentence was:

> *This text has been prepared with the needs of those students who are interested in the application of statistical methods to economic problems constantly in mind.*

> *(Crum and Patton, 1925, p. iii)*

I had to read that sentence three times before I realized that the last few words ("constantly in mind") did not refer to the students but rather to the authors. That is, the sentence meant to say something like this:

> *While preparing this text, the authors constantly kept in mind the needs of those students who are interested in the application of statistical methods to economic problems.*

Possibly that first sentence was a one-off rhetorical-effect blunder; possibly the rest of the book would be clearly written. I leave it to you to decide, based on the following additional quotations:

Their definition of correlation was given as:

> *It is a well-known fact that there exists some sort of interrelation between many of the simplest natural phenomena, and we have a somewhat less confident opinion that economic and other social phenomena are linked together in groups.... We are usually at a loss to determine the exact nature of the causes and effects involved in concomitant variations. The interest of the statistician is fixed less upon the question whether one phenomenon causes another than upon the discovery of a mutual relationship between the phenomena and upon the <u>measurement of the extent of this relationship</u>. Such a mutual relationship between two variable phenomena is <u>called correlation</u>* (p. 218; underlining added).

In that definition, correlation is defined as a relationship; but on the very next page, the definition morphs into the following mathematical statement:

> *Correlation is essentially an <u>average of the relationship</u> between the associated variates, for all the pairs of values; such relationship, in its simplest form, is a ratio between the two variables, each measured from an appropriately selected origin. If this average is highly typical—if it is highly representative of the individual relationship of every pair—correlation is said to be high, and otherwise, low* (pp. 219–220; underlining added).

Correlation-related text continued for 11 more pages, on which were described correlation tables, scatter diagrams, regression lines, and frequency distributions. Finally, *r* is defined:

> *The coefficient of correlation, r, is the arithmetic <u>average of the products</u>, one product for each of the N associated pairs, of the deviation of one variate, from the mean of such variates for all N cases, by the deviation of the associated variate, from the mean of the associated variates for all N cases, deviations being measured in standard units* (p. 231; underlining added).

That sentence should win some kind of award for being the most difficult-to-follow summary description of the correlation coefficient. Thankfully, in a footnote on that page, the formula for r is clearly given. What is not given is an explanation of the meaning of the correlation coefficient: How is it to be interpreted? How is it to be used? Is the correlation coefficient the "measurement of the extent of...relationship" that they'd mentioned in their first definition of correlation? The next 32 pages did not answer even one of those questions, although the pages did include instructions for how to calculate various types of correlation coefficients from various types of data.

In 1925, **Louis Leon Thurstone** was an associate professor of psychology at the University of Chicago when he published a textbook titled *The Fundamentals of Statistics*. It was "the result of seven years of teaching the fundamental principles of statistics and mental measurements to classes of about thirty graduate students annually" (Thurstone, 1925, p. ix). He described such students as having undergraduate degrees in "unmathematical subjects", among which he included "economics". Given the huge amount of research that had been completed during the prior three decades by Edgeworth and others on how to apply mathematical statistics to economics, it is unclear if Thurstone's description of that subject as being unmathematical is a reflection of his narrow breadth of knowledge or of his university's behind-the-times curriculum.

His book included four chapters on correlation, which discussed (respectively) correlation tables, the correlation coefficient, how to calculate the correlation coefficient, and rank correlation. In the chapter on correlation tables, he surprisingly used the words relation and relationship, rather than correlation and correlationship, when describing what is shown by a scatter diagram:

> *The scatter diagram is a chart for showing graphically the relation between two variables. The scatter diagram shows graphically not only the presence or absence of relationship, but it also enables one to judge by inspection the degree of relation between the two variables plotted.*
>
> *(Thurstone, 1925, p. 196)*

The closest thing to a discussion of what correlation is and how to interpret a measurement of it was given in the first paragraph in the chapter on the correlation coefficient; that entire paragraph was:

> *The <u>correlation coefficient</u> is a pure number, a constant which <u>indicates the degree of relation between two variables</u>. It varies from +1 to −1. When the relation is perfect and positive, the correlation coefficient is +1. When the relation is perfect but inverse, the correlation coefficient is −1. When there is no relation whatever between the two variables, the coefficient is zero. <u>Other values of the coefficient indicate intermediate degrees of relation</u>. Thus a coefficient of +.8 indicates that the points on the scatter diagram cluster rather closely about a diagonal line across the diagram, whereas a coefficient of +.3 indicates that*

the points scatter more from the diagonal tendency although the relation is still noticeable. The degree of relation between height and weight is approximately +.5. It is apparent, then, that <u>the correlation coefficient is only a numerical way of describing the scatter diagram,</u> although the diagram gives more information than can be found from the single numerical value of the coefficient. <u>When a great number of relations are being studied, the correlation coefficients serve as abbreviations or indices of the degree of relation from which the experienced statistician can visualize the diagram, more or less roughly.</u> If one has the option of seeing the scatter diagram and the correlation coefficient, one would of course choose the diagram because the coefficient can be found from the diagram but the diagram cannot be at all accurately constructed from the coefficient. The diagram gives more information than the coefficient, but when many relations are to be compared, <u>the coefficient serves as an objective and impartial measure of the degree of relation</u> (pp. 205–206; underlining added).

In other words, he recommends using the correlation coefficient as a screening tool. The problem with such a recommendation is that it assumes that the magnitude of the correlation coefficient and that of the regression coefficient are correlated (i.e., high correlation indicates a relatively steep regression line), but unfortunately for Thurstone, that is not always true (see Figures 7.2 and 7.12; for a more detailed discussion, see Zorich, 2017).

In 1925, **Edmund E. Day** (previously mentioned) published a 459-page text-book titled *Statistical Analysis*. Given his 1918 battle with King over the interpretation of correlation (previously discussed), it is not surprising that his chapter on correlation was titled "The Meaning of Correlation".

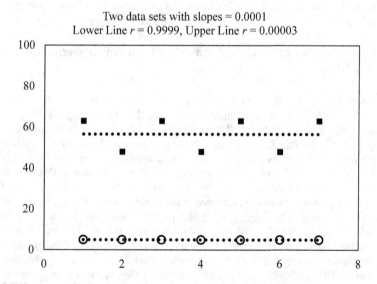

Two data sets with slopes = 0.0001
Lower Line r = 0.9999, Upper Line r = 0.00003

FIGURE 7.12
This chart shows that a very low regression coefficient (nearly zero) can be associated with either a high or low correlation coefficient (0.9999 or 0.00003). (Adapted from Zorich (2017).)

Day was careful to provide eight pages of introduction before he defined correlation:

> *By correlation is meant, in brief, a definite tendency for two or more variables to vary together. The variables may move in the same or opposite directions, but if they are correlated they are never indifferent to one another,—they are either mutually attractive or mutually repellent. Correlation involves a one-to-one correspondence between paired variables.*
>
> (Day, 1925, pp. 188–189)

Apparently, he did not consider that to be clearly stated, because he continued in the very next sentence to appeal to a higher authority for the definition of correlation:

> *Bowley* [author of Elements of Statistics, previously discussed] *gives a clear statement of the condition: "When two quantities…."* (p. 189)

Day did not indicate from which edition of Bowley's he took that quote, only that it was from page 316; however, that exact quote is found on page 316 in each of Bowley's first three editions, dated 1901, 1902, and 1907, respectively. It is unknown why Day did not instead quote from Bowley's 1920 fourth edition; possibly Day preferred the original rather than the new "completely rewritten" text (Bowley, 1920, p. v).

With King likely in mind, Day continued:

> *It must not be supposed that the existence of correlation between any two variables proves any simple and direct causal connection between the two* (p. 189).

Day then described correlation "phases", a term which seems to have been unique to him:

> *Measurement of correlation is one thing; interpretation of correlation, quite another. In fact, the* explanation of observed correlations is one of the most difficult task in the whole field of statistical analysis. *The concept of correlation just given really has* two distinct phases. *In the* first *place* [sic], *it carries the idea of varying degrees of conformity to some clearly defined (functional) relationship between the paired variables; and* secondly, *it involves the idea of this definite functional relationship to which the observations of the paired variables tend in some measure to conform. Corresponding to these two phases of the concept, there are* two distinct lines of analysis. *The* first *is designed to determine the extent to which the observed values of the variables actually conform to some functional relationship between the two—this is the analysis of the* extent *of correlation. The* second *is directed towards the precise description of the functional relationship which best fits, or conforms to, the observed values of the two variables—this is the analysis of the* form *of the correlation. These two problems, as we shall see, are very closely related. In a measure* [,] *they are solved by the*

same line of analysis. This analysis will be considered in the following chapter [which is titled "The Measurement of Correlation"] (pp. 189–190; underlining added).

The first subchapter in "The Measurement of Correlation" was titled "The Pearsonian Coefficient", in which he demonstrated in detail the "most convenient" methods for calculating the linear correlation coefficient. His second subchapter was titled "The Line of Regression", the first paragraph in which is given here next (while reading it, please refer back to the underlined words in the just-given quote from his pages 189–190):

> Indices of correlation [such as the linear correlation coefficient] *show the* <u>*extent*</u> *to which to which the data appear to indicate a definite functional relationship between the paired variables. If there is evidence that such a relationship exists, the determination of the most probable relationship constitutes a* <u>*second phase*</u> *of the correlation analysis.... When the relationship between two correlated variables appears to take the* <u>*form*</u> *of a straight line,* <u>*the line*</u> *best expressing the relationship between the variables* <u>*may be readily obtained from the Pearsonian coefficient*</u> (pp. 199–200; underlining added).

The formula he gave for the slope of that line was the equivalent of this:

$$\text{Coefficient of regression} = (\text{Coefficient of correlation})(Sy/Sx).$$

Thus, as with Thurstone (previously discussed), his words and his formula *together* risk leading the novice student to internalize a mistaken idea, namely that a large correlation leads to a large regression, and similarly that small leads to small; the average student now mistakenly thinks that the absolute values of the correlation coefficient and the regression coefficient are themselves correlated.

At the end of his chapter, Day provided important information for the novice student; but he hid it where most students would not notice, namely in the second to the last paragraph in the chapter. There, he admitted that...

> It is not feasible in an elementary text to deal comprehensively with the difficult problem of <u>interpreting statistical coefficients</u> in the light of the general theory of probability. About all that can be done is to state certain general conclusions.... <u>In most cases the</u> [correlation] <u>coefficient assumes intermediate values, and conclusions as to the extent of correlation have to be stated with caution</u> (p. 209; underlining added).

That last sentence was footnoted to this text:

> It should be noted that values of indices of correlation are not to be thought of as percentages. A coefficient of .74 cannot be regarded as indicative of twice as much correlation as shown by a coefficient of .37... (p. 209n3).

That was a valuable warning. However, despite the fact that his chapter was titled The <u>Measurement</u> of Correlation, Day does not explain to the reader how the reader <u>should</u> think about or compare different values of correlation coefficients. Instead, he provided another warning, in the chapter's final paragraph:

> *There are few more difficult tasks for the statistician than the evaluation of correlation coefficients* (p. 210).

G. Irving Gavett's 1925 book titled *A First Course in Statistical Method* is important for its being possibly the first *elementary* statistics textbook that was written by a professor of mathematics rather than by a professor in another field such as business or economics. In his chapter on correlation, the definitions, descriptions, vocabulary, and cautions/warnings tended to be more abstract than those found in prior textbooks (the following are examples from his book):

- *Correlation: Two variables are said to be correlated if, when any value of the first variable be selected, it is found that the average of the associated values of the other variable seems to depend on the size of the selected value of the first variable. If, when the selected value of the first variable is small, the average of the associated values of the other variable is small also, increasing with increase in size of the selected value of the first variable, the correlation is direct or positive. If, on the other hand, the average of the associated values of the second variable is large, decreasing with increase in size of the selected value of the first variable, correlation is inverse or negative* (Gavett, 1925, p. 212).
- *Measure: The numerical measure of the tendency of the size of one variable apparently to affect the average size of the associated values of the other variable should be of the nature of a coefficient or ratio. For a perfect direct correlation, its value would be +1, and for perfect inverse correlation, −1. For no correlation, it would be zero. One of the most used coefficients of correlation is one devised by Karl Pearson and is known as Pearson's coefficient of correlation* (p. 214).

In a subchapter titled "The Quantity r as a Measure of Correlation" that was less than half a page long, he did not explain how r is a "measure" of correlation, but instead simply provided formulas and then repeated himself by claiming that r is "an excellent measure of correlation" (p. 228).

He ended his chapter on correlation with this advice:

> **Warning**: <u>Too much stress cannot be put on using the coefficient of correlation,</u> as well as other statistical constants, <u>with care</u>. Do not draw conclusions that are unwarranted…. <u>A high correlation does not prove anything</u>. <u>It suggests</u> the probability of a cause-and-effect relationship between the two variables. The statistician has no further concern. The investigator, dealing with these variables, takes the suggestion and searches for a reason. Frequently, the statistician and the investigator are the same individual. He must <u>be careful not to let his statistics run away with his reason</u> (pp. 249–250; underlining added).

"Don't let your statistics run away with your reason!" Maybe that is a motto we all should adopt.

It seems that Gavett never published a second edition; possibly the reason was that R. A. Fisher published his soon-to-be very famous textbook that same year (discussed here next).

Ronald Aylmer Fisher (1890–1962) graduated from Cambridge University in 1912 with a degree in astronomy. He then taught high-school mathematics and physics until 1919 when he was hired as a statistician for the Rothamsted Experimental Station to help with plant-breeding experiments there. While at Rothamsted, he invented the method of DOE (design of experiments) and ANOVA (analysis of variance). He was knighted in 1952 (Britannica. com-Fisher).

In 1925, he published a book titled *Statistical Methods for Research Workers*. Many years later, one historian wrote: "It is now twenty-five years since R. A. Fisher's *Statistical Methods for Research Workers* was first published. These twenty-five years have seen a complete revolution in the statistical methods employed in scientific research, a revolution which can be directly attributed to the ideas contained in this book" (Yates, 1951, p. 19). Another historian gave similarly high praise: "His ideas form much of the basis of the courses in statistics taught in many British and American universities. His students occupy many of the leading positions in statistical and genetic research. His books—notably the many editions of *Statistical Methods for Research Workers*...—have become classics" (MacKenzie, 1981, p. 183).

In the 1925 first edition of that book, Fisher does not mention r^2 as having meaning of its own, except in one sentence (shown here next); in it, ρ refers to the population correlation coefficient, and ρ^2 therefore can be interpreted as the population coefficient of determination (although he does not use that term):

> ...*of the total variance of y [,] the fraction $(1 - \rho^2)$ is independent of x, while the remaining fraction, ρ^2, is determined by, or calculable from, the value of x.*
>
> *(Fisher, 1925, p. 145)*

In the 1958 thirteenth and final edition of the same-titled book, he included that same sentence (Fisher, 1958, p. 182).

That sentence described ρ^2 as the "fraction" of total variance that is dependent on x. No similarly mathematically precise description was provided for the correlation coefficient. Yet that sentence was not intended to urge the reader to use r^2 (i.e., the sample statistic corresponding to ρ^2) rather than r as a measure of correlation strength, as evidenced by the fact that in neither the first nor final edition is there any subsequent discussion or even mention of the meaning or interpretation of r^2 or ρ^2. Instead, Fisher focused on r, the correlation coefficient. For example, this next sentence likely led many of his readers to erroneously conclude that magnitudes of correlation coefficients can be compared to each other on a linear scale ("on a conventional scale", as he described it):

> ...*the only satisfactory estimate of the correlation, when the variates are nor-mally correlated, is found from the "product moment."...r is an estimate of the correlation ρ. Such an estimate is called the correlation coefficient, or the product moment correlation.... The correlation between A and B measures, on a conventional scale, the importance of the factors which (on a balance of like and unlike action) act alike in both A and B, as against the remaining factors which affect A and B independently.*

> (Fisher, 1925, pp. 146, 153–154; the wording is
> virtually the same in Fisher, 1958, pp. 183ff.)

I personally consider that description very confusing, compared to descriptions by most other contemporary authors.

Under the leadership of Fisher, the practice of reporting correlation as r rather than r^2 became established mid-20th-century practice, which was a continuation of the same practice that had been promoted in prior decades by Galton and his immediate successors.

To be called a correlation, Galton required that the variability in both data sets be linked by common causes (previously discussed). Fisher agreed:

> *The idea of regression is usually introduced in connection with the theory of correlation, but it is in reality a more general, and, in some respects, a simpler idea, and the regression coefficients are of interest and scientific importance in many classes of data where the correlation coefficient, if used at all, is an artificial concept of no real utility....it is seldom, with controlled experimental conditions, that it is desired to express our conclusion in the form of a correlation coefficient.*

> (Fisher, 1925, pp. 114, 138; underlining added)

That first edition text is essentially identical to what is found in the 13th edition (1958, pp. 129, 175).

In 1925, a mathematics professor at Harvard University (USA) bemoaned the fact that "The great trouble at present with the theory of correlation seems to be that there is no general agreement as to how large r^2 must be in order that we may safely conclude that there is a real connexion [sic] between the two sets of phenomena" (Coolidge, 1925, p. 149). Fisher helped to alleviate that great trouble by not only developing a t-test method for assessing the statistical significance of a correlation coefficient but also then promoting general agreement by including that method in his famous textbook (1925, pp. 157ff). Unfortunately, the t-table that he provided (his Table IV) did not include any t-values between "n"=30 and infinity; I say unfortunate because there is more than a 4% difference between the t-value for $n=30$ and $n=$infinity at alpha=5% (in Zorich, 2021c, more can be found about the history of textbook t-tables and about how $n>30$ came to be called a "large" sample).

The next section in his textbook described a way to transform the distribution of correlation coefficients to normality, by converting each r value into a normal-distribution z-table value. He provided an example distribution of correlation coefficients derived from samples of $n=8$, taken from populations that had true correlation coefficients of either 0.0 or 0.8. That example showed that $r=0$ can sometimes be "observed" (i.e., calculated from a sample) even when the true population value of r is relatively large. He then warned that…

> …*with higher correlations [,] small changes in r correspond to relatively large changes in z. In fact, measured on the z-scale, a correlation of 0.99 differs from a correlation 0.95 by more than a correlation 0.6 exceeds zero.*
>
> *(Fisher, 1925, p. 165; for clarity, a zero has been added here in front of each numerical correlation value—those zeros were not present in the original)*

A reader of that quote could easily conclude that r is not a very good *measure* of correlation, but Fisher did not say anything like that in his book.

In 1925, the end-of-year issue of the *Journal of the American Statistical Association* included a paper titled "An Automatic Correlation Calculating Machine". Its author was **Clark Leonard Hull** (1884–1952); he was an instructor in psychology at the University of Wisconsin who was well-known for his research on learning and for his work on the application of mathematics to psychological theory (Britannica.com-Hull). He described the machine as having been "built into a steel table and occupies a space 26 inches by 32 inches square"; a photo of the machine (shown here as Figure 7.13) was included in Hull's paper (Hull, 1925, pp. 527–528).

Hull's paper's entire first paragraph was:

> *The machine described in the following pages was designed to eliminate the drudgery and the persistent arithmetical errors from the calculation of the standard deviation and the Pearson product-moment coefficient of correlation. This has been accomplished by rendering the lengthy preliminary manipulation of the data quite automatic and mechanical, thus eliminating at once both the costliness and the fallibility of the human element (p. 522).*

That last sentence was revisited at the end of the paper:

> …*the machine eliminates almost absolutely the element of human error which has so long vexed the computer [i.e. a person who computes] of the product-moment coefficient of correlation (pp. 530–531).*

Prior to using the correlation machine…

> *The columns of original data are simply transferred to perforated paper data test strips by means of a special auxiliary machine. This operation may be compared roughly to running the data off on an ordinary listing adding machine (p. 524).*

PLATE II

Upper view of correlation calculating machine. K and L are rolls of perforated tape being fed into the multiplier and the multiplicand respectively. The answers are read off from the dial D.

FIGURE 7.13
This photo shows Hull's table-top "correlation calculating machine". (From Hull (1925, p. 528).)

A separate roll of paper strip had to be created for each of two paired columns of data (which were called columns A and B in this explanation), then...

> *All that is necessary for the operator to do in this typical operation is to place the strip containing the numbers of column A in one position, the strip containing the numbers of column B in the other position and then press the starter (S). The machine then, without any further attention from the operator, automatically multiplies each pair of numbers...adding up the products as it goes along. When the machine reaches the bottom of the columns, whatever their length, it stops automatically and the $\sum(A \times B)$ may be read from the dial (D) at the convenience of the operator (p. 527).*

In other words, the machine did *not* calculate correlation coefficients, despite the paper's title. On the third page of the paper, we discover that "The machine [is] essentially an automatic products-sum calculator". It calculated intermediate values that could then be used in short-cut formulas to calculate the correlation coefficient, formulas such as the one given on the second page of Hull's paper.

Although the last step in calculation of the correlation coefficient had to be performed by a human, the fact that the lengthy preliminary steps could now be semi-automated was a boon to scientists world-wide. In 1925, the paper's readers may have been anxious to build their own "correlation calculating machine"; Hull stated that his most recent version "can be duplicated for about $1200" (p. 531). It is curious that Hull does not provide or even reference instructions on how to construct one. Given that $1200 in 1925 is equivalent to about $21,053 in 2023 dollars (*https://www.officialdata.org/us/inflation*), and given that "a considerable amount of data has already been sent in" by many researchers (p. 531), Hull may have been hoping to become a high-tech entrepreneur, as evidenced by the last sentence in his paper:

> *If the demand for such a service* [i.e., calculation of correlation coefficients for huge data-sets]...*continues to increase, there will be established at Madison* [the city where he lived] *in the near future a national correlation bureau operating on a regular <u>commercial</u> basis* (p. 531; underlining added).

1926

In 1926, the first issue of a new journal titled *Industrial Psychology* was published. One of the premier articles was titled "A Correlation Machine". That paper was authored by **Stuart Carter Dodd** (1900–1975), and the machine had been developed while he was a student in psychology at Princeton University in New Jersey, USA (Wikipedia: Dodd).

As just discussed, C. L. Hull had previously published a paper on a similarly purposed "correlation calculation machine" at the University of Wisconsin, USA. Hull's paper was published at the end of 1925; Dodd's paper was published at the start of 1926.

Hull's paper did not mention Dodd's machine, but Dodd's paper discussed Hull's in detail, in its final half-page, as if that discussion were a last-minute addendum to an already completed paper. It is unknown when these two university professors of psychology first became aware of each other's machines; possibly they met in December 1924, when Hull "demonstrated" a "sufficiently perfected" model of his machine at the "Washington meeting of the American Psychological Association" (Hull, 1925, p. 522n1).

It is interesting that although Hull included only a page of statistical theory in his 10-page paper, Dodd included more than five pages of statistical theory in his 13-page paper. Dodd's theoretical discussion contained what looked to be a heretofore unpublished application of correlation:

> <u>The aim of research by correlation methods is to increase the accuracy of prediction</u> (or reduce the standard error of estimate) <u>by developing better measures and more accurate data which will yield higher correlations</u>. An example of the use of

correlation in industry may be taken from the employment office of a large corporation which is looking for the best applicants for clerks, foremen, or typists. The corporation would collect data on their employees of that type covering such variables as length of experience, age, amount of schooling, average previous earnings over some period, and the score on some aptitude test or questionnaire. They would then form a criterion on which to correlate these variables. The criterion would be some index of success in the job such as average annual wage. Those variables which correlated most highly with this criterion of ability on the job should be most carefully studied in scrutinizing an [future] applicant's qualification, for from them the most certain prediction of success may be obtained.

(Dodd 1926, pp. 46–47; underlining added)

Another of his uncommon but practical views was this:

...coefficients of correlation need to be corrected for the range on which they are calculated before they can be compared. The size of the coefficients will vary greatly with the range of the two variables from which they are derived. The range is conventionally expressed by the standard deviation or sigma, "σ", of each variable.... The coefficients of correlation derived from.... a longer range... would be much higher than coefficients from a shorter range (p. 49; underlining added).

Although surprisingly not stated in his paper, that final sentence can be derived from this well-known formula (previously discussed), but only in the very unlikely case where σ_y and *Slope* are identical in the two studies whose *r* values are being compared:

$$\text{Correlation coefficient} = \left(\text{Linear regression slope}\right)\left(\sigma_x/\sigma_y\right)$$

In contrast to Hull's paper, Dodd described his machine in such detail that the reader might be tempted to construct one, if not discouraged by Dodd's final comments:

We are at present going through the stages of experimenting to find the most mechanically durable, compact, and simple design.... The difficulties of getting a competent designer, financing, and patenting will delay the availability of the machine for some time (p. 57).

Hull and Dodd were not the only ones researching how to machine-calculate correlation coefficients; for example, in 1925, H. A. Wallace, C. F. Searle, and G. W. Snedecor at the then-named Iowa State College (USA) were making similar attempts using IBM punch-card equipment that they'd borrowed from a local insurance company (Johnson and Kotz, 1997, p. 339).

Why do people like to collect things such as stamps, coins, or books? Apparently **Percival M. Symonds** (1883–1960), a professor of education at

Columbia University (Intelltheory.com-Symonds), liked to collect formulas for the correlation coefficient. By 1926, he'd collected "52 variations of the product-moment (Pearson) coefficient of correlation"; he then decided to share them with the world. He published in the *Journal of Educational Psychology*, which was an unfortunate choice, given that most mathematicians, statisticians, and biologists would not be routine readers of that publication.

In his paper, the recommendations that he gave for which formula to use were based in part on whether sample size was large or small and on whether or not a "calculating machine" was available. He referenced commercially available forms that helped to organize and facilitate calculation, although some of them cost what he called a "high price". Per 100 forms, the prices he listed range from $2 (from the Stanford University Bookstore) to $6 (from a publisher in Chicago)—in 2023 dollars, those prices are $35 and $104, respectively (*https://www.officialdata.org/us/inflation*). He also referenced papers by Hull and Dodd on correlation calculating machines (previously discussed). It is unfortunate that one of the most widely known and practical formulas, first developed in the early 1890s, was given incorrectly in Symond's paper: The correct formula is $r = \sqrt{(b_x b_y)}$, but his published version was missing the square-root sign (note: this formula was discussed here previously, in chapter 1).

1927

By the mid-1920s, the number of statistics textbooks written by non-mathematicians (biologists, economists, etc.) was noticeably in decline, and the number authored by mathematicians was increasing dramatically. In 1927, one such mathematician was **Robert Wilbur Burgess**, who was then a senior statistician at Western Electric but who had previously been an assistant professor of mathematics at Brown University (Burgess, 1927, title page).

His textbook's editor was **John Wesley Young**, a professor of mathematics at Dartmouth College; Young wrote a half-page "Editor's Introduction", in which he spoke unkindly about those non-mathematician authors:

> *The more serious books on statistics which have appeared of late in this country* [USA] *fall essentially into two classes: those written primarily by mathematicians, which are largely theoretical and make considerable demands on the mathematical preparation of their readers; and those written by specialists in other fields intended primarily for use by other specialists in these fields. The latter are usually limited as to their methods and are often* <u>*unsound as to their mathematics*</u>*.*

(Burgess, 1927, page not numbered; underlining added)

In a chapter on correlation, Burgess first provided explanations and formulas for linear regression, and then explained…

> *These two equations, known as the regression equation of y on x and of x on y, respectively, summarize the relationship between x and y from two different points of view…. These equations, however, do not in this form summarize the degree of the relationship between x and y in a single figure…. <u>A single figure summarizing the degree of relationship</u>, however, has been derived….[using] standardized deviations; that is to say, to divide the actual deviations by the standard deviations…. This form of coefficient is therefore more general and can be understood apart from the context better than the unmodified regression coefficients. It is therefore used as <u>a general measure of degree of relationship</u> and is <u>known as the coefficient of correlation</u>.*
>
> (Burgess, 1927, pp. 204–205; underlining added)

Amazingly, his subsequent statements contradict those ones just given:

> *Because of the vagueness of the interpretation of r…[it] should not be relied on to give the complete and only analysis and summary of relationships in a given problem. Other schemes…are valuable supplements, and the use of the two regression equations, with attention to their degree of linearity, is to be recommended (p. 209).*

He summarized his chapter on correlation:

> *An important statistical problem is to determine and express <u>the extent to which the possession of more or less of one characteristic (x) by any individual in a group implies the possession of more or less of another characteristic (y)</u>. More briefly, the problem is to determine the relationship between two variables x and y in a group of individuals. The relationship may be shown [several ways, including…] By finding the average value of the second characteristic for each of a series of selected values (or ranges of values) of the first characteristic…. When the method of averages, [just given] above, is used, it is often advisable to fit an equation to these averages…. Analysis of the formula for this equation shows that the number which expresses <u>the essence of the relationship</u>…. is called the coefficient of correlation (pp. 227–228; underlining added).*

As shown by the just-given quotations, even a mathematician has difficulty explaining what correlation is and what it measures. However, his chapter did not mention the coefficient of determination nor even discuss it conceptually.

In 1927, mathematician **Henry Lewis Rietz** (previously mentioned) published a book titled *Mathematical Statistics*; this was a full 20 years after he wrote the "Correlation Theory" appendix for Davenport's 1907 book on plant and animal breeding (previously discussed).

In Rietz's book, his treatment of correlation is interesting because, in regard to the word correlation and its coefficient, he stated explicitly what others had been saying implicitly:

> [There are] *two fundamental ways of approach to the characterization of a distribution of correlated variables, although the two methods have much in common. The one may be called the "regression method," and the other the "correlation surface method."*

> *(Rietz, 1927, pp. 78–79)*

He presented "regression" as subservient to correlation, which was reflected in his choice of formula for *defining* the regression coefficient (p. 86); his formula was the equivalent of this:

$$\left(\text{Regression coefficient of } y \text{ on } x \right) = \left(\text{Correlation coefficient} \right)\left(\sigma_y / \sigma_x \right)$$

In other words, to calculate the regression coefficient, one needs to have first calculated the correlation coefficient.

The last words in his correlation chapter accurately predicted the future:

> *Although the many omissions make it fairly obvious that our discussion is not at all complete, it is hoped that enough has been said about* the theory of correlation *to indicate that this theory may be properly considered as constituting* an extensive branch in the methodology of science *that should be further improved and extended* (p. 113; underlining added).

1929

In 1929, **Helen Mary Walker** (1891–1983) obtained a Ph.D. from Teacher's College at Columbia University, where she had taught statistics since 1925. In later decades, she became the first female president of the American Statistical Association and then president of the American Educational Research Association (Amstat.org-Walker). Her 1929 book titled *Studies in the History of Statistical Method with Special Reference to Certain Educational Problems* was originally written "in partial fulfillment of the requirements for the degree of Doctor of Philosophy" (Walker, 1929, title-page); the book was so popular that a second printing was published two years later, but curiously without the "in partial..." tag-line. Her questionable claims in that book (regarding the writings of Bowditch and Galton) have been discussed here previously.

In that book's chapter titled "Correlation", she stated that Galton's December 1888 paper was the first use of the term "correlation" in a "technical sense" (p. 106); no explanation was given as how to distinguish technical and non-technical usage. That statement was followed immediately by what she claimed was Galton's paper's introductory sentence that focused on "Co-relation or correlation of stature" and "how to measure its degree" (p. 106); however, Galton had there used the word "structure" not "stature" (in her defense, "stature" was the variable most talked about in Galton's paper).

Chapter Summary

The period from 1900 to 1930 saw publication of the first generation of statistical textbooks; in them and in journal articles, correlation was explained in various creative ways that were sometimes inaccurate and/or in conflict with each other. After 1930, it seems that there was much less creativity and conflict, or possibly all the low-hanging fruit had already been picked and all the combatants had tired of fighting. However, I chose 1930 as the boundary between this chapter and the next for two reasons: First, it was the year of publication of the first book entirely devoted to the subject of correlation; second, starting in the 1930s, college departments, and instructors were beginning to have the word "Statistics" and even "Bio-Statistics" in their titles, as if the "science" of mathematical statistics had matured. S. Stigler, in his 1999 book titled *Statistics on the Table*, wrote that "I propose to advance and defend the claim that mathematical statistics began in 1933", which he considered a "point estimate" that deserved to be qualified by means of a "confidence interval"; he then explained that "I refer to the birth of mathematical statistics as a *discipline*" (Stigler 1999, pp. 157–158; italics in the original). It is interesting that other historians have used that same 1930 cut-off date: D. MacKenzie's 1981 book titled *Statistics in Britain: 1865–1930—The Social Construction of Scientific Knowledge*, and A. Hald's 1998 book titled *A History of Mathematical Statistics from 1750 to 1930*.

8

1930 to 2000

Introduction

This chapter covers more than twice as many decades as the prior one but contains less than half the pages. The 30 years of research that resulted in this book took me to many used-bookstores in the United States, United Kingdom, and Canada; in them, I found many more statistics books from the mid-to-late 20th century than I found from earlier in the century, and yet I found relatively few of those more recent texts that were interesting, humorous, or historically significant. This chapter discusses those few.

Almost all the authors discussed in this chapter can be said to have worked primarily in just three fields of study:

- *Economics*: Mordecai Ezekiel, James Smith, W. F. C. Nelson, F. C. Mills, Albert Waugh, Morris Blair.
- *Mathematics*: Burton Camp, Harold Davis, George Snedecor, Victor Goedicke, Henry Alder, Carol Ash.
- *Statistics*: Alexander Tschuprow, William Cochran, Alan Treloar, Leonard Tippett, Maurice Kendall, Albert Bowker, Marty Schmidt, Bart Holland.

When starting the research for this post-1930 era, I had assumed that I would discover a trend of ever-improving presentations of correlation history, theory, and practice. Although I did find many improvements, I also found so many examples of erroneous presentations that I began to wonder if the quality of the average presentation may have worsened rather than improved. As we shall see in this chapter, the types of errors are numerous; the following are examples:

- Karl Pearson is said to have invented the coefficient of correlation.
- Francis Galton's coefficient of correlation is equated with his coefficient of reversion.
- Charles Darwin is said to have published in 1868 a "landmark" step in the history of mathematical correlation.
- r is denominated as a percentage and also as a non-percentage number between 0 and 100.

DOI: 10.1201/9781003527893-8

- r is said to equal the ratio of the number of causes that X and Y have in common to the total number of causes.
- r is claimed to be a reliable screening tool when searching for datasets in which X has a large effect on Y.
- The angle between the regression lines of Y on X and X on Y is presented as a measure of the strength of correlation.
- r^2, the coefficient of determination, is ignored or depreciated.
- Highly correlated curvilinear plots are labeled as showing "no correlation."

The details of how the improvements and the mistakes were worded in publications are the subject of this chapter.

1930

In 1930, the first ever entire book on correlation was published; the title was *Methods of Correlation Analysis*; and the author was **Mordecai Joseph Brill Ezekiel** (1899–1974), who had in 1924 co-invented an improved version of the correlation ratio (as previously discussed). In 1930, he was not only the economic adviser to the US Secretary of Agriculture but also the vice-president of the American Statistical Association (Ezekiel, 1930, title page). His book's first edition's Preface explained its purpose:

> ...the aim throughout has been to show how the various methods may be employed in *practical* research.... It is hoped that this presentation will assist research workers in many fields to appreciate both the possibilities and the limitations of correlation analysis....
>
> (Ezekiel, 1930, p. v; underlining added)

The practical nature of the book was reflected in the fact that eight of its 23 chapter titles began with the word "Determining," six began with "Measuring," and one began with the word "Practical."

Unfortunately, his explanation of the meaning of the correlation coefficient was misleading:

> ...the coefficient of correlation...is simply a measure of how large the variation in the estimated values is, in *proportion* to the variation in the original values. The coefficient of correlation thus measures the *proportion* of the variation in one variable which is associated with another variable, and therefore is a *measure* of the relative importance of the concomitance of variation in the two factors (p. 119; underlining added).

I say "unfortunately" because he'd fallen into the trap that had ensnared many authors before him, and would ensnare many after him, namely the trap of re-phrasing Galton's description of the correlation coefficient. In 1888, 1889, and 1890, Galton had repeatedly stated that *r* was a measure of correlation. Any scientific measurement depends for its accuracy on a defined scale; the scale may be linear (e.g., as used by a balance to measure mass) or logarithmic (as used by the Richter Scale to measure the intensity of earthquakes) or some other mathematically valid scale. What mathematically valid scale does *r* use? None, as many authors have stated implicitly or explicitly (see list in previous discussion of S. G. Wright, 1921), including Ezekiel himself later in that same 1930 book:

> Although the coefficient of correlation was the earliest measure used, it can be seen that it may be misinterpreted.... If instead the coefficient of determination is used.... Since this [the coefficient of determination] *is the most direct and unequivocal way of stating the proportion of the variance in the dependent factor which is associated with the independent factor, it may be used in preference to the other methods....* Particularly in linear correlation, there are three constants which summarize nearly all that a correlation analysis reveals. First, the standard error of estimate.... Second, the coefficient of determination.... Finally, the coefficient of regression....
>
> (1930, pp. 120, 136; *underlining added;* notice that the correlation coefficient was not among those "three constants")

His title for the chapter in which he discussed those three constants was "Three Measures of Correlation—The Meaning and Use for Each," but the correlation coefficient was not one of them (p. 136; underlining added).

In Ezekiel's second edition (1941, pp. 137–138), he repeated his first edition's unfortunately worded explanation of the correlation coefficient's meaning. However, in his third edition, he eliminated that explanation and did not replace it with another (Ezekiel and Fox, 1959, pp. 127–128).

In all three editions, he summarized his chapter titled "Measuring Accuracy of Estimate and Degree of Correlation" without even mentioning the correlation coefficient; the text of that summary is virtually identical in all three editions (on pages 124, 144, and 133, respectively):

> *Summary: This chapter has pointed out that the closeness of relation between two variables may be measured either by the absolute closeness with which values of one may be estimated from known values of the other or on the basis of the proportion of the* variance *in one which can be explained by, or estimated from, the accompanying values of the other. The accuracy of estimate is measured by the* standard error of estimate, *which indicates the reliability of values of the dependent variable estimated from observed values of the independent variable.* The relative closeness of the relation is best measured by the coefficient of determination, in the case of linear relationship, or by the index of determination, in the case of curvilinear relationship. *These measures show the proportion of the*

> variance in the dependent variable which is associated with differences in the other variable. *In the case of variables causally related, they measure the proportion of the variance in one which can be said to be "caused by" variations in the other.*
>
> (Ezekiel and Fox, 1959, p. 133; underlining added)

An interesting fact is that the title and subject matter of the 1959 third edition expanded to include the topic of regression: *Methods of Correlation and Regression Analysis*. One section of that revised edition was curiously titled "Uses and <u>Philosophy</u> of Correlation and Regression Analysis" (p. 434; underlining added). That new edition also provided a history of correlation:

> **Brief History.** <u>The methods of correlation</u> and regression analysis were <u>first developed by students of heredity, notably Karl Pearson</u>. The professional journal in this field, <u>Biometrika, contains the original papers</u> establishing the method, and many studies using it in the field of heredity. These include such studies as the relation of the stature of children to that of their parents. The very term "regression" itself comes from this initial use. When it was found that very tall or very short parents tended to have children who were on the average less tall or short, this was described as a tendency to "regress toward the mean," and the line describing this was called "the regression line." (pp. 434–435; underlining added)

Sadly, Ezekiel's "Brief History" did not mention Francis Galton, whose name does not appear anywhere in any edition of the book, not even in a footnote, nor in any end-of-chapter publication reference, nor in the end-of-book "Author Index."

1931

Burton Howard Camp (1880–1980) taught mathematics at MIT and Harvard in his early 20s prior to being named Associate Professor of Mathematics at Wesleyan University (USA), where he remained for the next four decades (Wikipedia: Camp). In 1931, he published his *The Mathematical Part of Elementary Statistics: A Textbook for College Students*. He introduced his work as "an elementary textbook" (Camp, 1931, p. iii), a claim that is not supported by the complicated wording of his first attempt to define the correlation coefficient:

> The coefficient of correlation is denoted by r, and is defined as the first product moment about the general mean point in terms of the σ's as units: [he then gave a formula for r].
>
> (Camp, 1931, p. 137)

His second attempt at a definition (given 25 pages later!) is more supportive of his "elementary" claim:

> *Let the standard deviation be chosen as units. Then the coefficient of correlation measures the degree to which it is true that a change in one variable determines an equal change in the other. This is probably the best simple description of the character of the coefficient of correlation which can be given in words, without the aid of mathematical symbols* (p. 162).

Given that he considered it the "best," it is surprising that he immediately issued what he called a "re-statement" of it:

> *The coefficient of correlation measures the degree to which it is true that a relative change in one variable determines an equal relative change in the other. By a relative change is meant the ratio of the absolute change to the standard deviation* (p. 162).

One of his chapters is curiously titled "Regression, Interpretation of r"; it begins with the following paragraph:

> *Consider the general case of correlation, where N is large, and the data might be represented by dots spread over the [X,Y plot] paper. Suppose we wish to draw and to find the equation of that straight line which, on the whole, will come nearest to all these dots. We shall suppose the best-fitting line is that one which fits best in the sense of least squares, but even with this understanding there are at least three different possible points of view. Let δ be the distance between a dot and the line. We wish to make $\Sigma\delta2$ a minimum. The three cases that arise depend on whether:*
>
> > *Case (a) δ is measured parallel to the y-axis*
> > *Case (b) δ is measured parallel to the x-axis, or*
> > *Case (c) δ is measured perpendicular to the line.*
>
> *In Case (a), the line is called the "regression of Y on X"; in Case (b) it is called the "regression of X on Y"; in Case (c) it has no generally accepted name. We shall call it the "geometrically best-fitting line," because in geometry we usually prefer to think of the distance between a point and a line as measured perpendicular to the line* (p. 152).

Several pages later, he claimed:

> $|r|$ *measures the closeness with which the dots* [on an X,Y plot] *cluster about the geometrically best-fitting line; r^2* [measures] *the closeness with which they cluster about the regression lines (distances in the last case being measured parallel to the y- and x-axes, respectively)* (p. 161).

He then provided what he considered to be mathematical proofs for both of those claims. After first reading Camp's elegant claims and proofs, the reader might internalize the idea that two different X, Y data-sets can be compared using r-values, the idea being: Whichever data-set produces the larger r has data points that are more closely clustered about its best-fit line. However, a review of his proofs reveals that such a conclusion is valid only in the unlikely event that the slope of the lines are identical (see discussion of this general topic in Zorich, 2017).

One bewildering aspect of his book is the choice of data for some homework exercises. For example, apparently he had someone measure the length and breadth of 900 library books. Although such a data set is as good as any other to a mathematician, to the average student it might convey the idea that the correlation coefficient is a meaningless toy rather than a valuable tool.

1934

In 1934, **James G. Smith** was an associate professor of economics at Princeton University. He seems to also have been an historian, psychologist, and philosopher at heart, as evidenced by the contents of the statistics textbook that he authored that year. The book's title was *Elementary Statistics*, but its subtitle was *An Introduction to the Principles of Scientific Methods*. What is most interesting about his statistics book is that there are many non-statistical discussions and topics; for example, in his Preface, he wrote...

> It is with the idea of giving to the student a genetic [sic] treatment of the subject of scientific method that this textbook is written (p. v).
>
> The attempt has been made here to write <u>a textbook in statistics</u> as will be found desirable for elementary courses in liberal arts colleges, but it is hoped that the business schools will also find this text <u>useful as collateral material in connection with some standard statistics textbook</u>. In a real sense, <u>this is an "introduction" to scientific method</u>, making no claim of being a complete exposition of scientific method.

> (Smith, 1934, p. vi; underlining added)

Early in the book, he provided comfort to apprehensive students:

> There is a natural psychological reaction unfavorable to the study of statistics because symbols are used for various purposes. The uninitiated are mystified and frightened away from the subject on account of the symbolic presentation, simple as it may be in reality. It is important therefore to become familiar with the secret that the symbols used in statistics are really quite simple—and that there are not very many of them. Furthermore, they are easily learned and remembered, as soon as one has seen the real purpose they serve (p. 10).

His book included a formula-based definition and discussion of the combined concepts of linear and curvilinear correlation (pp. 385–387). He used S_y to represent what today is called the Standard Error of Estimate, which in effect is the standard deviation of the vertical distances that the plotted X, Y points are away from the regression line (linear or curvilinear). He used σ_y to represent the standard deviation of the Y values of those plotted points. He then stated that "…we can generalize to the effect that when S_y is less than σ_y *there is correlation. When* $S_y = \sigma_y$ *there is no correlation.*" From a graphical point of view, what he was saying is this:

- If the regression line is a horizontal straight line (slope=0), then there is *no* correlation (because that is the only possible way that S_y can equal σ_y).
- If the regression line has any other linear slope or any other shape, then there *is* correlation.

I fail to see how that generalized definition has any practical or pedagogical value; however, it does sound impressive, *philosophically*. To be fair, his book does use one sentence to explain how to interpret situations that lie between $S_y = \sigma_y$ and $S_y = 0$; but he fails to explicitly mention something that would be helpful to the typical novice student, namely that the only time that $S_y = 0$ is when *all* the plotted points fall *exactly* on top of the regression line.

It is disappointing that the history of correlation that Smith provided in his book was poorly written. A reader of it who is not aware of Galton's December 1888 paper would likely mistakenly conclude that it was Karl Pearson, not Galton, who had discovered the correlation coefficient and named it "r":

> *Between the years 1877 and 1889, Galton worked out a mathematical method by which he could give an exact measure of the relationship between (for example) heights of children and the average heights of their parents…. This is the famous law of regression to type…. The method Galton used was based upon the median and quartiles and has not been generally followed. In the 1890's another method, based upon the arithmetic mean and the standard deviation, was* <u>devised by Karl</u> <u>Pearson</u> *and his method has been widely adopted and is known as the Pearsonian* <u>coefficient of correlation (symbol r)</u>. *Nevertheless, it was Galton's work which led to Pearson's discovery, and Karl Pearson, a devoted disciple of Galton, says that Galton worked out much of the fundamental theory of correlation and was the first to define the measurement of correlation…* (p. 362, underlining added).

The last five of Smith's 26 chapters had nothing to do with statistical method; instead, they covered what he called "The Evolution of the Scientific Method." The chapter titles were:

- From Ancient Heritage to the Middle Ages.
- From Religious Sanction to the Natural Order.

- From Natural Order to Variability.
- Relativity, Indeterminacy, and the Dynamic View.
- Scientific Method and the Future.

Such a philosophical approach may not have appealed to all students and teachers; but on the inside cover of the hardcopy that I own, a proponent of his approach wrote the following words on August 20, 1942 (in this quote, I assume that Eldon had gifted the book to Nicky):

> *Nicky: Statistics is a peculiar tool for Science. Like a new friend [,] it needs thoughtful probing for real understanding, a realization of weak points as well as good; don't ask too much from it [,] for it will only give back honest answers in proportion to what you give it in study. But when well acquainted, like with an old friend [,] you will know what can be asked of it [;] with assurance[,] it won't fail you. Eldon*

In 1934, **Herbert Arkin** and **Raymond B. Colton**, who were instructors at the College of the City of New York, published an introductory statistics textbook that went through four editions in just 5 years. Its full title (including the "etc." at the end) was *An Outline of Statistical Methods as applied to Economics, Business, Education, Social and Physical Sciences, etc.* I don't have access to their first edition, but the Preface to their "revised and enlarged" second edition (1935) stated that their book helped to fill a "noticeable gap in educational literature." That Preface likened the book to a "manual" that "gives the distilled essence of material which might well require one or more large volumes for a full discussion…. To all statistical workers this little volume [224 pages] will be as indispensable as an adding machine" (Arkin and Colton, 1935; page not paginated). The book's Preface promised that "no formula is included that has not practical applications."

As far as I have been able to determine, this may have been the first *basic* textbook to include a discussion of and formula for the Coefficient of Determination (p. 87). Also interesting is that the authors chose to denominate the correlation coefficient as a percentage, both in text and in an example calculation:

> *The coefficient of correlation will have the same limits as the value outlined above; viz., zero to 100%…. r = …99.95%*

> (Arkin and Colton, 1935, p. 83)

They included that *description* of r as a percentage in their first through fourth editions, but in that fourth edition they removed the % sign from the end of their example calculation, i.e., "$r=…99.25\%$" became "$r=…99.25$" (1939, p. 80). Thirty-one years later, in their fifth edition, they finally converted their descriptive text to "*The coefficient of correlation has the same limits as the value outlined above, zero and 1*" and their example calculation to "$r=…0.9925$" (1970, p. 89). All three versions of the example calculation are shown here in Figure 8.1.

Different numerical values for the same correlation
coefficient in successive editions of the textbook titled
An Outline of Statistical Methods, by Arkin and Colton.

$$r = \sqrt{1 - (2.04)^2/(16.74)^2} = \mathbf{99.25}\% \quad \text{(2nd edition, 1935)}$$

$$r = \sqrt{1 - (2.04)^2/(16.74)^2} = \mathbf{99.25} \quad \text{(4th edition, 1939)}$$

$$r = \sqrt{1 - (2.04)^2/(16.74)^2} = \mathbf{0.9925} \quad \text{(5th edition, 1970)}$$

FIGURE 8.1

Arkin and Colton denominated the correlation coefficient differently in different editions: First in units of percent, then in non-percentage values up to 100, and finally in decimal fractions.

_____

1935

The English language has a finite number of words that might appropriately be included in the title of an introductory textbook on statistics. Even so, it is bewildering that **Harold T. Davis** and **W. F. C. Nelson** decided to title their 1935 book *Elements of Statistics*. I say bewildering because *Elements of Statistics* by A. L. Bowley (previously discussed) had since 1901 been one of the best-known of all statistical textbooks and was in 1935 in its fifth edition. In defense of Davis and Nelson, they did subtitle their book "With Application to Economic Data" whereas Bowley's book had no subtitle.

Davis was a professor of mathematics at Indiana University and Nelson was an economist working for the Cowles Commission for Research in Economics (Davis and Nelson, 1937, title page). I do not have a copy of their first edition, but the Preface to their 1937 second edition states that "no essential changes have been made in the text" compared to the first edition, except "to correct certain errors" and to include an introduction to the Student's *t*-test (p. ix).

Their second edition chapter titled "Elements of Correlation" began in a typical manner but ended strangely:

> We may then <u>define the theory of correlation</u> as the theory of the concomitant variation of two or more attributes of a group of individual entities, the attributes being measured with respect to each entity…. If these points [Yi vs. Xi], when… plotted, appear to lie approximately along some curve, then one may say that the two sets of numbers are correlated. If they group themselves about a straight line, then one is concerned with the case of a linear correlation, otherwise the correlation is said to be non-linear…. In order to have some way of arriving at a numerical measure of linear correlation, the <u>so-called correlation coefficient</u> has been devised.
>
> (Davis and Nelson, 1937, pp. 253–254; underlining added)

It is interesting that they described the correlation coefficient as being "so-called," as if they did not agree with its being called by that name. In these next set of quotations, they revealed their disregard and discomfort with *r* itself and their indecision as to how to explain it:

> It should be emphasized that in practically every case *it is advisable to make a* *scatter diagram before computing a correlation coefficient*. In this way, much tedious computation is avoided while much valuable information, especially with regard to the linearity or nonlinearity of relationship, is often gained (p. 265; underlining added).

They then continued to discuss linear correlation:

> The angles between the regression lines [Y on X, and X on Y] serve as a measure of the relative magnitude of the correlation coefficient (p. 279).

That was a method originally developed by Boas in 1894 (previously discussed). To use the word "measure" with such a method is invalid because there is no appropriate scale of units and no way to tie a given angle to a given *r* value. It would have been better to use the words "crude indicator" instead of "measure."

They continued:

> The correlation coefficient, for all its importance in the theory of statistics, *is* *rather a difficult constant to interpret*.... If in the two sequences of statistical values,
> X data: $X_1, X_2, X_3....... X_n$,
> Y data: $Y_1, Y_2, Y_3....... Y_n$,
> the X and Y sequences are affected by $m + n$ equally probable causes of which *m* are common to both, then the correlation coefficient is equal to
> $r = m/(m + n)$
> or, in other words, *the correlation coefficient is the ratio of the common causes* *to the total number of causes. This very beautiful interpretation of the correla-* *tion coefficient is not easily proved*... (p. 280; underlining added).

That may be "beautiful" but it is also fantastical; not only is $r = m/(m + n)$ "not easily proved" with biometric data, it is impossible to prove, because it is impossible to determine the value of either *m* or *n* in any real-life situation. Conversely, when dealing with controlled studies, e.g., amount of fertilizer applied vs. crop yield, the concept of *m* and *n* is not applicable because there are no "causes" that are "common" to crop yield and fertilizer amount. Thus, although their "ratio of common causes" explanation of correlation seems reasonable at first glance, it is invalid.

It is suspicious that such a ratio is conceptually identical to A. L. Bowley's explanation of the "Nature of r" in the 1920 edition of his textbook (previously discussed). It could be hypothesized that Davis and Nelson not only copied the title of Bowley's book but also used some of its contents without mentioning those facts either in text, reference, or footnote. Or is it possible that mathematics professor Davis and research economist Nelson were simply unaware of Bowley's famous textbook, the first edition of which had been published 34 years previously, and the fifth edition of which had been in print for almost a decade?

Their summary was Galton-like, and not just because of the hyphen between "co" and "variation":

> *Simple correlation is a measurement of the amount of <u>co-variation</u> between two series, and may indicate the degree to which one element affects another, or the degree to which the two are affected by <u>common causes</u>* (p. 293; underlining added).

1937

Austin Bradford Hill (1897–1991) was an English epidemiologist and statistician who pioneered the randomized clinical trial, and who was one of the first scientists to demonstrate the statistical correlation between cigarette smoking and lung cancer (ASU.edu-Hill).

Apparently, he was a jokester: During a lecture, he pointed out an audience member whose wife had just given birth to twins; Hill then humorously claimed that they baptized one twin but kept the other for a control (Armitage, 1995, p. 143). In the Preface to a revised edition of one of his books, he thanked friends who had proof-read the manuscript; he then said "for the faults that remain [in my book] I trust sincerely that the reader may hold them largely responsible" (Hill, 1961, p. vi).

In the 1930's, Hill authored a series of articles on how to apply mathematical statistics to medical research data. They were published in the journal *The Lancet*, whose Editor in 1937 was "happy to accede to the many requests we have received for the reissue of these articles in book form" (Hill, 1939, p. iv). As far as I can determine, at least 11 more editions of that book were published (the 12th and last was published the year that Hill died). The book's title was *Principles of Medical Statistics*, which, half a century after its first edition, was put on an historical pedestal:

> *Before Hill's Lancet articles...the effect of statistics on medical research and practice was minimal.... After 1945 statistics was gradually introduced into the medical curriculum* [at most medical schools]....

> (Armitage, 1995, p. 150)

I don't have access to the 1937 first edition, but the Preface to the 1939 second edition of *Principles* said that the "principal change" vs. the first edition was to add a chapter (unrelated to correlation). Its second-edition's Definitions appendix was divided into five subsections. It is interesting that he included the definitions for Regression Coefficient, Regression Equation, and Scatter Diagram in the subsection titled Correlation, as if those topics were subordinate to correlation. No definition of correlation was given; the closest he came was in his definition of the...

> *Correlation Coefficient.—A measure of the degree of association found between any two characteristics in a series of observations (on the assumption that the relationship between the two characteristics is adequately described by a straight line).*

> *(Hill, 1939, p. 184)*

Hill also described *r* as being "a measure of the amount of relationship" (p. 103). Of the 18 chapters in the second edition, two of them concern correlation. According to one historian, the reason for such an large emphasis on correlation in the late 1930s was that at that time and for many decades thereafter...

> *Epidemiologists studying the aetiology of disease must rely for the most part on the interpretation of associations between disease measures and possible causative factors.*

> *(Armitage, 1995, p. 151)*

Hill demonstrated correlation using an *X, Y* scatter plot of *X* = "Mean Weekly [outside] Temperature" for 26 weeks, vs. *Y* = "Number of Deaths Registered" from either bronchitis or pneumonia in an unspecified population. The weekly number of deaths ranged from 280 at 35°F to 60 at 46°F; the plotted points appeared to be arranged linearly; he determined that a best-fit straight line through them had a slope of –20 deaths per °F and that the correlation coefficient was –0.90 (pp. 98, 102, 104).

He speculated that if that data had been elaborated into age-groups, such as ≤5 years and ≥65 years, then "we might calculate two such [correlation] coefficients...and thus determine in which of [those] two age-groups are deaths from these causes more closely associated with temperature level. We can also pass beyond the coefficient of correlation and find the equation to the straight line...drawn through the [*X, Y* plotted] points" (p. 103). His explanation and demonstration for how to calculate the *regression* coefficient for that line required the user to *first* calculate the *correlation* coefficient.

A problem with Hill's approach is that the casual reader is led to believe that when screening data sets, the most important statistic is not the regression coefficient (i.e., an indication of the size of the effect) but the correlation coefficient (i.e., an indication of the consistency of the effect). The problem for the reader is that in some cases, an X may have very little effect on a Y (i.e., the regression coefficient is relatively small), and yet their correlation coefficient may be large, or vice versa. His introduction to correlation does not make that clear, and in fact it leads the reader to believe the opposite.

It is interesting that the 20-page exposition on correlation in his 1955 sixth edition seems at first glance to be identical to that in his 1939 second edition. However, he had made slight changes, some of which incorporate the wisdom of his accumulated years of experience; for example:

> In practice, we have <u>first, then,</u> to answer this question: could the value of the [correlation] coefficient we have reached have arisen quite easily by chance....

> (Hill, 1955, p. 154; underlining added—those two underlined words were not present in the 1939 second edition's otherwise-identical copy of that sentence on page 107)

1938

We previously discussed **F. C. Mills**'s 1924 paper in which he introduced his own version of a coefficient of non-linear correlation. In that same year, he published a book titled *Statistical Methods Applied to Economics and Business*. I don't have access to that edition, but I do own a copy of his mammoth (736-page) second edition (1938), which he Prefaced by explaining that it had been much revised and expanded:

> In preparing the present edition of Statistical Methods [,] account has been taken of the more important of the recent developments that have a bearing on the economic and business applications of statistics.... In the chapters added to this edition [,] I have sought to exemplify economic applications of the newer methods of analysis.... In these sections I have drawn heavily on the path-breaking work of R. A. Fisher.

> (Mills, 1938, p. viii)

Mills considered r to be a measure of confidence ("The greater the value of r, the greater the confidence that may be placed in the [regression] equation as an expression of a relation which is approximated in a high percentage of cases." p. 336). Such a view was similar in approach to that taken in 1919 by Persons, who considered r to be a measure of probability (previously discussed).

It is interesting that Mills agreed with Ezekiel (previously discussed) that there are three values that help measure correlation:

> *The <u>measurement</u> of relationship in a given case is <u>completed</u> when we have secured the <u>three measures</u> described. [1] The equation of average relationship [i.e., the linear regression formula] is an expression of the underlying law connecting the two variables, if such a law may be assumed. [2] The standard error of estimate measures the variation, in absolute terms, about the line of relationship. [3] <u>The coefficient of correlation is an abstract measure of the degree to which the average relationship actually holds in practice</u>* (p. 337; underlining and bracketed numbers added).

That was the only succinct definition of the correlation coefficient that he had provided so far. He placed a slightly larger version of it in a "Summary of Procedure" section, 30 pages later:

> *This coefficient [r] is an abstract measure of the degree of relationship between the two variables, in so far as this relationship may be described by a straight line* (p. 367).

Both Mills and Ezekiel include the regression coefficient and the standard error of the estimate in their list of three; Ezekiel also included the coefficient of determination but not the correlation coefficient, whereas Mills did the opposite. Ezekiel was able to mathematically justify his choice by *proving* that his choice was a measurement, but Mills did not even try to do so.

Also interesting is that although Mills made much use of the mathematical symbol r^2, he never mentioned the term Coefficient of Determination; nor did he explain that r^2 had been shown by others to be a much better *measure* of correlation than r (e.g., by Spearman, Wright, and so on, previously discussed). Instead, Mills described r^2 only as an intermediate value that simplified calculation of the correlation coefficient and other statistics (pp. 339, 349, 350, 354, 367, 371).

Albert Edmund Waugh joined the University of Connecticut in 1924 as an instructor in agricultural economics, in which department he subsequently became an assistant professor and then an associate professor; in 1945 he was named dean of the College of Liberal Arts and Sciences, and named Provost in 1950 (N.Y. Times: Waugh).

In 1938, he authored a book titled *Elements of Statistical Method*; thus it was one of a dwindling number of basic statistical textbooks that were written by non-mathematicians. In it, he confusedly explained the difference between data that are "related" vs. data that exhibit "correlation":

> *[W]here a knowledge of the value of one variable helps us in estimating the value of another variable, we say that the two variables are "related."*
>
> (Waugh, 1938, p. 230)

> *[W]e should say that there is covariance (since the values of the two variables tend to vary together) or correlation (since there seems to be some relationship between the variables)* (p. 238).

However, he did clearly and succinctly explain that...

> *There is no necessary relation between their values except that their signs are always the same* (p. 251n2; "their" refers to the regression coefficient and the correlation coefficient).

As we have seen already (e.g., E. E. Day) and will see again (e.g., G. W. Snedecor), many statistics books mistakenly imply that there actually *is* a dependable relationship between the magnitudes of the regression and correlation coefficients.

1939

Alexander A. Tschuprow was given an Honorary Fellowship in England's Royal Statistical Society, despite his being a *Russian* mathematical statistician. In 1939, a statistics book he'd written was translated into English and then published in London; the book's title was *Principles of the Mathematical Theory of Correlation*. As far as I have determined, this was only the second English-language book published solely on the topic of correlation (the first was by Ezekiel in 1930, previously discussed).

He had lofty goals for his book, as he explained in his Preface:

> *The purpose of this book differs from other works on correlation, inasmuch as its intention is to provide a logical foundation for the theory of correlation and not a guide to the practical application to its methods.... The present treatise is an attempt to work out the doctrines of the modern theory of correlation into a homogeneous and comprehensive system from this point of view.*

> *(Tschuprow, 1939, p. vii)*

He sought to achieve his just-stated purpose by explaining correlation from the viewpoint of both mathematicians and non-mathematicians, the two of whom he viewed as being opponents in warring camps in regard to correlation. His purpose was also served by his eloquent and lengthy explanations of not only theory, formulas, plots, and tables but also by his use of many detailed examples. One such example is humorous, given that even today, Russia is infamous for its consumption of liquor:

> *No other country possesses such reliable and ample statistics of the consumption of spirits as Russia during the time of the State monopoly of the sale of spirits. I have attempted in my Seminar to turn this splendid material to the best scientific account* (p. 151; the word "seminar" appears here because, at least in part, this book is a compilation of his seminar lectures on correlation).

In his opinion, what purpose does correlation serve? In a word, objectivity rather than subjectivity. He shared that view in his book's last chapter, which was titled "Object and Value of Correlation Measurement." No textbook or paper that I've read that was published prior to 1939 contained such clearly worded justification for correlation analysis; here are example sentences from that chapter:

> *What are the advantages of these 'mathematical' methods of inquiry over non-mathematical ones? First of all: in the more precise framing of judgement...* (p. 145).
>
> *Two [non-mathematical] investigators with the same series of numbers before them will often come to contradictory conclusions, and in such cases... each of them will think he is in the right and reject the other's judgement as subjective. The verification by means of 'mathematical' methods is then the only means of deciding the controversy* (p. 146).
>
> *The non-mathematician also forms a notion of whether values of one variable increase or decrease on the average with the growth of the values of another one, and is even able to gain an idea of the rate of increase or decrease when the form of the regression curve does not deviate too significantly from linearity. However, he works with rather vague notions and with still vaguer ideas of the suppositions on which the method he is employing depends; his quantitative judgements suffer by uncertainty and inevitable subjectivity and he is not in a position to attach due consideration to the disturbing influence of chance fluctuations; either he is too confident or, disillusioned, he begins to be too cautious in his conclusions. The mathematical statistician, on the contrary, is in a position to make a more precise estimate of the reliability of his conclusions by the computation of the relevant standard error* (p. 147).

On the last page of his book, he provided a valuable warning:

> *A routine-like mechanical reliance on ready-made prescriptions leads, even when the most complicated formulae are employed and the most precise calculations are carried out, to an unproductive waste of time and energy and to the accumulation of numerical values which are but little likely to enrich our essential knowledge* (p. 158).

George Waddel Snedecor (1881–1974) founded the Iowa State University Statistical Laboratory in 1933, the first such lab in the United States. Such efforts lead to the establishment of the Department of Statistics at Iowa State University. His 1937 book, *Statistical Methods Applied to Experiments in Agriculture and Biology*, went through a total of eight editions (later editions were titled only *Statistical Methods*); they are considered by some to be the most highly regarded statistics textbooks ever published. (Encyclopediaofmath. org-Snedecor)

I don't have access to any edition prior to the fourth. However, the following quote from the first edition is found in J. F. Kenney's *Mathematics of Statistics*:

The point of interest here is that r [the correlation coefficient] *is the geometric mean of the two regression coefficients* [for *Y* on *X*, and for *X* on *Y*]. *In ordinary units of measurements, therefore, r is an average of the two regression coefficients used in (i) estimating y from x and (ii) estimating x from y. This serves to clarify the relationship of the <u>two coefficients, correlation and regression</u>, in measuring relationship. <u>The latter is the appropriate one if one variable, y, may be designated as dependent on the other, x. Values of y may be partly controlled or caused by x</u>, as when the available amounts of some glandular secretion cause differences in the size of organisms. Or, y may be subsequent to x, as weight gain in nutrition experiments follows the measurement of initial weight. <u>In such cases, the regression of y on x is usually the statistic that furnishes the information desired</u>. It is then appropriate to attempt to estimate the value of y from a knowledge of the corresponding value of x. <u>Correlation, on the other hand, is the appropriate measure of the relation between two variates like statures of husband and wife.</u> The two heights are known to be associated through some complex of social and biological causes, but <u>neither may be looked upon as a consequence of the other</u>. In this sense <u>correlation is a two-way average of relationship, while regression is directional</u>. Of course, there are many variables whose relationship may be studied by means of either correlation or regression, or both. <u>It is necessary only to keep clearly in mind the character of the relation being considered.</u>*

(Kenney, 1939, Part 1, p. 167; underlining added; this text is essentially identical to what is found in Snedecor's fourth edition, 1946, p. 143)

That view of correlation is not unique to Snedecor; we have seen aspects of it in other textbooks and in Galton's 1889 explanations of "relation" vs. "correlation"; however, in my opinion, that explanation by Snedecor is unsurpassed in its clarity and practicality.

Snedecor in his textbooks used six dot-plots to show the differences between high, intermediate, and low correlation. He plotted both *X vs. Y*, and *Y vs. X* (i.e., two linear regression lines) on the same chart; in his textual discussion, he explained that the angle that those two lines made with each other is an indicator of the strength of correlation (that explanation is similar to what was said by other authors that we have discussed, starting with Boas in 1894). Unlike those other authors, he provided the raw data so that readers could re-create and investigate the plots for themselves. He used the same *X* values and the same *Y* values in each of his six plots, but those values were paired differently for each different plot. When both *X* and *Y* were each sorted in order of magnitude and then paired, the plotted data-points lined up perfectly linearly with a relatively steep slope and $r=1.00$. When the *Y* values were randomly ordered, the data-points plotted (vs. the sorted *X* values) in a shotgun pattern with a horizontal linear regression line and $r=0.00$. When the *Y* values were only somewhat randomly ordered and then plotted vs. sorted *X*, the data points looked like an elongated cloud that loosely hugged the linear regression line; that line had a medium slope and $r=0.60$. Figures 8.2 and 8.3 show the results of my analysis of those plots.

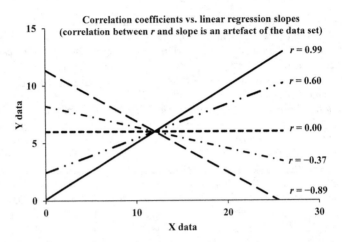

FIGURE 8.2
This chart summarizes five of the six plots provided in Snedecor, 1946, p. 140. It demonstrates that the data sets he chose caused an artificial correlation between the resulting correlation coefficients and the slopes of the linear regression lines.

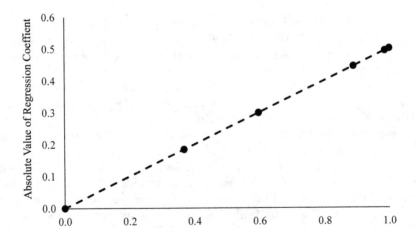

FIGURE 8.3
This chart shows an artificially perfect correlation between the correlation coefficients and the regression coefficients that resulted from six regression/correlation analyses that use the same data-set arranged differently for each analysis. (Derived from the data and plots in Snedecor (1946, p. 140); regression coefficients calculated using MS Excel.)

Such a perfectly straight line as seen in Figure 8.3 is to be expected, based upon the fact that...

- Sy, the standard deviation of the Y values, was identical in all six data sets.
- Sx, the standard deviation of the X values, was identical in all six data sets.
- A classic formula that connects the correlation coefficient to the regression coefficient is:

$$\text{Regression coefficient} = \left(\text{Correlation coefficient}\right)\left(Sy/Sx\right).$$

Snedecor's six plots promoted the mistaken notion that magnitudes of regression coefficients and correlation coefficients are themselves correlated in *real-life* situations. They certainly are correlated in the unreal data sets used in his examples, which he said he provided... "To help you acquire some experience of the nature of r..." (Snedecor, 1946, p. 140). However, in real life, researchers obtain measurements on many variables on the same subjects (e.g., when studying cancer rates in humans, variables are collected such as age, BMI, hours of exercise per week, number of cigarettes per day, glasses of alcohol per week, pounds of red meat per month, etc.). If such big-data is screened by searching for only those sets displaying a relatively large correlation coefficient (variable vs. cancer rate), valuable information could be lost if a large effect-size exhibits a small correlation coefficient.

By 1966, the five editions of his textbook had gone through a total of 18 printings; therefore, the just-discussed erroneous view of regression coefficient vs. correlation coefficient must have influenced thousands of future scientists.

His 1967 sixth edition was the first one to have a co-author, **William Gemmell Cochran**, who was then a statistics professor at Harvard University (Snedecor and Cochran, 1967, title page). Apparently, Cochran persuaded the octogenarian Snedecor to delete most of that large quotation just given ("The point of interest here is..."); it was replaced with an explanation of r^2 (i.e., the coefficient of determination, although that term was not used). In that replacement text, r^2 was said to be "another way of appraising the closeness of the relation between two variables," a way that is preferred because it is "the proportion of the variance of Y than can be attributed to its linear regression on X...." (Snedecor and Cochran, 1967, p. 176).

In the 1930s in Britain, the first professor of Statistics was appointed and the first university department of Statistics was formed (Mackenzie, 1981, p. 118). By the end of the decade, the subject of statistics was budding *sub*categories, as evidenced by the appointment of professors of *bio*statistics in the United States.

In 1939, **Alan E. Treloar**, an associate professor of biostatistics at the University of Minnesota, published a textbook titled *Elements of Statistical*

Reasoning; the text was based upon the author's prior 10 years of teaching "statistical methodology" to graduate students from various departments, whom he had helped "to secure an understanding of the principles of statistical reasoning" (Treloar, 1939, p. vii).

The textbook was subtly different from others available at the time, which seemed to have surprised at least one reviewer who concluded that...

> *The purpose of this book <u>seems</u> to be to develop fundamental statistical concepts for those who wish to reason carefully [,] <u>rather than</u> to provide a compendium of statistical techniques.*
>
> (Rider, 1941, p. 677; underlining added)

Despite Rider's comment, Treloar did offer practical advice in addition to explanation of theory. For example:

- *Biologists would do well to learn that this coefficient taken alone does not fully describe any association. Before embarking on discussions of such matters, one may well study further the trendlines, and the variation about them, which characterize the normal surface* (Treloar, 1939, p. 105).
- *The slope b of each regression line...is known as the <u>regression coefficient</u>. It <u>is given</u> in each case <u>by the product of the correlation coefficient and the appropriate quotient</u> of the standard deviations of the two variables, that of the dependent variable forming the numerator* (p. 115; underlining added). [The formula he described is one that had been known for many years, namely $b = r(Sy/Sx)$.]

In his textbook's introductory discussion of correlation, Treloar provided a chart that plotted the paired heights of more than 1000 husbands and wives (see Figure 8.4, which is his Figure 17). His Figure 21 (not shown here) fitted an ellipse around those points, with a long axis occurring approximately where the dashed line is placed here in Figure 8.5. His textual discussion gives the impression that he formed his conclusion (given here next) based solely on his visual analysis of those two figures (i.e., in his textbook, neither linear regression analysis nor a correlation coefficient was provided for this data); in his words:

> *It must be <u>quite clear</u> from Fig. 17 [Figure 8.4 here] that, <u>far</u> from there being a "law of opposites" governing human matings, <u>there is rather a tendency</u> for marriage to take place between individuals of <u>similar stature</u> (p. 87, underlining added).*

I assume that most readers would agree with Treloar's cautious conclusion (just given) that Figure 8.4's pattern of data points and Figure 8.5's ellipse-axis slope indicate that there is at least a tendency for humans to choose mates who have height similar to their own. That conclusion is a very much weaker version of

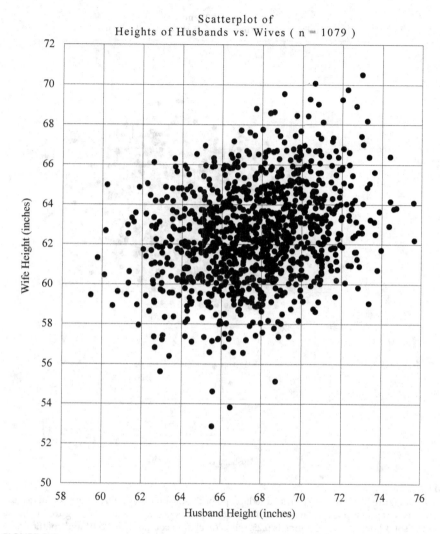

FIGURE 8.4
This chart shows heights of wives vs. that of their husbands. (Reconstructed from Figure 17 and the data in a correlation matrix table labeled as Figure 18, in Treloar (1939, pp. 86–87).)

the one given decades earlier by Karl Pearson in the paper from which Treloar claimed to have obtained his height data. I assume that most readers would *not* agree with Pearson's confident conclusion (given here next). His textual discussion gives the impression that he formed his conclusion based solely on the fact that the data yielded a correlation coefficient of 0.2804 with a probable error of 0.0189 (i.e., neither scatter-plot nor linear regression analysis of height data was provided in Pearson's paper). Pearson's conclusion was:

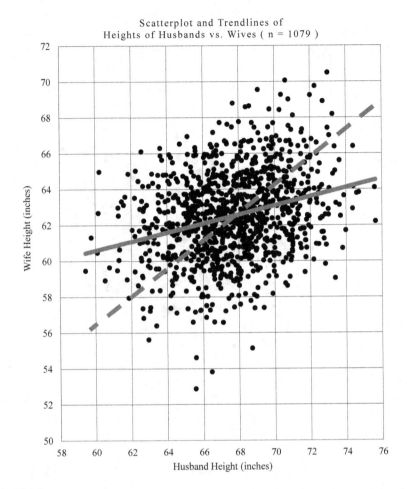

Scatterplot and Trendlines of
Heights of Husbands vs. Wives (n = 1079)

FIGURE 8.5
This is Figure 8.4 with trendlines added. Slope of the dashed line is 0.79; it approximates the long axis of an ellipse drawn around the plotted points (as in Treloar, 1939, Figure 21). Slope of the solid linear regression line is 0.25, with correlation coefficient=0.2804 and coefficient of determination=0.08.

> ...*there is a <u>very sensible resemblance</u> in size between husband and wife, which à priori I should have said was hardly conceivable.... We see at once* [in the table of correlation coefficients provided in his paper] *that between physical characters in the husband and wife of adult children* [,] *there is a correlation* [coefficient] *of upwards of 0.2, <u>a most remarkable degree of resemblance</u>.... <u>We could hardly want stronger evidence</u> of the existence of assortative mating in man, i.e. of the actuality of sexual selection.*
>
> (Pearson and Lee, 1903, p. 373; underlining and the "0" in 0.2 added; "assortative mating" refers to the practice of choosing a mate who is similar in appearance to oneself)

Pearson's correlation coefficient of 0.2804 translates to a coefficient of determination of 0.08; translated into words, the variation in the heights of husbands explains about 8% of the variation in heights of wives. To be fair, Pearson's paper was published the year *before* Spearman published a paper that showed that "not Galton's measure of correlation, but the square thereof" is the most accurate indicator of the degree of consistency of correlation (Spearman, 1904a, p. 75; previously discussed).

Treloar also liked to discuss history, especially that of Francis Galton. However, Treloar seems to have not read Galton's papers carefully, as evidenced by his confusing Galton's "coefficient of reversion" (i.e., the regression coefficient) with Galton's correlation coefficient; Treloar's statements equating the two (see next) are most definitely incorrect, except in cases where the standard deviations of paired variates are identical—he did not make that clear anywhere in his book.

> When Galton first arrived at an equivalent form of the quantity which we have symbolized as r, he named it the "*coefficient of reversion*." .... Its magnitude provided a measure of the extent to which offspring tended to regress towards mediocrity, and that was the immediate problem. The *symbol r for this measure of reversion* was selected by him and has been retained from that time. For some years it was *known as "Galton's coefficient,"* but *now the more general term "coefficient of correlation" is commonly applied to it* (p. 95; underlining added).

His claim about r having been known as "Galton's coefficient" is suspect; I've never encountered it in any other textbook or journal. Possibly Treloar meant to say "Galton's function," in which case he would have been correctly referencing the term coined by Weldon in 1893 (see discussion of the history of nomenclature, in Chapter 1 here).

Treloar focused on the error about the linear regression line; for its measurement, he used $\sqrt{(1-r^2)}$, which he said was "often now called the *coefficient of alienation*" (p. 126; he may have obtained the term "coefficient of alienation" from T. L. Kelley, e.g., from Kelley, 1923, p. 173). Galton had used that identical mathematical expression in his December 1888 paper (Galton, 1888c, p. 144); there, he had assigned it the letter *f*, explaining that it was "the Q value of the distribution of any system of x values, as x_1, x_2, x_3, &c., round the mean of all of them...." (where Q was his term for the probable error). In other words, Galton considered $\sqrt{(1-r^2)}$ to be the probable error not of the raw data but of the deviations from the linear regression line; it was his estimate of what we today call the standard error of estimate. Treloar subtracted that coefficient of alienation from unity (i.e., $1-\sqrt{(1-r^2)}$) and called the result a "prediction index" that "may be used as an index of the prediction value of r_{xy}" where r_{xy} was his symbol for the linear correlation coefficient for paired variates x and y (p. 126). However, although r^2 is a proportion and therefore $1-r^2$ is a proportion, the square root of a proportion is not a proportion; that is, neither his $\sqrt{(1-r^2)}$ nor his $1-\sqrt{(1-r^2)}$ is a proportion. Therefore, his two new terms cannot validly be said to "measure" anything, in the same way that it can be said that the correlation coefficient (r) does not measure anything (as previously discussed). It would have been more useful to have defined the coefficient of alienation as $(1-r^2)$; however, then his prediction index would have been $1-(1-r^2)$, which simplifies to r^2, the coefficient of determination.

The reader might expect that Treloar would explain how to rigorously interpret intermediate values of the correlation coefficient. Instead, he expressed regret that he could not provide a wholly satisfactory explanation:

> It is sometimes taken as a basis for condemnation of the correlation coefficient that _a difference of 0.1_ (or any fixed increment) _between two values of r means an increasing difference in intensity of association_, from the prediction point of view, _as one passes from low to high values of r. This deficiency is regrettable_, but _no wholly satisfactory way of avoiding it_ by using a function of r, such as _r²_ or the prediction index just considered, _has yet found general appeal_. It should be understood clearly that _the correlation coefficient, as a measure of intensity of the association, must be interpreted with care_. Spurious interpretation has its roots in lack of understanding by the interpreter and does not originate in the statistic itself.
>
> (Treloar, 1939, p. 127, underlining added)

In 1942, Treloar published a book that focused solely on correlation; its title was *Correlation Analysis*. It discussed and presented much on correlation as a technique (e.g., formulas and data handling) but very little on correlation as a concept. His reason for such a narrow focus was that "It is assumed that the reader has already had an introductory and well disciplined training in statistical reasoning…. These matters are discussed adequately for present purposes in the author's book 'Elements of Statistical Reasoning….'" (Treloar, 1942, p. 1)

1940

In the late 20th century, a series of books appeared under the generic title of "XXX for Dummies" and another series appeared with the title "The Complete Idiot's Guide to XXX," where I have used XXX to represent a topic such as Statistics. Those books remind me of a similar book by **Donald E. Church** that was published in 1940 and titled *Speed Methods of Statistics for Use in Business*; in its two chapters on correlation, we find only text—no tables, no data, no drawings, no charts, no plots, and no formulas. He was an instructor in the College of Commerce at Ohio University (Church, 1940, title page), and I wonder what his lectures on business statistics were like.

Maybe there was something strange about the early 1940s. Another surprisingly text-only book was published in 1940; it was edited by **J. Huxley**, was titled *The New Systematics*, and included 22 papers by different authors. In more modern times, the title would have been *The New Taxonomy*, because the book was an evaluation of the taxonomic systems used throughout all fields of biology. Taxonomy then and now is based primarily upon similarities and differences in anatomical structures or cytological tissues; however, in the 566 pages of Huxley's book, there was not one drawing or photo of a plant, animal, tissue, or single-cell organism (to be fair, there was a sketch of a sea-shell on page 403). Similarly, in 1942, **Adriance S. Foster** published the first edition of his *Practical Plant Anatomy*, which contained not a single photo or drawing in all its 142 pages, not even in the chapters on cytology; to be fair, its Preface did explain that "This [book] is in no sense to be regarded as a substitute for collateral reading in the standard texts in plant anatomy" but rather is "intended for use in the laboratory," as "a guide which will both direct was well as orient the student" (Foster, 1942, p. vii).

In the Preface to Church's *Speed Methods of Statistics* book, he explained that...

> The rapid succession of changes in business conditions has created a pressing demand for more extensive analyses of statistical data. *The simple methods presented in this book* sharply reduce the time and cost of such work and *avoid the mathematical* difficulties that usually are involved in customary statistical methods. *The use of technical terms has been avoided wherever possible* on the assumption that many readers may not be technically trained statisticians or, in fact, may be dealing with statistical analysis for the first time.
>
> *(Church, 1940, p. iii; underlining added)*

Regarding correlation, Church had this to say:

> The solutions of some business problems hinge upon the character and extent of the relationships that exist between two or more factors. Measurement of such relationships may lead to the discovery of the causes of the fluctuations and may suggest means by which undesirable situations may be improved.... Statistical analyses involving only two factors are said to be "simple" (or simple correlation) while those involving the relationships between one factor and two or more others are said to be "multiple" (or multiple correlations) (p. 60).

He viewed correlation as somewhat subjective:

> The extent to which practical significance may be attached to a statistically computed correlation depends largely upon a *logical analysis* of the character of the

relationship. The existence of a high degree of correlation between two series, as judged solely by the statistical results, proves merely that the two series fluctuate in harmony with each other, but does not prove the existence of causal relationships…. While supplementary correlation analyses of several related factors often proves helpful in confirming suspected causal relationships, the ultimate test must be based upon a rational explanation of the influence of the causal factors (p. 61; underlining added).

What did he recommend?

The relationship between any two time series may be recognized graphically by comparing the shape of the two curves or by studying their dot formation on a scatter diagram…. One of the principal merits of the line charts as compared with scatter diagrams for correlation studies is the flexibility of line chart handling. The minor movements may be discounted readily by eye, if the study is primarily concerned with the larger cyclical swings. On the other hand, if interest is centered around short term movements, the longer term fluctuations may be discarded (pp. 62–63; underlining added).

Church also described a "Graphic Determination of the Error of Estimate" (p. 71), which was his way of determining the Standard Error of Estimate (*See*) using an eye-balled linear regression line through the dots on a scatter diagram. He recommended creating both Y-on-X and X-on-Y regression plots, and thereby obtaining two estimates of *See*. What is surprising is that he did not tell the reader how to estimate the correlation coefficient using the formula: $r = \sqrt{(b_1 b_2)}$, where b_1 and b_2 are the slopes of those two plots. How are we to interpret such an oversight? Was he so little versed in correlation theory that he was unaware of that relationship? If he was aware, it is puzzling that a graphical estimate of the relatively common statistic r was omitted but a graphical estimate of the comparatively obscure statistic *See* was included. Or did he simply want to avoid mentioning even a single formula in his textbook?

1941

Leonard Henry Caleb Tippett (1902–1985) studied statistics under Karl Pearson and R. A. Fisher; he then spent his entire career working for the British Cotton Industry Research Association, during which time he invented the random number table (Wikipedia: Tippett). Not surprisingly, his book titled *The Methods of Statistics: An Introduction Mainly for Experimentalists* was more of a practical applications manual than a college textbook. Editions one through four were published in 1931, 1937,

1941, and 1952, respectively. I have access to only the third and fourth editions; similar to what Cochran did with his textbook's title (previously discussed), he shortened the title in those later editions to just *The Methods of Statistics*.

The following quotations were all taken from the 1941 third edition (all underlining added); they provided practical advice for all statisticians, not just experimentalists:

- *Attempts are sometimes made to explain correlation in terms of chances, and to state that if the* [correlation] *coefficient for two characters is 0.6 (say), and knowing one of them we use the regression formula to guess at the other, we shall be right six times in ten trials. Such an explanation is nonsense...* (pp. 160–161).
- *The uses (and abuses) of the coefficient of correlation are many...* (p. 161).
- *A rather more dangerous use of this* [correlation coefficient] *constant, however, is as evidence of causation. If two quantities are associated (as shown by the correlation coefficient)* [,] *the inference often made is that one is a cause and the other an effect. Such an inference is often erroneous when dealing with quantities susceptible to such close control that the correlation coefficient is unity, but it is particularly unsafe when there are uncontrolled variations and the relationship is not exact. Often two quantities are both affected in the same way by a third so that they appear to be related, when actually neither if altered independently would have any effect on the other.... Care, common sense, imagination, and a technical knowledge of the subject to which it is applied are particularly necessary in this use of correlation* (p. 162).
- *When considered as the expression of the relationship between two quantities...the correlation coefficient merely measures the importance of the variations in one quantity associated with the other (the independent variable) relative to all other variations.... It is thus not a physical quantity in the way the regression is* (pp. 162–163).
- *A low correlation coefficient does not necessarily mean that x is incapable of having an important influence on y; it may mean that an insufficiently large range of x has been tried* (p. 163).
- *To sum up;* [sic] *there are three important constants that express the properties of a correlation table: (1) The correlation coefficient that measures the importance of the variation in y associated with x relative to the total variation, (2) The regression coefficient that measures the average amount of increase or decrease in y per unit increase in x, and (3) The residual variance that measures the scatter of values of y about the regression line. For different populations, these constants are independent in that a high value of one constant does not necessarily mean a high or low value of either of the others* (p. 164).

Those "three important constants" are reminiscent of value-statements made by earlier authors that we've discussed, namely M. Ezekiel and F. Mills. Table 8.1 summarizes in modern terms what each author claimed:

TABLE 8.1

Correlation-Related Coefficients Considered Most Important by Three Influential Authors

	The Most Important Values Derived from Linear Correlation Analysis				
	Correlation Coefficient	Coefficient of Determination	Regression Coefficient	Standard Error of Estimate	Standard Variance of Estimate
	r	r^2	b	*See*	*See²*
Ezekiel (1930)		X	X	X	
Mills (1938)	X		X	X	
Tippett (1941)	X		X		X

1942

Everett Franklin Lindquist (1901–1978) received a Ph.D. in Education in 1927 from what is now called the University of Iowa; and he was subsequently employed there as a faculty member until he retired, 42 years later. His accomplishments include having designed and developed the ACT and other standardized tests, updated versions of which are still used to evaluate college-readiness of high-school students (Stateuniversity.com-Lindquist). In 1938, he authored an introductory statistics textbook titled *A First Course in Statistics: Their Use and Interpretation in Education and Psychology*; I have access only to the 1942 second edition. In one way, it was similar to Church's book (previously discussed): Church's contained not even one formula for computing statistics; similarly, the preface to Lindquist's book stated that "These materials [that is, these chapters] stress as much as possible the uses and interpretation of statistics, and <u>minimize as much as possible the mathematical theory of statistics and the mechanics of computation</u>" (Lindquist, 1942, p. iv; underlining added).

Lindquist's book used 75 of its 227 pages to discuss correlation; in those pages, the word "relation" and "correlation" were used almost synonymously. For example, in a chapter on "Correlation Theory," he wrote:

- *When measures of each of two traits are secured for each individual in a given group, it may frequently be noted that the two measures for any individual tend to have roughly the same relative position in their respective distributions.... When this is true, we say that the two traits (or measures) are "<u>positively related</u>" for the group in question, or that they show a "<u>positive correlation.</u>"* (Lindquist, 1942, p. 153; underlining added)
- *Whenever the relationship between measures of two variables is such that the means of the rows and the means of the columns on the*

> *scatter-diagram each tend to lie along a straight line, we say that these*
> *variables are "rectilinearly" related, or that they represent an instance*
> *of rectilinear correlation.* (p. 156; underlining added)

Despite his emphasis on correlation, Lindquist hesitated to explain its meaning:

> [The correlation coefficient]...*may not be considered as directly proportional*
> *to the degree of relationship. A coefficient of correlation of .80, for example, may*
> *not be said to represent exactly twice as close a relationship as one of .40, even*
> *though both are established over the same range of talent. To be able to make such*
> *a statement, we would have to be able to describe, independently of r, just exactly*
> *what we mean by closeness of or degree of relationship, and no such description*
> *or definition that is generally acceptable has yet been proposed. Because of our*
> *inability to define "degree of relationship," we are unable to state in general how r*
> *changes in value for given changes in that degree* (p. 198; underlining added).

He not only hesitated, he abandoned all hope:

> *It is therefore recommended that the beginning student in statistics make no*
> *attempt to arrive at any absolute interpretation of r.... When comparing r's of*
> *different magnitude, he should avoid trying to estimate "how much" closer rela-*
> *tionship is in one case than another, but should be content with the knowledge*
> *that there is a difference of some indeterminate amount.... If he wishes to secure*
> *a more definite notion of what an r of a given magnitude really means, he can do*
> *no better than to study the distribution of the tally marks on the scatter-diagram*
> *from which it is computed* (p. 202; underlining added).

Those were surprising statements, in 1942, given that explanations of r^2 and the coefficient of determination had been discussed by other authors, starting in 1904 (as previously discussed). His *First Course* book did not include any such discussion, and neither did his previously published book titled *Statistical Analysis in Educational Research.*

1943

In 1943, **Maurice G. Kendall** published Volume I of his *The Advanced Theory of Statistics.* He was then a "Fellow and Member of the Council of the Royal Statistical Society" and "Statistician to the Chamber of Shipping of the United Kingdom"; Volume II was published in 1946, when he'd become "An Honorary Secretary of the Royal Statistical Society" and also "Fellow of the Institute of Mathematical Statistics" (per title pages). It is interesting that on the approximately 1000 pages of those two volumes, the word correlation and related parts of speech appeared an astounding ≈ 870 times in text, titles, and bibliography; those appearances were throughout almost all the chapters in both volumes rather than solely in Volume I's few chapters on correlation. His explanation of correlation was similar to that provided by most of his contemporaries, however his use of equations and formulas in those explanations far exceeded that in other textbooks.

1944

In 1944, Sewall Wright (previously discussed as the inventor of the term Coefficient of Determination) must have been quite pleased with professor/economist **Morris Myers Blair**, who had just published a book titled *Elementary Statistics with General Applications.* I say pleased because Blair's chapter on correlation was titled "Correlation and Determination." Early in that chapter, he explained:

> Up to this time [in this textbook] *we have learned to measure...the relationship between two related frequency distributions by means of regression lines....* There is, however, need for another type of measurement, a ratio, or relative number, or percentage statement of the relationship between the variables.... Correlation is the square root of a percentage, and is, therefore, quite misleading to the beginner in statistics. *A more easily understood, and in many respects a better measure, is* the coefficient of determination, which is a true percentage of the portion of one variable that is associated with another. *Determination is the square of correlation,* and is coming into general use as the more accurate and easily understood measure.
>
> (Blair 1944, pp. 264–265; underlining added)

That sounds very clear. But in his next series of paragraphs, Blair may have confused some of his readers:

> *The relationship between determination and correlation may be illustrated* [by comparing values of the coefficient of determination and the correlation coefficient—thus, he equated the word correlation to the correlation coefficient and the word determination to the coefficient of determination].... *Correlation,* or *determination, is a method of measuring the similarity of the change in these two variables* [the ones that he used were sunlight and temperature].... Correlation and determination are measures *of the degree of association in the movements of two or more variables* (pp. 265–266; underlining added).

His readers may have wondered how a percentage and the square-root of a percentage can *both* be considered measures of correlation. Readers may have sought clarification in the book's "Appendix of Technical Terms," where they would have found the following (on p. 637; underlining added):

- *COEFFICIENT OF CORRELATION* = <u>a measure</u> *of the amount of variation in a* <u>dependent</u> *variable which is associated with variation in one or more* <u>independent</u> *variables* <u>expressed as the square root of a percentage</u>. *Complete or perfect correlation is designated as 1.00.*
- *COEFFICIENT OF TOTAL DETERMINATION* = <u>a measure</u> *of the amount of variation in a* <u>dependent</u> *variable which is associated with one or more* <u>independent</u> *variables* <u>expressed as a percentage</u>.

Paired together, I consider those definitions to be confusing.

It is interesting that both of those definitions involve dependent and independent variables—the reader may have wondered whether the methods of correlation and determination should be used when the concept of cause and effect does not apply. For example, in December 1888, Galton (as previously discussed) defined the correlation coefficient using a plot of human height vs. arm-length; essential to *his* definition of correlation was the fact that neither variable could be considered the dependent or independent variable. Blair's definitions could be viewed as the opposite of Galton's.

Blair was inconsistent in how he portrayed numerical values of determination. Sometimes he described them as decimal fractions and other times as percentages; for example, on his page 268, he said that "the determination, r^2, is 57.2%," but on page 277 he showed "$r^2 = \ldots 0.90$." In one table (see Figure 8.6, here) he confusedly used both methods.

Like many other authors (previously discussed; see also Zorich, 2017), Blair told his readers that there is a *dependable* relationship between the size of the correlation coefficient and how closely X, Y points cluster around the line of best fit:

> The closer the plotted points of data come to the regression line, the higher is the correlation. The more widely they are scattered, the lower the correlation…. The wider the data scatters about the line, the smaller the correlation (pp. 266, 270).

Coefficient of Correlation $r = \sqrt{\%}$	Coefficient of Determination $r^2 = \%$
1.00	1.00
0.90	0.81
0.80	0.64
0.70	0.49
0.60	0.36
0.50	0.25
0.40	0.16
0.30	0.09
0.20	0.04
0.10	0.01

FIGURE 8.6

This table shows r^2 and r defined as a percentage and square-root of a percentage, respectively; confusedly, the r^2 values shown here are decimal fractions rather than percentages. (This is a reconstructed copy of the table that appears in Blair (1944, p. 265).)

He forgot to mention that such a rule is true only when those lines all have the same slope.

He offered some well-written practical advice about how to achieve the most accurate correlation coefficient:

> *The size of the coefficient of correlation for any given set of data depends to a considerable extent on the degree to which the underlying regression equation and line measure the true relation between the variables. If the relation between the variables is truly a straight line, a larger coefficient of correlation will be obtained if a straight regression line is used. If, however, the relation between the variables is actually curvilinear, a larger coefficient of correlation will be obtained if the correct curvilinear regression equation is chosen. It is impossible to get a coefficient of correlation which will measure the full amount of relationship in the data unless the regression equation which best measures that relationship is chosen (p. 556).*

1953

In 1953, **Victor Goedicke** was an associate professor of mathematics at Ohio University (USA), when he published a textbook titled *An Introduction to the Theory of Statistics*. In it, he assigned the letter "D" to the coefficient of determination (p. 159) and then *defined* the correlation coefficient as "the square root of D" (p. 160). He also provided the following useful advice about r and D:

> *The student will perhaps feel that it is wasteful and unnecessary to master two separate statistical terms which measure, in different ways, exactly the same thing, namely, closeness of relationship.... The nature of the relationship between x and y is described by the line of best fit, and the strength of the relationship is described by D. But r is a composite quantity which describes the strength of the relationship by its absolute value, and part, but not all, of the nature of the relationship by its plus or minus sign. It is suggested that you regard r as an intermediate mathematical step and D as the final objective. In short, r should be studied because it is widely used in statistical reports and because it is a useful mathematical tool for a variety of purposes; but it is recommended that when you read a report containing a value of r, you should mentally square it to obtain D for the purpose of interpreting the results. It is sometimes useful to think of D somewhat loosely as the "percentage causation," although it is important to notice that we know nothing about the nature of the causation from the size of D or r. Variations in x may be causing variations in y, or vice versa, or both may be caused by variations in a third variable which was not measured by the investigator, or x and y may have varied together by chance.*
>
> (Goedicke, 1953, pp. 160–161; underlining added)

It is interesting that the final chapter in his book was titled "Statistics and Common Sense":

> In this chapter we shall point out a few of the pitfalls against which you should
> guard, both in performing statistical analyses of your own and in interpreting
> the results of others.... a correlation coefficient measures the degree to which
> two variables are related, but...it does not provide any direct information about
> the nature of the causal relationship. Failure to remember this leads to many a
> statistical absurdity (pp. 248, 253).

He must have been disappointed that his symbol "D" for the coefficient of
determination was not adopted by the wider scientific community.

1955

Many statistics textbooks claimed to have been written with practicality in
mind, but the 1955 *Handbook of Industrial Statistics*, may have had the most
practical *sounding* title. Its authors were **Albert H. Bowker** and **Gerald J.
Lieberman**, both of whom were statistics professors at Stanford University.
Their *Handbook* was less than 200 pages long (pages 774–959 in the *Handbook
of Industrial Engineering and Management*); they had very little to say in it about
correlation (their entire discussion of correlation comprised only six sentences), and what they did say was not flattering:

> A measure of the degree of association between two variables is the correlation
> coefficient.... In engineering applications, the correlation coefficient does not
> play a very important role. The correlation coefficient can be derived from the
> slope of the fitted least squares line.... Consequently it is clear that the correla-
> tion coefficient does not contain any additional information.
>
> (Bowker and Lieberman, 1955, p. 895)

Years later they published a college textbook titled *Engineering Statistics*.
I have access to the 1972 second edition, which was about three times as long
as their *Handbook*; its discussion of correlation (18 sentences) was also three
times as long as that in the *Handbook*. It offered a view of correlation that was
essentially identical to the one in their *Handbook*:

> The sample correlation coefficient can be derived from the slope of the fitted least
> squares line.... Consequently, it is clear that the sample correlation coefficient
> does not contain any additional information (1972, p. 363).

It is therefore not surprising that none of the book's chapter titles contained the
word correlation, and that the one sub-chapter that was titled "Correlation"
was buried at the end of a chapter titled "Fitting Straight Lines."

Their treatment of correlation did their students a disservice. It would not
have cost the authors much in the way of paper and ink to have included

an explanation that the reason that their disparaging statement was true is only because engineers almost always deal in *Relational* Correlation (where the variable of interest is dependent upon one or more independent variables). A few more sentences could have explained that *r* or *r*² would actually be valuable in an engineering study that involved two or more variables that were *not* obviously dependent one upon the other, i.e., in *Co-Relational* Correlation—for example: The strength of the user-formed seal on one side of a 4-sided sterile-barrier pouch vs. the strength of the seals that had been formed by the pouch manufacturer on the other three sides of the pouch.

1967

In 1967, **F. H. Lange** (a professor at the University of Rostock, Germany) published an English translation of his second edition (1962) German textbook on correlation; the English title was *Correlation Techniques*. The English translation of the Preface to the first German edition was included in that second edition; it explained:

> In the last decade the statistical approach has penetrated to a surprising extent into many branches of communication engineering.... The statistical approach has been considerably extended by means of...the so-called correlation function, which has led to numerous and very diverse application in electronics. The correlation concept is now at the centre of serious scientific work, not only on communications, but also on acoustics, optics, control engineering, physiology and radio astronomy, and more recently on radio reception systems.... The present book.... discusses the fundamentals and the applications of linear correlation analysis [,] whenever possible from the point of view of engineering methods.

> (Lange, 1967, p. 11—note that this is the Preface's page 11; the text of the book itself has another page 11)

With such a purpose, it is not surprising that the book's subtitle was: *Foundations and Applications of Correlation Analysis in Modern Communications, Measurement and Control*. The "Preface to the English Edition," which was written by Lange himself, explained:

> ...the function of the present book has somewhat changed. Instead of being a monograph on a highly specialised subject, it can now be regarded as a text for last-year students in radio, measurement and control engineering (p. 7).

His book focused on correlation of radio-frequency curves; such curves are superficially similar to the correlation curves of late-19th-century economists. His "definitions" of correlation are all mathematically formulaic—the reader fails to find a useful one in words.

1970

In 1970, a 48-page pamphlet titled *Correlation and Regression as Related to Statistics* was authored by a committee called "**The Schools Statistics Panel**," for use in British secondary schools. The book explained correlation first and then regression. To explain correlation, five example *X, Y* data-sets were presented in *X, Y* scatter plots:

> *In these five examples in which comparisons have been made between two variables, we have seen some kind of association or relationship between the variables in four cases and no relationship in the other. This relationship between variables is called correlation. We shall try to measure the degree (or amount) of relationship between two variables. The degree of relationship is measured by a number which is called the coefficient of correlation, and is denoted by r. For a perfect direct (positive) relationship...r has the value of +1. When no relationship exists between the variables, r has the value of zero.*

> *(Panel, 1970, p. 12)*

The four data-sets that exhibited a large effect of *X* upon *Y* (i.e., the lines had a relatively large slopes) also produced a large correlation coefficient, and the one set that exhibited a tiny effect (very small slope) produced a tiny correlation coefficient. Such examples accompanied by the just-given text must have resulted in tens of thousands of students internalizing the mistaken idea that a large correlation coefficient is always accompanied by a relatively large regression coefficient, and that a small correlation coefficient is always accompanied by a relatively small regression coefficient. Such a mistake in textbook presentation is not unprecedented (e.g., see similar mistakes by Persons in 1910 and Snedecor in 1946, both previously discussed; see also Zorich, 2017).

1977

I had the good fortune to be introduced to mathematical statistics by a professor who assigned the sixth and last edition (1977) of *Introduction to Probability and Statistics*, the authors of which were UC Davis professors **Henry L. Alder** (1922–2002) and **Edward B. Roessler**. After more than 40 years, I still consider it to be one of the best written statistics textbooks, primarily for its thorough and clear explanations of difficult topics. In 1976, Alder had received the UC Davis Award for Distinguished Teaching (MAA.org-Alder). The following is an example of text from that sixth edition:

> *...the <u>coefficient of determination is the proportion of the total variation</u> (or variance) in Y <u>that can be explained by the linear relationship</u> existing between X and Y. When multiplied by 100, the proportion is converted to a percent. Thus, the correlation coefficient of 0.952.... indicates that $(0.952)^2 \cdot 100 = 90.6\%$ of the variation...is due to the linear relationship...the rest of the variation is due to unexplained factors.... Note that <u>this interpretation in terms of percentages applies only to the variance of the Y's, not to the standard deviation of the Y's.</u>*

> *(Alder and Roessler, 1977, p. 231; underlining added)*

1979

In 1975, **Marty J. Schmidt** published the first edition of his book titled *Understanding and Using Statistics: Basic Concepts.* I have access to the second edition (1979), the Preface of which explained that "All chapters...have been updated; many have been rewritten" (p. vi). The second edition is interesting for two reasons:

- Unlike virtually all other contemporary authors, he used the word "co-related" as had Galton:

> *Two variables are <u>related</u> when changes in the value of one are systematically <u>related</u> to changes in the value of the other.... Correlation methods offer a means of determining whether or not such <u>relations</u> exist, that is, whether variables are "<u>co-related</u>" or correlated.*

> *(Schmidt, 1979, pp. 146–147; underlining added)*

- He claimed that *some* correlation coefficients have absolute values greater than 1.00:

> *<u>Most</u> correlation coefficients range in value from −1.00 to +1.00* (p. 151; underlining added).

A reader who skims the book and arrives at that last sentence is bound to wonder "How can there be correlation coefficients greater than +1 or less than −1?" However, after examining the previous and subsequent several paragraphs in the book, the reader discovers that Schmidt used the term "correlation coefficients" in reference to a wide range of types of coefficients of correlation, not just Pearson's product-moment version. His definition for them was:

> *Statistics designed to specify precisely the direction and degree of relations are known as correlation coefficients* (p. 151).

Some of *those* types of correlation coefficients can have absolute values greater than 1.00.

His scatter-plot examples of data-sets with varying correlation coefficients were an improvement over those found in most other textbooks, in that his plots look like they all might have approximately the same slope, i.e., the same regression coefficient (see Figure 8.7 here). Therefore, the student-reader intuitively correctly concludes that (in similar situations) the magnitude of the correlation coefficient tends to be inversely related to the amount of scatter about its corresponding linear regression line. Scatter-plots in almost all other textbooks previously discussed here lead the student to conclude incorrectly that the magnitude of the correlation coefficient is positively related to the magnitude of the regression coefficient.

Unfortunately, his fourth edition (2010) confusedly seemed to define correlation in terms of linearity:

> One of the simplest examples of a relation is when an increase in the values of one variable corresponds to a tendency of the values of the other variable to increase. Alternatively, the values of the second variable might decrease.... we need to know the strength of this tendency.... Relations between variables can also be characterized by degree. The degree of the relation refers to the extent that observed values adhere to the designated relation. <u>The strongest degree is when one variable is linearly related to another</u>, i.e., the rate at which values of the one variable increase (or decrease) is proportional to the rate of increase for the other variable. This means that if we make a scatter plot for pairs of values (one for each variable), then the plot will be a line.... In this chapter we will quantify the degree of a relationship by the correlation, sometimes called the correlation coefficient.

(Schmidt, 2010, pp. 256–257; underlining added)

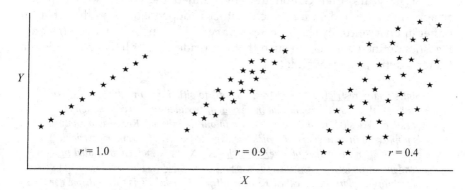

FIGURE 8.7
These are scatter-plots with similar slopes (i.e., similar regression coefficients) but correlation coefficients of varying magnitudes. The inverse relationship between *r* and the amount of scatter is obvious. (Reconstructed from Schmidt (1979, p. 150).)

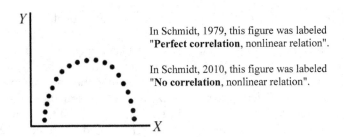

FIGURE 8.8
As seen in the labeling of this chart, Schmidt's description of curvilinear correlation changed drastically from 1979 to 2010. (Derived from Schmidt (1979, p. 154) and Schmidt (2010, p. 264). This image is from the 2010 edition, but the 1979 image is virtually identical.)

He subsequently acknowledged that there are "various correlation coefficients" but that "the most useful correlation coefficient [is] the Pearson product-moment correlation coefficient," which he correctly stated is "appropriate if the underlying relation between X and Y is essentially linear" (pp. 260–261, 264). Amazingly, the scatter-plot that his second edition labeled as "Perfect correlation, nonlinear relation" was in his fourth edition re-labeled as "No correlation, nonlinear relation" (see Figure 8.8 here); that re-labeling had to have misled no small number of readers.

1986

Almost 100 years after Galton initially defined the correlation coefficient as applying to X, Y data for which neither is dependent upon the other but rather upon a mutually shared cause, **Gary L. Tietjen**'s *A Topical Dictionary of Statistics* defined correlation such that the reader might likely conclude that Galton's definition is invalid:

> *Statisticians restrict the word* [regression] *to situations in which the dependent variable is random and the independent variables are fixed, mathematical variables. We shall now be more precise in our definitions. Regression analysis is applicable in situations in which the expected value of a random variable Y depends upon the values of other variables* X_1, X_2..., X_p, *which are called independent variables.... Y is called the dependent variable.... The proportion of* [Y] *variability explained...is called R^2 or Multiple R^2, while R is the multiple correlation coefficient.... That is widely viewed as a measure of the strength of the relationship between the dependent variable Y and the independent variables.*
>
> *(Tietjen, 1986, pp. 48, 52; underlining added)*

1988

Exactly 100 years after Galton presented his original paper on "Co-relations," **J. L. Rogers** and **W. Nicewander** published their list of "Thirteen Ways to Look at the Correlation Coefficient"; they added some historical discussion, which included this unelaborated, single-sentence claim:

> Galton's cousin, Charles Darwin, used the concept of correlation in 1868 by noting that "all the parts of the organisation are to a certain extent connected or correlated together." (p. 60)

That Darwin quote is from his *The Variation of Animals and Plants under Domestication* (1868, vol. 2, p. 319). Rogers and Nicewander gave only part of Darwin's sentence. The full sentence is...

> All parts of the organisation are to a certain extent connected or correlated together; but the connexion may be so slight that it hardly exists, as with compound animals or the buds on the same tree. [Compound animals are composed of a number of individuals each of whom are independently performing vital functions but yet are organically connected in a united colony.]

As discussed here in Chapter 2, that and similar uses by Darwin of the word correlation were non-mathematical Observational Correlations. I suspect that Rogers and Nicewander's claim has been misinterpreted by many readers to mean that Darwin was talking about mathematical Relational or Co-Relational Correlation.

To confuse the reader even more, Rogers and Nicewander's paper included a table (p. 61) titled "Landmarks in the History of Correlation and Regression," which listed only ten dates. One of them was 1868, because of the just-given inappropriate quote by Darwin. Another was 1985, which was listed as the "Centennial of regression <u>and correlation</u>" (underlining added)—I find that humorous: How can 1985 be the centennial of mathematical correlation, given that it was discovered in 1888?

In the USA, there is a humorous urban myth about patents: Supposedly, at the end of the 19th century, the Commissioner of the U.S. Patent Office claimed that everything that could be invented had already been invented (Wikipedia: Duell). Similarly, as my review of 20th-century statistics books came to an end, I expected to find no new ways to explain correlation. To my surprise, I found an 1988 statistics book by **George W. Bohrnstedt** and **David Knoke**, who used Venn diagrams; also, I found such diagrams in the correlation section of a 2000 statistics book by **Frederick J. Gravetter** and **Larry B. Wallnau.**

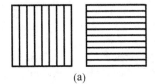 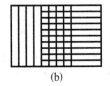

(a)   (b)   (c)

Venn Diagrams showing 0.00, 0.50, and 1.00 Correlation

FIGURE 8.9
This shows modified Venn diagrams that represent different degrees of correlation.

Both of those books used classic Venn diagrams involving overlapping circles. Because it is difficult to visualize the percentage overlap of circles, I am here (in Figure 8.9) using rectangles in my discussion of their explanations.

In both books, correlation was said to be quantified by the amount that the diagrams representing X and Y overlapped. To represent the absence of any correlation, the diagrams for X and Y were shown as not overlapping (as in Figure 8.9a). To represent perfect correlation, the diagrams for X and Y were shown as overlapping completely (as in Figure 8.9c). To represent moderate correlation, the diagrams for X and Y were shown as partially overlapping (as in Figure 8.9b, which shows 50% overlap).

Those two books contradicted each other in regard to the deeper meaning of such overlap. Bohrnstedt and Knoke (p. 367) labeled their equivalent of Figure 8.9b as representing $r=0.50$, which of course encouraged the reader to adopt the erroneous view that the magnitude of the correlation coefficient can be interpreted on a linear scale, i.e., as a proportion. On the other hand, Gravetter and Wallnau (p. 540) labeled their equivalent of Figure 8.9b as $r^2=0.50$, i.e., $r=0.71$, which helped the reader adopt the correct view, namely, that r^2 is a proportion and r is not.

1993

In 1993, **Carol Ash** was a professor at the University of Illinois (Urbana, USA) when she published her *The Probability Tutoring Book: An Intuitive Course for Engineers and Scientists (and Everyone Else!)* Her book is the only one in my bibliography whose title contains an exclamation point, the presence of which I interpret to mean that she felt passionately about teaching this subject; I would have enjoyed taking a course in probability from her.

Curiously, her book also discussed correlation, albeit on only two of its 465 pages; however, her discussion provided a fresh perspective on correlation and a new use for the correlation coefficient. In a chapter on Expectation, in a subchapter titled "Correlation," in a section titled "Definition," she provided Pearson's product-moment formula and then explained that "correlation is just the covariance 'normalized' to be in the range [–1,1]" (p. 235; bracketed text is in the original). The next section was titled "An Application of Correlation—The Least Squares Estimate"; in it, she provided formulas for determining the two coefficients of the "line of regression" (i.e., *a* and *b* in the formula $Y = a + bX$). Her two formulas required the user to have *already* calculated the correlation coefficient (p. 236).

1998

In 1998, **Bart K. Holland** was a professor in the Division of Biostatistics in the Department of Preventive Medicine at the New Jersey Medical School (Holland, 1998, title page). Apparently, his experience with clinicians led him to conclude that the best way to teach them statistical probability was to avoid mathematics entirely; and therefore his book on probability was titled *Probability without Equations: Concepts for Clinicians*. True to its title, there is not a single equation or formula in the entire book. Interestingly, similar to Carol Ash's book on probability (previously discussed), it used three of its 103 pages to discuss correlation.

Holland's explanation of correlation was given from the viewpoint of a probability "P value" (what we today call a "p-value" in significance testing), and is therefore unique among all the publications that I've researched:

> Sometimes in medical research what's needed is not a prediction but <u>a measure of association</u>. For example, in obstetrics, the size of a certain fetal bone, determined by ultrasonography, may be useful as a measure of fetal age. In general, as one variable increases, the other increases, so we say that there is a correlation, in this case a positive correlation. A correlation can also be negative of course, meaning that as one increases, the other decreases in an inverse association. The two main questions we're interested in are, <u>Is the association beyond what one might expect by chance?</u> If it is, how "tight," or perfect, is this association? <u>These questions are answered by calculating P values</u> based on a bivariate normal distribution. That is, we assume both variables to be normally distributed: [sic] each observation in the sample consists jointly of one value from one of the normal distributions and the corresponding value from the other distribution. The mathematics of the situation permits the calculation of the probability of obtaining a particular set of such pairs of corresponding values. <u>The summary statistic relating the observations to the P values is called the correlation coefficient, or r</u>... (pp. 46–47; underlining added).

Persons's 1919 paper (previously discussed) also described *r* as a probability statement about whether or not there is correlation; however, his paper did not mention p-values or anything like them.

Summary of This and the Previous Chapters

At the beginning of the 20th century, we saw the first ever textbooks on mathematical statistics; they included much discussion of correlation. By mid-century, at least two textbooks had been published *solely* on the topic of correlation.

What is the correct interpretation of correlation and how should it be use? That question was uncivilly argued about in print during the 1910s and was not answered to everyone's satisfaction even by the end of the century. No consensus was reached as to how to teach correlation in college textbooks: Some described correlation as a subset of regression, some described regression as a subset of correlation; some emphasized the correlation coefficient and others emphasized the coefficient of determination.

Ingenious new approaches were introduced to help explain correlation, such as Venn diagrams and p-values.

9

21st Century and beyond

Having read this book, you may wonder if it is possible to rigorously define the word "correlation." Such a task may seem as difficult as trying to rigorously define the word "vegetable"—is it a root (e.g. carrot), a leaf (lettuce), a stem (asparagus), a fruit (tomato), a fungus (mushroom), or an aggregate of flowers (broccoli)? Or maybe it is impossible—for example, I recall being amazed when, in the 1970s, my little sister first told me that her favorite rock-music band had just released a "bad" new song, by which she meant that it was a "good" new song.

Such difficulty may in part be the reason that some recent authors have cruelly criticized correlation; for example:

- In 1998, the *Encyclopedia of Biostatistics* was published in six large volumes. It touched on many applications of correlation but yet cautioned the reader against them:

    The high profile assumed by the concept of correlation during the early part of the twentieth century has now largely vanished. This is partly due to the emergence of more penetrating methods of statistical analysis. In particular, the emphasis has gradually moved away from an index measuring a degree of association, to an attempt to describe more explicitly the nature of that association. In a word, the emphasis has move away from correlation toward regression.

    (Armitage and Colton, 1998, p. 975; underlining added).

- In a 2001 book titled *Applying Regression & Correlation* we find:

    A correlation is just a number that represents a special case of a regression line....

    (Miles and Shevlin, 2001, p. 20)

- In a 2010 paper in the "Staying Current" section of a journal titled *Advances in Physiology Education*, the biostatistician author stated:

    Correlation can help provide us with evidence that study of the nature of the relationship between x and y may be warranted.... Unlike correlation, however, regression can estimate the nature of different kinds of relationships between two variables. Because of this versatility, regression is a far more useful technique.

    (Curran-Everett, 2010, pp. 186, 191; underlining added)

The authors of a late-20th-century textbook (*Applied Linear Statistical Models*, by Neter et al. (1996)) avoided using the word correlation in their four-page introduction to what a century earlier would have been called correlation (pp. 80–84). Instead, that introduction repeatedly used the term "degree of linear association"; its only use of the word correlation was in labeling r as the "correlation coefficient." In those same pages, the authors defined r as the square-root of the coefficient of determination and then explained that r "does not have...a clear-cut operational interpretation" whereas "r^2 indicates the proportionate reduction in the variability of Y attained by the use of information about X" (p. 82). In contrast to that introductory section, the word correlation was used many times in a later chapter titled "Normal Correlation Models"; the very first sentence in that chapter stated that... "The purpose of this chapter [Chapter 15] is to indicate the relation between regression models and their uses, discussed in Chapters 1–14, and normal correlation models" (p. 631). A century earlier, that ratio of chapters might likely have been the reverse, i.e. 14 chapters on correlation and only one on regression.

Another recently published textbook uses most of the introductory pages in its chapter on correlation to warn the reader about the inherent human weakness for confusing correlation with causation (Vidakovic, 2017, pp. 649–650). The battle against that weakness has been waged for more than a century, and it will probably still be going on a century from now.

At the end of the 20th century, Stephen Stigler published his book titled *Statistics on the Table: The History of Statistical Concepts and Methods*, which comprised 22 chapters with titles such as "Jevons as Statistician," "Galton and Identification by Fingerprints," and "Regression toward the Mean" (Stigler, 1999). The word "correlation" or "co-relation" does not appear in any chapter title. Under the word "correlation," the book's Index lists only eight pages (6, 20, 21, 89n, 104, 105, 182, and 340); those pages touch briefly on serial correlation, spurious correlation, the correlation coefficient, and the history of the word correlation. If that book had been written at the start of the 20th century, it would have included at least one large chapter on correlation.

Joseph Henrich's 2020 book titled *The WEIRDest People in the World* contains many X, Y plots that are labeled with correlation values, but he does not clearly state what most of those values represent, as if it is not important. For example, its first linear-regression plot was labeled with "$R^2 = 0.56$" (on its page 44), whereas what was labeled simply as "Correlation" on the plots on its page 201 is not R^2 but rather a type of r value called "Spearman correlations" (as explained only in end-note #6 on page 542). The book contains many other X, Y plots that include correlation values (e.g., in Figure 7.4.A on page 242, we find "Correlation = −0.59," and in Figure B.1.A on page 499, we find "Correlation = 0.23"), but the reader is nowhere told how to interpret those numbers; that is, are they $_sR^2$ or r values or some other statistic?

The correlation coefficient has fallen out of favor with some major statistical software programs (e.g., *JMP 15*, *Minitab 17*, and *Excel 2019*). In them, the standard output of X, Y linear-regression analysis does not include the cor-relation coefficient but rather the coefficient of determination.

It is clear that the initial 20th century's tidal wave of interest in correla-tion has receded, a process that, in 1967, Snedecor and Cochran claimed had begun in the 1920s:

> *Over the last forty years, investigators have tended to increase their use of regression techniques and decrease their use of correlation techniques. Several reasons can be suggested.* [The primary reason is that...] *The correlation coef-ficient r merely estimates the degree of closeness of linear relationship between X and Y, and the meaning of this concept is not easy to grasp. To ask whether the relation between Y and X is close or loose may be sufficient in an early stage of research. But more often the interesting questions are: How much does Y change for a given change in X? What is the shape of the curve connecting Y and X? How accurately can Y be predicted from X? These questions are handled by regression techniques.*

> *(Snedecor and Cochran, 1967, p. 188)*

It is important to note that Snedecor and Cochran were there focusing on Relational Correlation, not Co-Relational Correlation.

It is interesting that correlation's diminished reputation did not stop the *Journal of the American Statistical Association* from recently publishing an arti-cle titled "A New Coefficient of Correlation"; the author was a statistics pro-fessor at Stanford University (Chatterjee, 2021).

My hope for the future is that statistics textbooks and instructors will pro-mote the use of the signed coefficient of determination, instead of the correla-tion coefficient; as previously discussed, $_sR^2$ is a valid measurement, whereas r is not. Using r to measure correlation is like using square-root-year to mea-sure age. One of my grand-nieces proudly tells me that she is 4 years old, and another grand-niece is 9 years old; I suspect that they would be disappointed if their ages were reported as two and three (i.e., the square roots of four and nine). Using the correlation coefficient to measure the strength of correlation is as nonsensical as using square-root-year to measure the age of children.

If I could go back in time, I'd name the regression coefficient the "relation coefficient" and name the correlation coefficient the "co-relation coefficient." Those proposed names match up better not only with the historical record but also with the purpose of the respective coefficients, as can be seen in most chapters in this book.

I am happy to report that as of 4pm Central Daylight Time (USA) on August 30, 2021, the "#1 Best Seller [book] in Statistics" on Amazon.com was the 13th edition of Mario F. Triola's *Elementary Statistics*. I say "happy," because that book uses several pages to clearly instruct the reader to *not* use the correla-tion coefficient when *measuring* correlation, but rather to use its square, the coefficient of determination (r^2) (Triola, 2021, Kindle Edition, section 10.3).

Nobody knows the future, with the possible exception of "super-forecasters," who find correlation-related analysis to be "a valuable tool" (Tetlock and Gardner, 2015, p. 101). Unfortunately, I predict that many introductory statistics textbooks and instructors will continue to confuse students in regard to correlation, and that confused students will become confused engineers and researchers who will pen confusing reports and papers. I hope that the readers of this book will do better.

The history of the word correlation and its coefficient is in effect the history of the *conceptual* development of mathematical correlation. Its history seems to be similar to that of the calculus. In the mid-17th century, Newton and Leibniz formulated the concepts and methods that we now call calculus, but those concepts were not immediately universally accepted as being mathematically valid. Are the results of calculus calculations exact or approximate, albeit extremely precise approximations? If A has a limiting value of B, how can we say that $A = B$ since A never gets to B? Similarly, how can the limit of $1/X$ as X goes to zero have any meaning since $1/0$ is undefined? These and other questions were not answered satisfactorily until the late 19th century, when a "rigorous formulation" foundation for the calculus was finally built (Boyer, 1949, Chapter VII). Thus it took about 200 years to develop the supporting concepts, formulas, and words needed to achieve the present day virtually universal agreement on how to interpret a derivative and an integral. Correlation is much simpler than calculus, but it too may take centuries to achieve universal agreement on the meaning of the word correlation and on how its coefficient should be used and interpreted; I hope that this book will have played a part in achieving that goal. I hope that someone someday writes another history of correlation, a history that covers the years from 2000 to when that goal will have been achieved.

Bibliography

Alder, H. L., and Roessler, E. B. 1977. *Introduction to Probability and Statistics*, 6th ed. San Francisco, CA: W. H. Freeman.

Amstat.org-Walker: *https://magazine.amstat.org/blog/2016/12/01/sih-hwalker/*

Anonymous. 1839. "The Silurian System". *Q. Rev.*, 64(127): 102–119.

Anonymous. 1901. "Editorial". *Biometrika*, 1(1): 1.

Anonymous. 1911. "Obituary: Sir Francis Galton, D.C.L., D.Sc., F.R.S.". *J. R. Stat. Soc.*, 74(3): 314–20.

Arkin, H., and Colton, R. R. 1935. *An Outline of Statistical Methods as Applied to Economics, Business, Education, Social and Physical Sciences, etc.*, 2nd ed.; 1939 4th ed.; 1970 5th ed. New York: Barnes & Noble, Inc.

Armitage, P. 1995. "Before and after Bradford Hill: Some Trends in Medical Statistics". *J. R. Stat. Soc. A*, 158(1): 143–153.

Armitage, P., and Colton, T. (eds.) 1998. *Encyclopedia of Biostatistics*. Chichester: John Wiley & Sons.

ASA. 1926. "Constitution and By-laws of the American Statistical Association". *J. Am. Stat. Assoc.*, 21(sup 1): 3–5.

ASA. 1941. *Index to the Journal of the American Statistical Association, Volumes* 1-34, 1888–1939. Washington, DC: American Statistical Association.

Ash, C. 1993. *The Probability Tutoring Book: An Intuitive Course for Engineers and Scientists (and Everyone Else!)*. Piscataway, NJ: IEEE Press.

ASU.edu-Hill: *https://embryo.asu.edu/pages/austin-bradford-hill-1897-1991*

Athenaeum-History: *https://theathenaeum.co.uk/the-club/our-history/*

Bailey, W. B. 1911. "Reviews & Notes: An Introduction to the Theory of Statistics. By G. Udney Yule". *Pub. Am. Stat. Assoc.*, 12: 765.

Baldwin, M. 2015. *Making Nature: The History of a Scientific Journal*. Chicago, IL: University of Chicago Press.

Baldwin, M. 2020. "The Business of Being an Editor: Norman Lockyer, Macmillan and Company, and the Editorship of *Nature*, 1869–1919". *Centaurus*, 2020: 1–14. https://doi.org/10.1111/1600-0498.12274.

Bateson, W. 1889. "On Some Variations of *Cardium edule*, Apparently Correlated to the Conditions of Life". *Phil. Trans. R. Soc. London B.*, 180: 297–330.

Bell, A. G. 1885. "Is There a Correlation between Defects of the Senses?" *Science*, 5(106): 127–129.

Bergson, H. 1910. *L'Évolution Créatrice*, sixième édition. Paris: Felix Alcan.

Bergson, H. 1911. *Creative Evolution*. A. Mitchell's translation of *L'Évolution Créatrice*. London: MacMillan.

Bertillon, A. 1893. *Identification Anthropométrique: Instructions Signalétiques*, Nouvell édition, Melun: Imprimerie Adminstrative.

Bertillon, A. 1896. *Signaletic Instructions, Including the Theory and Practice of Anthropometrical Identification*. Translation by: McLaughry, R. W. (ed.). Chicago, IL: Werner Co.

Blair, M. M. 1944. *Elementary Statistics with General Applications*. New York: Henry Holt.

Boas, F. 1893. "Remarks on the Theory of Anthropometry". *Pub. Am. Stat. Assoc.*, 3: 569–575.

Boas, F. 1894. "The Correlation of Anatomical or Physiological Measurements". *Am. Anthropol.*, 7: 313–324.

Boas, F. 1921. "The Coefficient of Correlation". *Q. Pub. Am. Stat. Assoc.* 17: 683–688.

Bohrnstedt, G. W., and Knoke, D. 1988. *Statistics for Social Data Analysis*, 2nd ed. Itasca: F. E. Peacock Publishers.

Bowditch, H. P. 1877. "The Growth of Children". Boston: Albert J. Wright, State Printer ("From the Eighth Annual Report of the State Board of Health of Massachusetts"). Additionally, a "reprint" of this was published in 1894 by the American Statistical Association (Boston) in a book titled *Papers on Anthropometry* (no editor's name was given).

Bowditch, H. P. 1891. *The Growth of Children, Studied by Galton's Method of Percentile Grades.* (The only publication information provided in this stand-alone pamphlet was this: "Reprinted from the Twenty-second Annual Report of the State Board of Health of Massachusetts. Boston, 1891").

Bowker, A. H., and Lieberman, G. J. 1955. *Handbook of Industrial Statistics*. Englewood Cliffs, NJ: Prentice-Hall.

Bowker, A. H., and Lieberman, G. J. 1972. *Engineering Statistics*, 2nd ed. Englewood Cliffs, NJ: Prentice-Hall Inc.

Bowley, A. L. 1901. *Elements of Statistics*; 1902. 2nd ed.; 1907. 3rd ed.; 1920. 4th ed.; 1926. 5th ed. London: P. S. King & Son.

Bowley, A. L. 1937. *Elements of Statistics*, 6th ed. London: Staples Press (reprinted 1948).

Boyer, C. B. 1949. *The History of the Calculus and Its Conceptual Development*. New York: Dover Publications.

Bravais, A. 1846. "Analyse mathématique sur les probabilités des erreurs de situation d'un point", *Mémoires présentés par divers savants à l'Académie royale des sciences de l'Institut de France*, 9: 255–332.

Bray, C. 1866. *On Force, Its Mental and Moral Correlates*. London: Longmans, Green, Reader, and Dyer.

Britannica.com-Arithmometer: *https://www.britannica.com/technology/Arithmometer*

Britannica.com-Bergson: *https://www.britannica.com/biography/Henri-Bergson*

Britannica.com-Boas: *https://www.britannica.com/biography/Franz-Boas*

Britannica.com-Bravais: *https://www.britannica.com/biography/Auguste-Bravais*

Britannica.com-Bray: *https://www.britannica.com/biography/George-Eliot#ref260686*

Britannica.com-Fisher: *https://www.britannica.com/biography/Ronald-Aylmer-Fisher*

Britannica.com-Hull: *https://www.britannica.com/biography/Clark-L-Hull*

Britannica.com-Huxley: *https://www.britannica.com/biography/Thomas-Henry-Huxley*

Britannica.com-Owen: *https://www.britannica.com/science/fossil*

Britannica.com-Spencer: *https://www.britannica.com/biography/Herbert-Spencer*

Britannica.com-Twenties: *https://www.britannica.com/topic/Roaring-Twenties*

Britannica.com-Uniformitarianism: *https://www.britannica.com/science/uniformitarianism*

Britannica.com-Wright: *https://www.britannica.com/biography/Sewall-Wright*

Brooks, C. F. 1932. "William Gardner Reed", *Bul. Am. Meteor. Soc.* 13(5): 90–91.

Brown, W. 1911. *The Essentials of Mental Measurement*. Cambridge: University Press.

Brown, W., and Thomson, G. H. 1921. *The Essentials of Mental Measurement*. Cambridge: University Press.

Bulmer, M. 2003. *Francis Galton: Pioneer of Heredity and Biometry*. Baltimore, MD: John Hopkins University Press.

Burgess, R. W. 1927. *Introduction to the Mathematics of Statistics*. Boston, MA: Houghton Mifflin Co.

Camp, B. H. 1931. *The Mathematical Part of Elementary Statistics: A Textbook for College Students*. Boston, MA: D. C. Heath & Co.

Chatterjee, S. 2021. "A New Coefficient of Correlation". *J. Am. Stat. Assoc.*, 116(536): 2009–2022.

Chiang, C. L. 2003. *Statistical Methods of Analysis*. Singapore: World Scientific Publishing.

Church, D. E. 1940. *Speed Methods of Statistics for Use in Business*. New York: Ronald Press.

Columbia.edu-Mendel: *https://statmodeling.stat.columbia.edu/2012/08/08/gregor-mendels-suspicious-data/*

Coolidge, J. L., 1925. *An Introduction to Mathematical Probability*. London: Oxford University Press.

Cornell.edu-Day: *https://president.cornell.edu/the-presidency/edmund-ezra-day/*

Cox, D. R. 2001. "*Biometrika*: The First 100 Years". *Biometrika*, 88(1): 3–11.

Crow, E. L., et al. 1960. *Statistics Manual*. New York: Dover Publications.

Crum, F. S. 1901. [A Review of] "A Study of Municipal Growth". *Pub. Am. Stat. Assoc.*, 7(56): 84–86.

Crum, W. L., and Patton, A. C. 1925. *An Introduction to the Methods of Economic Statistics*. New York: McGraw-Hill.

Csiszar, A. 2018. *The Scientific Journal: Authorship and the Politics of Knowledge in the Nineteenth Century*. Chicago, IL: University of Chicago Press.

Cummings, J. 1920. "The Conception of Statistics as a Technique". *Q. Pub. Am. Stat. Assoc.*, 17: 164–176.

Curran-Everett, D. 2010. "Explorations in Statistics: Correlation". *Adv. Physiol. Educ.* 34: 186–191.

Cuvier, G. 1812. *Recherches sur les ossemens fossiles de quadrupèdes*, vol. 1. Paris: Chez Deterville.

Cuvier, G. 1825a. *Discours sur les révolutions de la surface du globe, et sur les changemens qu'elles ont produits dans le régne animal*, 3rd ed. Paris: Dufour and D'Ocagne.

Cuvier, G. 1825b. *Discourse on the Revolutionary Upheavals on the Surface of the Globe and on the Changes Which They Have Produced in the Animal Kingdom*. Translated 2009 by Ian Johnston. Downloaded from https://web.viu.ca/johnstoi/cuvier/cuvierweb.pdf.

Cuvier, G. 1831. *A Discourse on the Revolutions of the Surface of the Globe and the Changes Thereby Produced in the Animal Kingdom*. Philadelphia, PA: Carey & Lea.

Darwin, C. 1845. *Journal of Researches...*, 2nd ed. Republished 1939 as *The Voyage of the Beagle* by New York: P. F. Collier & Son.

Darwin, C. 1859. *On the Origin of Species...*. London: John Murray.

Darwin, C. 1868. *The Variation of Animals and Plants under Domestication*. London: John Murray.

Darwin, C. 1869. *On the Origin of Species...*, 5th ed. Re-published 1871 by New York: D. Appleton & Co.

Darwin, C. 1871. *The Descent of Man*, 1st ed. Re-published 1873 in two volumes by New York: D. Appleton & Co.

Darwin, C. 1872. *The Origin of Species...*, 6th ed. Re-published 1991 by New York: Prometheus Books.

Darwin, C. 1876. *Recollections of the Development of My Mind and Character* (Re-published 1958 as *The Autobiography of Charles Darwin, 1809–1882, with Original Omissions Restored,* edited by N. Barlow). New York: W. W. Norton.

Darwin, F., and Seward, A. C. (eds.) 1903. *More Letters of Charles Darwin,* vol. 1. New York: D. Appleton & Co.

Daston, L. 1988. *Classical Probability in the Enlightenment.* Princeton, NJ: Princeton University Press.

Davenport, C. B. 1899. *Statistical Methods with Special Reference to Biological Variation.* New York: John Wiley & Sons.

Davenport, C. B. 1900. "A History of the Development of the Quantitative Study of Variation". *Science,* 12(310): 864–870.

Davenport, C. B. 1904. *Statistical Methods with Special Reference to Biological Variation,* 2nd ed. New York: John Wiley & Sons.

Davenport, C. B. 1914. *Statistical Methods with Special Reference to Biological Variation,* 3rd ed. New York: John Wiley & Sons.

Davenport, C. B., and Bullard, C. 1896. "A Contribution to the Quantitative Study of Correlated Variation and the Comparative Variability of the Sexes". *Proc. Am. Acad.,* 32: 87–97.

Davenport, E., and Rietz, H. L. 1907. *Principles of Breeding: A Treatise on Thremmatology.* New York: Ginn & Co.

Davis, H. T., and Nelson, W. F. C. 1937. *Elements of Statistics, with Applications to Economic Data,* 2nd ed. Bloomington, IN: Principia Press.

Day, E. E. 1918. "A Note on King's Article on 'The Correlation of Historical Economic Variables and the Misuse of Coefficients in This Connection'". *Q. Pub. Am. Stat. Assoc.,* 16(122): 115–118.

Day, E. E. 1925. *Statistical Analysis.* New York: MacMillan Company.

de Beer, G. 1965. *Charles Darwin: A Scientific Biography.* Garden City: Doubleday & Co.

de Beaumont, E. 1865. "Memoir of August Bravais". C. A. Alexander (translator). Washington: Government Printing Office. *Annual Report of the Board of Regents of the Smithsonian Institution, Showing the Operations, Expenditures, and Conditions of the Institution for the Year 1869.* 1871, pp. 145–168.

Desrosières, A. 1998. *The Politics of Large Numbers: A History of Statistical Reasoning.* Cambridge: Harvard University Press.

Dictionary: Contradistinction: https://dictionary.cambridge.org/dictionary/english/contradistinction.

Dodd, S. C. 1926. "A Correlation Machine". *Industrial Psychology,* 1: 46–58.

Edgeworth, F. Y. 1881. *Mathematical Psychics: An Essay on the Application of Mathematics to the Moral Sciences.* London: C. Kegan Paul & Co.

Edgeworth, F. Y. 1892. "Correlated Averages". *London Edinburgh Dublin Philos. Mag. J. Sci.,* 34: 190–204.

Edgeworth, F. Y. 1893. "Statistical Correlation between Social Phenomena". *J. R. Stat. Soc.,* 56: 670–675.

Edgeworth, F. Y. 1917. "Review of *Scope and Methods of Statistics,* by H. Westergaard". *J. R. Stat. Soc.,* 80: 546–551 (reprinted in *Edgeworth on Chance, Economic Hazard, and Statistics,* by P. Mirowski (ed.). 1994 by Lanham: Rowman & Littlefield Publishers).

Elderton, W. P. 1906. *Frequency-Curves and Correlation.* London: Charles & Edwin Layton.

Elderton, W. P. 1927. *Frequency Curves and Correlation*, 2nd ed. London: Charles & Edwin Layton.

Elderton, W. P., and Elderton, E. M. 1910. *Primer of Statistics*, 2nd ed. London: Adam & Charles Black.

Elderton, W. P., and Elderton, E. M. 1912. *Primer of Statistics*, 3rd ed. London: Adam & Charles Black (reprint 1914).

Encyclopediaofmath.org-Bowley: *https://encyclopediaofmath.org/wiki/Bowley,_Arthur_Lyon*

Encyclopediaofmath.org-Snedecor: *https://encyclopediaofmath.org/wiki/Snedecor,_George_Waddel*

Enebuske, C. J. 1893. "An Anthropometrical Study of the Effects of Gymnastic Training on American Women". *Pub. Am. Stat. Assoc.*, 3: 600–610.

Ezekiel, M. 1924. "A Method of Handling Curvilinear Correlation for Any Number of Variables". *Pub. Am. Stat. Assoc.*, 19: 431–453.

Ezekiel, M. 1930. *Methods of Correlation Analysis*. 1941, 2nd ed. New York: John Wiley & Sons.

Ezekiel, M., and Fox, K. A. 1959. *Methods of Correlation and Regression Analysis* (in Effect, This Is the 3rd ed. of *Methods of Correlation Analysis*). New York: John Wiley & Sons.

Fancher, R. E. 1989. "Galton on Examinations: An Unpublished Step in the Invention of Correlation". *ISIS*, 80: 446–455.

Fasten, N. 1930, *Origin through Evolution*. New York: F. S. Crofts & Co.

Fisher, I. 1917. "The 'Ratio' Chart for Plotting Statistics". *Q. Pub. Am. Stat. Assoc.*, 15(118): 577–601.

Fisher, R. A. 1925. *Statistical Methods for Research Workers*. Edinburgh: Oliver and Boyd.

Fisher, R. A. 1958. *Statistical Methods for Research Workers*, 13th ed. (Re-printed 1963). New York: Hafner.

Fitzharris, L. 2017. *The Butchering Art*. New York: Scientific American/Farrar, Straus, and Giroux.

Forrest, D. W. 1974. *Francis Galton: The Life and Work of a Victorian Genius*. New York: Taplinger Publishing.

Fosdick, R. B. 1915. "Passing of the Bertillon System of Identification". *J. Criminal Law Criminol.*, 6(3): 363–369.

Foster, A. S., 1942. *Practical Plant Anatomy*. New York: D. Van Nostrand Co.

Freund, J. E. 1960. *Modern Elementary Statistics*, 2nd ed. Englewood Cliffs, NJ: Prentice-Hall.

G. H. D. 1912. "Sir Francis Galton, 1822–1911". In "Obituary Notices of Fellows Deceased". *Proc. R. Soc. London B*, 84: i–xxvii.

Gallant, R. A. 1972. *Charles Darwin: The Making of a Scientist*. Garden City: Doubleday & Co.

Galton, F. 1856. *The Art of Travel...*, 2nd ed. London: John Murray.

Galton, F. 1861. "Zanzibar". *Mission Field*, 6: 121–130.

Galton, F. 1863. "A Development of the Theory of Cyclones". *Proc. R. Soc. London*, 12: 385–386.

Galton, F. 1865. "The First Steps towards the Domestication of Animals". *Trans. Ethnol. Soc. London*, 3: 122–138.

Galton, F. 1866. "On an Error in the Usual Method of Obtaining Meteorological Statistics". *Athenaeum*, 2027: 274.

Galton, F. 1867. *The Art of Travel; or, Shifts and Contrivances Available in Wild Countries*, 4th ed. London: John Murray.

Galton, F. 1869. *Hereditary Genius: An Inquiry into Its Laws and Consequences*. London: Macmillan & Co.

Galton, F. 1870a. "Barometric Predictions of Weather". *Rep. Br. Ass. Adv. Sci.*, 40: 31–33.

Galton, F. 1870b. "Barometric Predictions of Weather". *Nature*, 2(October 20): 501–503.

Galton, F. 1871. "Experiments in Pangenesis…". *Proc. R. Soc. London*, 19: 393–410.

Galton, F. 1872a. "Description of the Trace Computer, Designed by Mr. Galton". *Rep. Meteorol. Committee R. Soc.*, 1871: 24–28.

Galton, F. 1872b. "Geography: Opening Address by the President, Francis Galton, F.R.S.". *Nature*, 6: 343–345.

Galton, F. 1872c. "On Blood-Relationship". *Proc. R. Soc. London*, 20: 394–402.

Galton, F. 1872d. "Statistical Inquiries into the Efficacy of Prayer". *Fortnightly Rev.*, 68: 125–135.

Galton, F. 1873. "Hereditary Improvement". *Fraser's Mag.*, 7: 116–130.

Galton, F. 1874a. "On Men of Science, Their Nature and Their Nurture". *Proc. R. Inst. Great Britain*, 7: 227–236.

Galton, F. 1874b. "Men of Science, Their Nature and Their Nurture". *Nature*, 9: 344–345.

Galton, F. 1874c. *English Men of Science: Their Nature and Nurture*. London: Macmillan & Co.

Galton, F. 1874d. "On a Proposed Statistical Scale". *Nature*, 9: 342–343.

Galton, F. 1877a. "Typical Laws of Heredity". *Proc. R. Inst. Great Britain*, 8: 282–301.

Galton, F. 1877b. "Typical Laws of Heredity". *Nature*, 15: 492–495, 512–514, 532–533.

Galton, F. 1877c. *Address to the Anthropological Department of the British Association, Plymouth, 1877*. London: William Clowes & Sons.

Galton, F. 1879. "The Geometric Mean, in Vital and Social Statistics". *Proc. R. Soc. London*, 29: 365–367.

Galton, F. 1880a. "Mental Imagery". *Fortnightly R.*, 28: 312–324.

Galton, F. 1880b. "Visualised Numerals". *Nature*, 21: 252–256.

Galton, F. 1881a. "The Anthropometric Laboratory". *Fortnightly Rev.*, 31: 332–338.

Galton, F. 1881b. "Isochronic Postal Charts". *Rep. Br. Ass. Adv. Sci.*, 51: 740–741.

Galton, F. 1883. *Inquiries into Human Faculty and Its Development*. New York: Macmillan & Co.

Galton, F. 1884a. *Records of Family Faculties*. London: Macmillan & Co.

Galton, F. 1884b. "The Cost of Anthropometric Measurements" [Letter to Editor]. *Nature*, 31: 150.

Galton, F. 1885a. [President's] *Address to the Section of Anthropology of the British Association*. London: Spottiswoode.

Galton, F. 1885b. *On the Anthropometric Laboratory at the Late International Health Exhibit*. London: Harrison and Sons.

Galton, F. 1885c. "On the Anthropometric Laboratory at the Late International Health Exhibition". *J. Anthro. Inst. Great Britain Ireland*, 14: 205–221.

Galton, F. 1885d. "Some Results of the Anthropometric Laboratory". *J. Anthro. Inst. Great Britain Ireland*, 14: 275–287.

Galton, F. 1886a. "Family Likeness in Stature". *Proc. R. Soc. London*, 40: 42–63.

Galton, F. 1886b. "President's Address". *J. Anthro. Inst. Great Britain Ireland*, 15: 489–99.

Galton, F. 1886c. "Regression towards Mediocrity in Hereditary Stature". *J. Anthro. Inst. Great Britain Ireland*, 15: 246–263. This paper is sometimes referenced as 1885, the year it was presented orally.

Galton, F. 1887a. "Notes on Permanent Colour Types in Mosaic". *J. Anthro. Inst. Great Britain Ireland*, 16: 145–147.

Galton, F. 1887b. "Thought without Words". *Nature*, 36: 28–29.

Galton, F. 1888a. "President's Address". *J. Anthro. Inst. Great Britain Ireland*, 17: 346–354.

Galton, F. 1888b. "Personal Identification and Description", *Proc. R. Inst.*, 12: 346–360.

Galton, F. 1888c. "Co-relations and Their Measurement, Chiefly from Anthropometric Data". *Proc. R. Soc. London*, 45: 135–145.

Galton, F. 1889a. *Natural Inheritance*. Facsimile Published 1997. Placitas: Genetics Heritage Press.

Galton, F. 1889b. "Correlations and Their Measurement, Chiefly from Anthropometric Data". *Nature*, 39: 238.

Galton, F. 1889c. "Tables of Observations". *J. Anthro. Inst. Great Britain Ireland*, 18: 420–430.

Galton, F. 1889d. "Human Variety". *Nature*, 39: 296–300.

Galton, F. 1889e. "President's Address...January 22nd, 1889". *J. Anthro. Inst. Great Britain Ireland*, 18: 401–419.

Galton, F. 1889f. *President's Address...January 22nd, 1889*. London: Harrison and Sons.

Galton, F. 1890a. "Kinship and Correlation". *North Am. Rev.*, 150(401): 419–431.

Galton, F. 1890b. *Anthropometric Laboratory: Notes and Memoirs*. London: Richard Clay & Sons.

Galton, F. 1891. "Method of Indexing Finger Marks". *Nature*, 44: 141.

Galton, F. 1892a. *Finger Prints*. London: Macmillan & Co.

Galton, F. 1892b. *Finger Prints*. Facsimile Published 1965 by New York: Da Capo Press.

Galton, F. 1893. "The Just-Perceptible Difference". *Proc. R. Inst.*, 14: 13–26.

Galton, F. 1894. "The Part of Religion in Human Evolution". *Natl. Rev.*, 23: 755–763.

Galton, F. 1895. *Fingerprint Directories*. London: Macmillan & Co.

Galton, F. 1897. "Note to the Memoir by Professor Karl Pearson, F. R. S., on Spurious Correlation". *Proc. R. Soc. London*, 60: 498–502.

Galton, F. 1900. "Identification Offices in India and Egypt". *Nineteenth Century*, 48: 118–26.

Galton, F. 1901a. "The Possible Improvement of the Human Breed under the Existing Conditions of Law and Sentiment". *Nature*, 64: 659–665.

Galton, F. 1901b. "Biometry". *Biometrika*, 1(1): 7–10.

Galton, F. 1907. *Probability, the Foundation of Eugenics*. Oxford: Clarendon Press.

Galton, F. 1908a. "Suggestions for Improving the Literary Style of Scientific Memoirs". *Trans. R. Soc. Lit.*, 28(Part II): 113–129 (including the "Discussion").

Galton, F. 1908b. *Memories of My Life*. London: Methuen & Co.

Galton, F. 1909. *Essays in Eugenics*. London: The Eugenics Education Society (Reprinted 1996 by Scott-Townsend Publishers, Washington, DC, USA).

Galton, F., and Mahomed, F. A. 1882. "An Inquiry into the Physiognomy of Phthisis by the Method of 'Composite Portraiture'". *Guy's Hospital Rep.*, 25: 475–493 + plates.

Garnett, J. C. M. 1919, "On Certain Independent Factors in Mental Measurements". *Proc. R. Soc. London A*, 96(675): 91–111.

Garson, J. G. 1900. "The Metric System of Identification of Criminals, as Used in Great Britain and Ireland". *J. Anthro. Inst. Great Britain Ireland*, 30: 161–198.

Gavett, G. I. 1925. *A First Course in Statistical Method*. New York: McGraw-Hill Book Co.

Gehlke, C. E. 1917. "On the Correlation between the Vote for Suffrage and the Vote on the Liquor Question: A Preliminary Study," *Q. Pub. Am. Stat. Assoc.*, 15(117): 524–532.

Giffen, R. 1913. *Statistics.* London: Macmillan & Co. (Its title page states that it was "written about the years 1898–1900" and edited/published in 1913 by H. Higgs and G. U. Yule).

Gillham, N. W. 2001. *A Life of Sir Francis Galton: From African Exploration to the Birth of Eugenics.* Oxford: University Press.

Goedicke, V. 1953. *Introduction to the Theory of Statistics.* New York: Harper & Brothers.

Goldenweiser, E. A. 1916. "Classification and Limitations of Statistical Graphics". *Q. Pub. Am. Stat. Assoc.*, 15: 205–209.

Goodrich, E. S. 1919. *The Evolution of Living Organisms*, 2nd ed. London: T.C. & E. C. Jack Ltd.

Gravetter, F. J., and Wallnau, L. B. 2000. *Statistics for the Behavioral Sciences*, 5th ed. Belmont, CA: Wadsworth.

Gray, J. 1911. "Sir Francis Galton, M. A., D. C. L., F. R. S. Born, February 16, 1822; Died, January 17, 1911". *Man*, 11: 33–34.

Grove, W. R. 1846. *On the Correlation of Physical Forces.* London: Samuel Highley [Note: The title of this 1st edition started with "On"; all subsequent edition titles started with "The".]

Grove, W. R. 1850. *The Correlation of Physical Forces*, 2nd ed. London: Samuel Highley.

Grove, W. R. 1855. *The Correlation of Physical Forces*, 3rd ed. London: Longman, Brown, Green, & Longmans.

Grove, W. R. 1862. *The Correlation of Physical Forces*, 4th ed. London: Longman, Green, Longman, Roberts, & Green.

Grove, W. R. 1867. *The Correlation of Physical Forces*, 5th ed. London: Longmans, Green, & Co.

Grove, W. R. 1874a. *The Correlation of Physical Forces*, 6th ed. London: Longmans, Green, & Co.

Grove, W. R. 1874b. *The Correlation of Physical Forces*, 6th ed. London: Longmans, Green, & Co. This is Galton's personal copy that is held by University College London's Special Collections Library, Galton Lab Misc. 37, Barcode # UCL0132620.

H'Doubler, F. T. 1910. "A Formula for Drawing Two Correlated Curves So as to Make the Resemblance as Close as Possible". *Pub. Am. Stat. Assoc.*, 12: 165–167.

Hacking, I. 1975. *The Emergence of Probability.* Cambridge: University Press.

Hacking, I. 1984. "Trial by Number". *Science*, 5(9): 69–70.

Hacking, I. 1990. *The Taming of Chance.* Cambridge: University Press.

Hald, A. 2007. *A History of Parametric Statistical Inference from Bernoulli to Fisher, 1713–1935.* New York: Springer.

Hallock, E. B. 1898. *Suggestions for Primary and Intermediate Lessons on the Human Body: A Study of its Structure and Needs Correlated with Nature Study.* New York: E. L. Kellogg & Co.

Harris, J. A. 1916. "An Outline of the Current Progress in the Theory of Correlation and Contingency". *Am. Nat.*, 50: 53–64.

Harris, J. A. 1917. "Reed on the Coefficient of Correlation". *Q. Pub. Am. Stat. Assoc.*, 15: 803–805.

Hartwell, E. M. 1893. "A Preliminary Report on Anthropometry in the United States". *Pub. Am. Stat. Assoc.*, 3: 554–568.

Harvard.edu-Bowditch: *https://hollisarchives.lib.harvard.edu/repositories/14/resources/7971*

Hayakawa, S. I., and Hayakawa, A. R. 1990. *Language in Thought and Action*, 5th ed. San Diego: Harcourt, Brace, Jovanovich.

Healey, J. F. 1984. *Statistics: A Tool for Social Research*. Belmont, CA: Wadsworth Publishing.

Henrich, J. 2020. *The WEIRDest People in the World*. New York: Farrar, Straus and Giroux.

Hetwebsite.net-Persons: *https://hetwebsite.net/het/profiles/persons.htm*

Hibben, J. G. 1896. *Inductive Logic*. New York: Charles Scribner's Sons.

Hill, A. B. 1939 *Principles of Medical Statistics*, 2nd ed. London: The Lancet Limited.

Hill, A. B. 1955. *Principles of Medical Statistics*, 6th ed. London: The Lancet Limited.

Hill, A. B. 1961. *Principles of Medical Statistics*, 7th ed. London: The Lancet Limited.

Hilts, V. L. 1973. "Statistics and Social Science", in Giere, R. N., and Westfall, R. S. (eds.), *Foundations of Scientific Method: The Nineteenth Century*. Bloomington, IN: Indiana University Press.

Hilts, V. L. 1975. "A Guide to Francis Galton's *English Men of Science*". *Trans. Am. Philos. Soc.*, 65(5): 1–85.

Himmelfarb, G. 1968. *Darwin and the Darwinian Revolution*. New York: W. W. Norton & Co.

Holland, B. K. 1998. *Probability without Equations: Concepts for Clinicians*. Baltimore, MD: John Hopkins University Press.

Holmes, G. K. 1891. "A Plea for the Average". *Pub. Am. Stat. Assoc.*, 2: 421–426.

Holmes, G. K. 1892. "Note from Mr. Francis Galton to Mr. George K. Holmes on the Subject of Distribution". *Pub. Am. Stat. Assoc.*, 3: 271–273.

Holzinger, K. J. 1923. "A Combination form for Calculating the Correlation Coefficient and Ratios". *J. Am. Stat. Assoc.*, 18: 623–627.

Hooker, R. H. 1908. "An Elementary Explanation of Correlation: Illustrated by Rainfall and Depth of Water in a Well". *Q. J. R. Meteorol. Soc.*, 34: 277–292.

Hull, C. L. 1925. "An Automatic Correlation Calculating Machine". *Pub. Am. Stat. Assoc.*, 20: 522–531.

Hunt, J. 1896. *Religious Thought in England in the Nineteenth Century*. London: Gibbings & Co.

Huntington, E. V. 1919. "Mathematics and Statistics, with an Elementary Account of the Correlation Coefficient and the Correlation Ratio". *Am. Math. Monthly*, 26: 421–435.

Huxley, J. (ed.) 1940. *The New Systematics*. London: Oxford University Press.

Huxley, T. H. 1863. *Evidence as to Man's Place in Nature*. London: Williams & Norgate.

Huxley, T. H. 1898, 1899, 1901, 1902. *The Scientific Memoirs of Thomas Henry Huxley*, vol. 1, 2, 3, 4. London: Macmillan.

Intelltheory.com-Symonds: *https://intelltheory.com/intelli/percival-symonds/*

Jerome, H. 1924. *Statistical Method*. New York: Harper & Brothers.

Jevons, W. S. 1874a. *The Principles of Science: A Treatise on Logic and Scientific Method*. London: Macmillan

Jevons, W. S. 1874b. *The Principles of Science: A Treatise on Logic and Scientific Method*, vol. 2. London: Macmillan. This is Galton's personal copy that is held by University College London's Special Collections Library, item # UCL0072831.

Jevons, W. S. 1883. *The Principles of Science: A Treatise on Logic and Scientific Method*, 2nd ed. London: Macmillan.

Johnson, N. L., and Kotz, S. 1997. *Leading Personalities in Statistical Sciences from the Seventeenth Century to the Present*. New York: John Wiley & Sons.

Jones, C. E. 1918. "A Genealogical Study of Population". *Q. Pub. Am. Stat. Assoc.*, 16: 201–219.

Jordan, D. S., and Kellogg, V. L. 1907 (Reprinted 1919). *Evolution and Animal Life*. New York: D. Appleton & Co.

Kelley, T. L. 1923. *Statistical Method*. New York: Macmillan Company.

Kendall, M. G. 1943. *The Advanced Theory of Statistics*, vol. I. London: Charles Griffin & Co.

Kendall, M. G. 1946. *The Advanced Theory of Statistics*, vol. II. London: Charles Griffin & Co.

Kendall, M. G. 1962. *Rank Correlation Methods*, 3rd ed. London: Charles Griffin & Co.

Kendall, M. G., and Buckland, W. R. 1960. *A Dictionary of Statistical Terms*. New York: Hafner Publishing Co.

Kendall, M., and Plackett, R. L. 1977. *Studies in the History of Statistics and Probability*, vol. II. New York: MacMillan Publishing Co.

Kenney, J. F. 1939. *Mathematics of Statistics*, Part 1 & Part 2. New York: D. Van Nostrand Co.

Keynes, M. (ed.) 1993. *Sir Francis Galton FRS: The Legacy of His Ideas*. London: Macmillan Press.

King, W. I. 1912. *The Elements of Statistical Method*. New York: Macmillan Company.

King, W. I. 1917. "The Correlation of Historical Economic Variables and the Misuse of Coefficients in this Connection". *Q. Pub. Am. Stat. Assoc.*, 15: 847–853.

King, W. I. 1918. "Reply to Dr. Day's Criticism of My Article on Correlation Coefficients". *Q. Pub. Am. Stat. Assoc.*, 16(123): 171–173.

Kline, M. 1953. *Mathematics in Western Culture*. London: Oxford University Press (paperback edition, 1980).

Kolln, M. 1996. *Rhetorical Grammar: Grammatical Choices, Rhetorical Effects*, 2nd ed. Boston: Allyn & Bacon.

Koren, J. 1915. "Some Statistical Ideals". *Q. Pub. Am. Stat. Assoc.*, 14(109): 351–357.

Koren, J. (ed.) 1918. *The History of Statistics: Their Development and Progress in Many Countries*. New York: Macmillan.

Kozelka, R. M. 1961. *Elements of Statistical Inference*. Reading: Addison-Wesley Publishing.

Kurtz, A. K., and Edgerton, H. A. 1939. *Statistical Dictionary of Terms and Symbols*. New York: John Wiley & Sons.

La Faye, D. (ed.) 2011. *Jane Austen's Letters*, 4th ed. Oxford: University Press.

Lange, F. H. 1967. *Correlation Techniques*. London: Iliffe Books Ltd.

Le Conte, J. 1873. "Correlation of Vital with Chemical and Physical Forces". *College Courant*, 13(23): 266–267.

Lee, S. (ed.) 1901. *Dictionary of National Biography*, Supplement vol. II. London: Smith, Elder, & Co.

Lindquist, E. F. 1940. *Statistical Analysis in Educational Research*. Boston: Houghton Mifflin.

Lindquist, E. F. 1942. *A First Course in Statistics: Their Use and Interpretation in Education and Psychology*, 2nd ed. Boston, MA: Houghton Mifflin Co.

Littledale, R. F. 1876. "The Rationale of Prayer", in Tyndall, J., Galton, F., et al. (eds.), *The Prayer-Gauge Debate*. Boston, MA: Congregational Publishing Society, 37–82.

Lyell, C. 1830, 1832, 1833. *Principles of Geology*, 1st ed., vols. 1, 2, 3. London: John Murray.

Lyell, C. 1873. *Geological Evidences of the Antiquity of Man*, 4th ed. London: John Murray.

Lyell, C. 1875. *Principles of Geology*, 12th ed. (2 vols). London: John Murray.

MAA.org-Alder: *https://www.math.ucdavis.edu/application/files/4516/0996/2718/1999.pdf*

MAA.org-Huntington: *https://mathshistory.st-andrews.ac.uk/Biographies/Huntington/*

MAA.org-Todhunter: *https://old.maa.org/press/periodicals/convergence/mathematical-treasure-todhunters-algebra-for-beginners*

MacKenzie, D. A. 1981. *Statistics in Britain: 1865–1930 – The Social Construction of Scientific Knowledge*. Edinburgh: Edinburgh Univ. Press.

Magee, J. D. 1912. "The Degree of Correspondence between Two Series of Index Numbers". *Q. Pub. Am. Stat. Assoc.*, 13(98): 174–181.

Magee, J. D. 1913. *Money and Prices: A Statistical Study of Price Movements*. Chicago, IL: University of Chicago.

Magee, J. D. 1914. "Note". *Q. Pub. Am. Stat. Assoc.*, 14: 345–346.

Matheson, R. E. 1889. "The Mechanism of Statistics". *J. Stat. Soc. Inquiry Soc. Ireland*, IX: Appendix, 3–25.

Maxwell, J. 1874. "Grove's 'Correlation of Physical Forces'". *Nature*, 10: 302–304.

Mayo-Smith, R. 1899. *Statistics and Economics*. New York: MacMillan Company.

Meadows, A. J. 2008. *Science and Controversy: A Biography of Sir Normal Lockyer*, 2nd ed. London: Macmillan

Miles, J., and Shevlin, M. 2001. *Applying Regression & Correlation*. Los Angeles: Sage Publications.

Mills, F. C. 1924. "The Measurement of Correlation and the Problem of Estimation". *J. Am. Stat. Assoc.*, 19: 273–300

Mills, F. C. 1938. *Statistical Methods Applied to Economics and Business*, 2nd ed. New York: Henry Holt.

Morus, I. R. 2017. *William Robert Grove: Victorian Gentleman of Science*. Cardiff: University of Wales Press.

Mulhall, M. G. 1892. *The Dictionary of Statistics*. London: George Routledge & Sons.

Murray, J. A. H. 1893. *A New English Dictionary on Historical Principles*, vol. II. C. Oxford: Clarendon Press.

Murray, J. A. H. 1914. *A New English Dictionary on Historical Principles*, vol. VIII. Q, R, S–Sh. Oxford: Clarendon Press.

Murray, J. A. H. 1933. *A New English Dictionary on Historical Principles*, Supplement. Oxford: Clarendon Press.

Neter, J., et al. 1996. *Applied Linear Statistical Models*, 4th ed. Chicago, IL: Irwin.

N.Y. Times: Waugh: *https://www.nytimes.com/1985/03/08/nyregion/albert-e-waugh.html*

Onions, C. T., et al. (eds.) 1937. *Oxford Universal Dictionary on Historical Principles*. Oxford: Oxford University Press.

Ostle, B. 1963. *Statistics in Research*, 2nd ed. Ames, IA: Iowa State University Press.

OSU.edu-Darwin: *https://news.osu.edu/darwins-children-represent-highs-and-lows-of-famous-scientists-personal-life/*

Owen, R. 1843. *Lectures on the Comparative Anatomy and Physiology of the Invertebrate Animals, Delivered at the Royal College of Surgeons, in 1843*. London: Longman, Brown, Green, and Longmans.

Owen, R. 1866. *On the Anatomy of Vertebrates*, vol. 1. London: Longmans, Green, & Co.

Panel, The Schools Statistics. 1970. *Correlation and Regression as Related to Statistics*. London: W. Foulsham.

Pearson, E. S. (ed.) 1956. *Karl Pearson's Early Statistical Papers*. Cambridge: University Press.

Pearson, E. S. 1978. *The History of Statistics...Lectures by Karl Pearson*. London: Charles Griffin & Co.

Pearson, E. S., and Kendall, M. G. (eds.) 1970. *Studies in the History of Statistics and Probability*. Darien: Hafner Pub. Co.

Pearson, K. 1892. *The Grammar of Science*. London: Walter Scott.

Pearson, K. 1895. "Note on Regression and Inheritance in the Case of Two Parents". *Proc. R. Soc. London*, 58: 240–242.

Pearson, K. 1896a. "Mathematical Contributions to the Theory of Evolution. – III. Regression, Heredity, and Panmixia". *Phil. Tran. R. Soc. London*, 187: 253–318.

Pearson, K. 1896b. "Contributions to the Mathematical Theory of Evolution. Note on Reproductive Selection". *Proc. R. Soc. London*, 59: 300–305.

Pearson, K. 1897. "Mathematical Contributions to the Theory of Evolution. – On a Form of Spurious Correlation which May Arise When Indices Are Used in the Measurement of Organs". *Proc. R. Soc. London*, 60: 489–498.

Pearson, K. 1900, *The Grammar of Science*. 2nd ed. London: Adam & Charles Black.

Pearson, K. 1911, *The Grammar of Science*. 3rd ed. Part 1. London: Adam & Charles Black.

Pearson, K. 1905. *Mathematical Contributions to the Theory of Evolution. – XIV: On the General Theory of Skew Correlation and Non-Linear Regression*. Draper's Company Research Memoirs, Biometric Series II (Reprinted in E. S. Pearson, 1956, pp. 477–528).

Pearson, K. 1906. "Walter Frank Raphael Weldon. 1860–1906". *Biometrika*, 5(1): 1–52.

Pearson, K. 1914. *The Life, Letters and Labours of Francis Galton*, vol. I: *Birth 1822 to Marriage 1853*. Cambridge: University Press.

Pearson, K. 1920. "Notes on the History of Correlation". *Biometrika*, 13(1): 25–45.

Pearson, K. 1924. *The Life, Letters and Labours of Francis Galton*, vol. II: *Researches of Middle Life*. Cambridge: University Press.

Pearson, K. 1930a. *The Life, Letters and Labours of Francis Galton*, vol. IIIA: *Correlation, Personal Identification and Eugenics*. Cambridge: University Press.

Pearson, K. 1930b. *The Life, Letters and Labours of Francis Galton*, vol. IIIB: *Characterisation, Especially by Letters*. Cambridge: University Press.

Pearson, K. 1934. Text from Speech Given by Pearson, Provided in Porter, 1986, 298–299.

Pearson, K., and Filon, L. N. G. 1898. "Mathematical Contributions to the Theory of Evolution. – IV. On the Probable Errors of Frequency Constants and on the Influence of Random Selection on Variation and Correlation", in Republished 1948 by E. S. Pearson (ed.), *Karl Pearson's Early Statistical Papers*. Cambridge: Cambridge University Press, pp. 179–261.

Pearson, K., and Lee, A. 1897. "On the Distribution of Frequency (Variation and Correlation) of the Barometric Height at Divers Stations". *Phil. Trans. A*, 190: 423–469.

Pearson, K., and Lee, A. 1903. "On the Laws of Inheritance in Man: I. Inheritance of Physical Characters". *Biometrika*, 2(4): 357–462.

Persons, W. M. 1910. "The Correlation of Economic Statistics". *Pub. Am. Stat. Assoc.*, 12: 287–322.

Persons, W. M. 1914a. "Review: Money and Prices...". *Q. Pub. Am. Stat. Assoc.*, 14: 79–82.

Persons, W. M. 1914b. "Rejoinder". *Q. Pub. Am. Stat. Assoc.*, 14: 347–350.

Persons, W. M. 1916. "Comment on Westergaard's 'Scope and Method of Statistics'". *Q. Pub. Am. Stat. Assoc.*, 15: 277–278.

Persons, W. M. 1919. "An Index of General Business Conditions". *Rev. Econ. Stat.*, 1: 111–205.

Pidgin, C. F. 1889. "The Mechanism of Statistics" [review of Matheson's paper]. *Pub. Am. Stat. Assoc.*, 1: 477–480.

Pidgin, C. F. 1890. "How to Make Statistics Popular". *Pub. Am. Stat. Assoc.*, 2: 107–115.

Porter, T. M. 1986. *The Rise of Statistical Thinking: 1820–1900*. Princeton, NJ: Princeton University Press.

PSU.edu-Rothrock: *https://pabook.libraries.psu.edu/literary-cultural-heritage-map-pa/bios/Rothrock__Joseph_Trimble*

Raverat, G. 1953. *Period Piece*. New York: W. W. Norton & Co.

Reed, W. G. 1917. "The Coefficient of Correlation". *Q. Pub. Am. Stat. Assoc.*, 15: 670–684.

Rider, P. R. 1941. "Alan E. Treloar, Elements of Statistical Reasoning". *Bull. Am. Math. Soc.* 47(9): 677. *https://projecteuclid.org/euclid.bams/1183503842*.

Rietz, H. L. 1911. "On the Theory of Correlation with Special Reference to Certain Significant Loci on the Plane of Distribution in the Case of Normal Correlation". *Ann. Math.*, 13: 187–199.

Rietz, H. L. 1919. "On Functional Relations for Which the Coefficient of Correlation Is Zero". *Q. Pub. Am. Stat. Assoc.*, 16: 472–476.

Rietz, H. L. 1927. *Mathematical Statistics*. La Salle: Open Court Publishing.

Rogers, J. L., and Nicewander, W. 1988. "Thirteen Ways to Look at the Correlation Coefficient". *Am. Stat.*, 42(1): 59–66.

Rothrock, J. T. 1873. "The Conservation and Correlation of Vital Force". *Am. Nat.*, 7(6): 332–340.

Schmidt, M. J. 1979. *Understanding and Using Statistics: Basic Concepts*, 2nd ed. Lexington, KY: D. C. Heath & Co.

Schmidt, M. J. 2010. *Understanding and Using Statistics: Basic Concepts*, 4th ed., "Edited by Dr. Donald E. Meyers". Redding: BVT Publishing.

Secrist, H. 1917. *An Introduction to Statistical Methods: A Textbook for College Students, a Manual for Statisticians and Business Executives*. New York: Macmillan Company.

Secrist, H. 1918. "Statistical Units as Standards". *Q. Pub. Am. Stat. Assoc.*, 16: 886–898.

Secrist, H. 1920a. *Readings and Problems in Statistical Methods*. New York: Macmillan Co.

Secrist, H. 1920b. *Statistics in Business: Their Analysis, Charting and Use*. New York: McGraw-Hill.

Secrist, H. 1925. *An Introduction to Statistical Methods: A Textbook for College Students, a Manual for Statisticians and Business Executives*, 2nd ed. New York: Macmillan Co. (this is a 1936 reprint of the 1925 edition).

Selkirk, K. E. 1981. *Rediguide 32: Correlation and Regression*. Nottingham: Nottingham University.

Simpson, J. A., and Weiner, E. S. C. (eds.) 1989. *The Oxford English Dictionary*, 2nd ed. Oxford: Clarendon Press.

Singer, C. 1990. *A History of Scientific Ideas*, 3rd ed. New York: Dorset Press.

Smith, J. G. 1934. *Elementary Statistics: An Introduction to the Principles of Scientific Methods*. New York: Henry Holt & Co.

Snedecor, G. W. 1946. *Statistical Methods Applied to Experiments in Agriculture and Biology*, 4th ed. Ames: Iowa State College Press.

Snedecor, G. W., and Cochran, W. G. 1967. *Statistical Methods*, 6th ed. Ames, IA: Iowa State University Press.

Snedecor, G. W., and Cochran, W. G. 1989. *Statistical Methods*, 8th ed. Ames, IA: Iowa State University Press.

Somervell, D. C. 1929. *English Thought in the Nineteenth Century*. New York: David McKay Co. (Reprinted 1962).

Spearman, C. 1904a. "The Proof and Measurement of Association between Two Things". *Am. J. Psychol.*, 15(1): 72–101.

Spearman, C. 1904b. "'General Intelligence,' Objectively Determined and Measured". *Am. J. Psychol.*, 15(2): 201–292.

Spencer, H. 1852. (Anonymously) "The Development Hypothesis". *Leader*, 1852: 280–281.

Spencer, H. 1862. *First Principles*. London: Williams & Norgate.

Spencer, H. 1865. *Principles of Biology*, vol. 1. London: Williams & Norgate.

Spencer, H. 1867. *Principles of Biology*, vol. 2. London: Williams & Norgate.

Spencer, H. 1901. *Essays: Scientific, Political, & Speculative*, vol. 1. London: Williams & Norgate.

Spencer, H. 1904a. *First Principles*, 6th ed. London: Watts & Co.

Spencer, H. 1904b. *An Autobiography*. New York: D. Appleton & Co.

Spiegel, M. R. 1961. *Schaum's Outline of Theory and Problems of Statistics*. New York: McGraw-Hill.

Stateuniversity.com-Lindquist: *https://education.stateuniversity.com/pages/2183/Lindquist-E-F-1901-1978.html*

Stevens, G. B. 1911. *The Pauline Theology: A Study of the Origin and Correlation of the Doctrinal Teachings of the Apostle Paul*. New York: Charles Scribner's Sons.

Stigler, S. M. (ed.) 1980. *American Contributions to Mathematical Statistics in the Nineteenth Century*. New York: Arno Press.

Stigler, S. M. 1986. *The History of Statistics: The Measurement of Uncertainty before 1900*. Cambridge: Belknap Press.

Stigler, S. M. 1989. "Francis Galton's Account of the Invention of Correlation". *Stat. Sci.*, 4(2): 73–79 (Note: the header of the printed published page incorrectly gives the pagination as 73–86).

Stigler, S. M. 1999. *Statistics on the Table: The History of Statistical Concepts and Methods*. Cambridge: Harvard University Press.

Symonds, P. M. 1926. "Variations of the Product-Moment (Pearson) Coefficient of Correlation". *J. Educ. Psychol.*, 17: 458–469.

Taleb, N. N. 2004. *Fooled by Randomness: The Hidden Role of Chance in Life and in the Markets*, 2nd ed. New York: Random House.

Tappenden, H. J. 1962. "Memoirs: Sir Wiliam Palin Elderton, K.B.E., Ph.D. (Oslo)". *J. Inst. Actuaries*, 88(2): 245–251. https://doi.org/10.1017/S0020268100015055.

Tetlock, P. E., and Garner, D. 2015. *Super Forecasting: The Art and Science of Prediction*. New York: Broadway Books.

Thomas, A. B. 1898. *The First School Year: A Course of Study with Selection of Lesson Material, Arranged by Months, and Correlated for Use in the First School Year*. California: State Normal School.

Thompson, H. 1894. "On Correlation of Certain External Parts of *Palaemon serratus*". *Proc. R. Soc. London*, 55: 234–240.

Thurstone, L. L. 1925. *The Fundamentals of Statistics*. New York: Macmillan Co.

Tietjen, G. L. 1986. *A Topical Dictionary of Statistics*. New York: Chapman & Hall.

Tippett, L. H. C. 1941. The *Methods of Statistics: An Introduction Mainly for Experimentalists*, 3rd ed. London: Williams & Norgate.

Tippett, L. H. C. 1952. *The Methods of Statistics*, 4th ed. New York: John Wiley & Sons.

Tobias, P. A, and Trindade, D. C. 2012. *Applied Reliability*, 3rd ed. Boca Raton: CRC Press.

Todd, R. B. (ed.) 1859. *The Cyclopaedia of Anatomy and Physiology*. London: Longman, Brown, Green, Longmans, & Roberts (Note: publication was accomplished in stages starting in 1835 and ending in 1859; the entirety was re-published in five volumes in 1859; it is the pagination of that final edition that is used in references here).

Todhunter, I. 1865. *A History of the Mathematical Theory of Probability*. Reprinted 1949 by New York: Chelsea Publishing Company.

Treloar, A. E. 1939. *Elements of Statistical Reasoning*. New York: John Wiley & Sons.

Treloar, A. E. 1942. *Correlation Analysis*. Minneapolis, MN: Burgess Publishing.

Triola, M. F. 2021. *Elementary Statistics*, 13th ed. Kindle Edition by Amazon.com.

Troup, C. E., Griffiths, A., and Macnaghten, M. L. 1894. *Identification of Habitual Criminals*. London: Eyre & Spottiswoode.

Trouton, F. T. 1889. "A Correction" [Letter to Editor]. *Nature*, 39: 412.

Tschuprow, A. A. 1939. *Principles of the Mathematical Theory of Correlation*, translated by M. Kantorowitsch. London: William Hodge.

UNESCO: https://www.unesco.org/en/education/digital/artificial-intelligence

Uoregon.edu-King: *https://scua.uoregon.edu/agents/people/1024*

Vidakovic, B. 2017. *Engineering Statistics*. Hoboken, NJ: John Wiley & Sons.

Vigna, S. 2014. "A Weighted Correlation Index for Ranking with Ties". urn:arXiv:1404.3325. *https://arxiv.org/abs/1404.3325*.

Vorzimmer, P. J. 1971. *Charles Darwin: The Years of Controversy*. Philadelphia, PA: Temple University Press.

Walcott, C. D. 1891. *Correlation Papers: Cambrian*. Washington D.C.: Government Printing Office.

Walker, H. M. 1929 & 1931. *Studies in the History of Statistical Method with Special Reference to Certain Educational Problems*. Baltimore, MD: Williams & Wilkins Co.

Ward, L. F. 1891. "Principles and Methods of Geologic Correlation by Means of Fossil Plants". Science, 18:282.

Warner, F. 1893. "Results of an Inquiry as to the Physical and Mental Condition of Fifty Thousand Children Seen in One Hundred and Six Schools". *J. R. Stat. Soc.*, 56: 71–100.

Watkins, G. P. 1915. "Theory of Statistical Tabulation". *Q. Pub. Am. Stat. Assoc.* 14: 742–757.

Waugh, A. E. 1938. *Elements of Statistical Method*. New York: McGraw Hill Co.

Waugh, F. G. 1894. *Members of the Athenaeum Club from Its Foundation*. London: Athenaeum. Reprinted by the British Library in 2011as *Members of the Athenaeum Club, 1824–1887*.

Weldon, W. F. R. 1890. "The Variations Occurring in Certain Decapod Crustacea. – I. *Crangon vulgaris*." *Proc. R. Soc. London*, 47: 445–453.

Weldon, W. F. R. 1892. "Certain Correlated Variations in *Crangon vulgaris*". *Proc. R. Soc. London*, 51: 2–21.

Weldon, W. F. R. 1893. "On certain Correlated Variations in *Carcinus moenas*". *Proc. R. Soc. London*, 54: 318–329.

Wells, H. G. 1896. *The Island of Dr. Moreau*. New York: Stone & Kimball.

Wells, H. G. 1901. *The First Men in the Moon*. London: George Newnes, Limited.

West, C. J. 1915. "The Value to Economics of Formal Statistical Methods". *Q. Pub. Am. Stat. Assoc.*, 14: 618–628.

West, C. J. 1918. *Introduction to Mathematical Statistics*. Columbus: R. G. Adams & Co.

Westergaard, H. 1916. "Scope and Method of Statistics". *Q. Pub. Am. Stat. Assoc.*, 15: 229–276.

Westergaard, H. 1932. *Contributions to the History of Statistics*. London: P. S. King & Sons.

Wheeler, D. J., and Chambers, D. S. 1992. *Understanding Statistical Process Control*, 2nd ed. Knoxville: SPC Press.

Whipple, G. M. 1907. "A Quick Method for Determining the Index of Correlation". *Am. J. Psych.*, 18(3): 322–325.

Wikipedia: Bowley: *https://en.wikipedia.org/wiki/Arthur_Lyon_Bowley*

Wikipedia: Camp: *https://en.wikipedia.org/wiki/Burton_Howard_Camp*

Wikipedia: Dodd: *https://en.wikipedia.org/wiki/Stuart_C._Dodd*

Wikipedia: Duell: *https://en.wikipedia.org/wiki/Charles_Holland_Duell*

Wikipedia: Goldenweiser: *https://en.wikipedia.org/wiki/Emanuel_Goldenweiser*

Wikipedia: Harris: *https://en.wikipedia.org/wiki/James_Arthur_Harris*

Wikipedia: Tippett: *https://en.wikipedia.org/wiki/L._H._C._Tippett*

Wikipedia: Weldon: *https://en.wikipedia.org/wiki/Raphael_Weldon*

Wikipedia: WeldonPrize: *https://en.wikipedia.org/wiki/Weldon_Memorial_Prize*

Willcox, W. F. 1916. "Introduction". *Q. Pub. Am. Stat. Assoc.*, 115: 225–228.

Willemsen, E. W. 1974. *Understanding Statistical Reasoning*. San Francisco, CA: W. H. Freeman & Co.

Williams, H. S. 1909. *The Story of Nineteenth-Century Science*. New York: Harper & Brothers Publishers.

Winslow, C. 1906. "Statistics of Heredity". *Pub. Am. Stat. Assoc.*, 10: 116–117.

Working, H. 1921. "A Use for Trigonometric Tables in Correlation". *Q. Pub. Am. Stat. Assoc.*, 17: 765–769.

Wright, S. 1921. "Correlation and Causation". *J. Agric. Res.*, 20: 557–585.

Yates, F. 1951. "The Influence of *Statistical Methods for Research Workers* on the Development of the Science of Statistics". *J. Am. Stat. Assoc.*, 46: 243, 19–34.

Yule, G. U. 1895. "On the Correlation of Total Pauperism with Proportion of Out-Relief: 1. All Ages". *Econ. J.*, 5(20): 603–611.

Yule, G. U. 1897a. "On the Significance of the Bravais' Formulae for Regression, &c., in the Case of Skew Correlation". *Proc. R. Soc. London*, 60: 477–489.

Yule, G. U. 1897b. "On the Theory of Correlation". *J. R. Stat. Soc.*, 60: 812–854.

Yule, G. U. 1909. "The Applications of the Method of Correlation to Social and Economic Statistics". *J. R. Stat. Soc.*, 72(4): 721–730.

Yule, G. U. 1910. "The Applications of the Method of Correlation to Social and Economic Statistics". *Bull. Inst. Int. Stat.*, XVIII: 537–551.

Yule, G. U. 1911. *An Introduction to the Theory of Statistics*. 1912, 2nd ed.; 1919, 5th ed. London: Charles Griffin & Company.

Yule, G. U. 1921. "On the Time-Correlation Problem, with Especial Reference to the Variate-Difference Correlation Method". *J. R. Stat. Soc.*, 84(4): 497–537.

Yule, G. U., and Kendall, M. G. 1937. *An Introduction to the Theory of Statistics*, 11th ed. London: Charles Griffin.

Yule, G. U., and Kendall, M. G. 1944. "Obituary: Reginald Hawthorn Hooker, M.A". *J. R. Stat. Soc.*, 107(1): 74–77.

Yule, G. U., and Kendall, M. G. 1950. *An Introduction to the Theory of Statistics*, 14th ed. London: Charles Griffin ("5th impression", 1968: per Preface, p. vi, this version is the same as the 14th edition 1st impression, except for "minor corrections" and "minor amendments").

Zorich, J. N. 2017. "Four Formulas for Teaching the Meaning of the Correlation Coefficient". *MathAMATYC Educator*, 8(3): 4–7.

Zorich, J. N. 2018. "Reasons for Teaching and Using the Signed Coefficient of Determination Instead of the Correlation Coefficient". *MathAMATYC Educator*, 9(3): 48–51.

Zorich, J. N. 2019. Personal email received 2019-04-12 from the archive department at Macmillan, which has owned *Nature* since 1869.

Zorich, J. N. 2020a. The approximately 400 pre-December-1888 files were downloaded from the Complete Works section of Galton.org on 2020-06-20; file by file, searches for words starting with "co-" (which found "co-relation", "co-relate", etc.) were conducted either electronically or by reading the entire file; an electronic search was conducted on a file only after it was converted to "Searchable PDF" format using "PDF Converter Professional 8.1 by Nuance Communications"; a small percentage of the files contained documents not authored by Galton (e.g. reviews of Galton's books), which therefore were not counted in this search.

Zorich, J. N. 2020b. The approximately 200 post-December-1888 files were downloaded from the Complete Works section of Galton.org on 2020-06-20; file by file, searches for words starting with "co-" (which found "co-relation", "co-relate", etc.) were conducted either electronically or by reading the entire file; an electronic search was conducted on a file only after it was converted to "Searchable PDF" format using "PDF Converter Professional 8.1 by Nuance Communications"; a small percentage of the files contained documents not authored by Galton (e.g. reviews of Galton's books), which therefore were not counted in this search; Galton's novel was not among these 200 files.

Zorich, J. N. 2021a. *Reliability and Distribution Plotting*™, v. 12.20, copyright 2019 by John N. Zorich.

Zorich, J. N. 2021b. Every article in the Publications of the American Statistical Association (and the *Quarterly Publications of the American Statistical Association*) that was published from 1888 through 1917 was either read or skimmed.

Zorich, J. N. 2021c. "Reasons for No Longer Teaching the Normal Approximation Confidence Interval". *MathAMATYC Educ.*, 12(2): 24–27.

Index

Note: **Bold** page numbers refer to tables and *italic* page numbers refer to figures.

Printed in the United States
by Baker & Taylor Publisher Services